观 赏 桃

ORNAMENTAL PEACHES

胡东燕　张佐双　著

中 国 林 业 出 版 社

China Forestry Publishing House

图书在版编目（CIP）数据

观赏桃：汉英对照 / 胡东燕，张佐双著.—北京：中国林业出版社，2010.5

ISBN 978-7-5038-5810-9

Ⅰ.①观… Ⅱ.①胡…②张… Ⅲ.①桃－花卉－观赏园艺－汉、英 Ⅳ.①S685.99

中国版本图书馆CIP数据核字（2010）第048122号

策划编辑：李　惟　陈英君

责任编辑：陈英君

出　　版：中国林业出版社

　　　　　（100009 北京西城区德内大街刘海胡同7号）

网　　址：www.cfph.com.cn

　　　　　E-mail：cfphz@public.bta.net.cn

电　　话：(010) 83227584

发　　行：新华书店北京发行所

制　　版：北京美光制版有限公司

印　　刷：北京画中画印刷有限公司

版　　次：2010年6月第1版

印　　次：2010年6月第1次

开　　本：787mm×1092mm　1 / 16

印　　张：18.75

字　　数：297千字

印　　数：1～3000册

定　　价：160.00元

观赏桃

序

——陈俊愉

胡东燕、张佐双两先生的新著《观赏桃花》书稿，在我这里已经放置一两个月了，我因既老又忙，一直未曾批阅。前两天佐双电询，这才翻阅了中英全文，然后写成这个小序。

在此序中，我提了三方面的看法，即：（1）关于书名；（2）关于本书的优点；（3）对本书的建议。

首先，关于书名。建议中名《观赏桃》，英名Ornamental Peaches，省却中名里的"花"字，又把中、英文统一了起来。既改变原来中名"观赏桃花"、英名"Ornamental Peaches"之不一致的问题，又合理地含括了观叶（如紫叶桃等）为主的品种在内，做到了名实相符、中英一致。

其次，关于本书优点。我认为这是一部优点显著的新著，因其：

系统而全面——从古代到现今、从本国到外国，从种质资源、分类与育种、栽培到应用，以及中华艺术中的桃花，无不包罗于本书之中。

新颖而时尚——如在观赏桃品种分类方面，一改过去分型的老习惯，改为与国际接轨的品种群分类——6个品种群分得合理，既合现况实际，又切中华传统名花"二元品种分类"之真谛——由简单到复杂、由低级至高级、由纯种至远缘杂种。

中文简明清晰，英文真实恰当，参考文献丰富全面，图（片）照（片）生动美观，做到了图文并茂，参证齐全。在国内新出书刊中确已达到了领先水平。

最后，对本书提出三点建议：

（1）进一步丰富观赏桃在我国栽培历史与发展部分的内容。

（2）在观赏桃育种选种趋势上考虑更为周全。

（3）突出观赏桃花大色艳叶色有变，抗性与适应性强的优势，育种目标少而精，重点突出，以选育出更精彩的新品种来。

在我要结束此一小序之际，谨此向著者、编者及有关同志致以衷心的祝贺，因为这是我国、也是世界上第一部观赏桃佳著，它将对全球桃业和广大人民做出多方面的贡献。

九三叟陈俊愉

序于北京林业大学梅菊斋中

2010.4.14

FOREWORD
—— Professor Junyu Chen

The manuscript of the new book *Ornamental Peaches* written by Dr. Dongyan Hu and Prof. Zuoshuang Zhang has been placed on my desk for one or two months. Since I am old and too busy these days, I have not ever read it yet until Zuoshuang called me a couple of days ago and asked about the foreword. So that I started to go through all Chinese and English versions. Then, here is the foreword for this monograph.

Three aspects are put forward in this foreword. 1. The title. 2. The merit of the book. 3. Suggestions.

First, about the title. I suggest using "ornamental peaches" instead of using "ornamental peach blossoms" in Chinese title. It not only can cover some purple leaved cultivars, but also can be in accord with the English title.

Second, about the merit of this book. I think it is a new work with remarkable merit. The reasons are as follows:

1. Systematic and comprehensive——From the ancient times to nowadays, from China to oversea countries, the monograph covered germplasm, classification, breeding, cultivation, application, and the Chinese artwork of ornamental peaches.

2. New and fashionable——For example, the classification of ornamental peach cultivars. Instead of the old classification systems, new idea of group, which is in line with international norms, has been put forward in this book. Six groups are classified reasonable, not only in accord with the reality, but also take the essence of the so-called "dual classification system for ornamental plants", which is from simple to complex, from low state of evolution to the higher level, from species to distant hybrids.

3. The Chinese version is concise, the English version is appropriate. The reference is plentiful and comprehensive. The illustrations and pictures are vivid and beautiful. Both pictures and texts are excellent with elaborate quotation, achieving the leading level among the new books and publications in China.

Last, I have three suggestions. 1. The introduction of history and cultivation of ornamental peaches in China can be in large. 2. The discussion on breeding trend can be more thorough. 3. The breeding target should focus on the advantages of ornamental peaches, which are big flower with bright color, combined with variation on leaves, and adaptable and strong adversity resistance. To breed more brilliant cultivars, the target should be fewer but focused, which can give top priority to what is the most important.

On the occasion of writing this foreword, please allow me to express my congratulations to the authors, editors, and all related people. This is the first great monograph on ornamental peaches in China, even in the world. It will contribute to the area on the peach and also to all the people in many aspects. .

Junyu Chen, 93 years old
Professor of the Beijing Forestry University
Academician of the Chinese Academy of Engineering
April 14, 2010

FOREWORD

——Ralph Scorza Ph.D.

There is little doubt as to the great admiration that the Chinese people have for the peach as a fruit producing tree and as a tree of great beauty in form and color. Being native to China and a subject of selection and breeding for centuries, the peach has been expanded through selection and breeding seemingly to the extremes of its genetic capacity, and yet there appears to be no end to the genetic variation that can be produced. Nowhere has this genetic plasticity been more apparent than in its effects on tree form, and flower color and form, and therefore its effects on the peach as an ornamental species. As the authors discuss in this work, the ornamental peach is the object and the co-celebrant of festivals cherished by the Chinese people. Yet, in the Western world the peach is generally considered as a fruit producing species. Numerous breeding programs, particularly in Europe and North America, but on other continents as well, have greatly expanded the fruit qualities from the early peaches carried throughout the world from China. A vast array of fruiting varieties has been produced that fit market demands in terms of flavor, ripening time, size, color and firmness. Peaches can now be enjoyed tree ripe from local farm stands or they can undertake journeys of weeks, over thousands of miles of ocean to be consumed in distant markets. Yet few of those who consume these delicious and healthful fruits consider the great value of the peach tree as an ornamental species. Indeed, few consumers in the West have even seen peach blossoms other than perhaps in Chinese artwork. The focus on peach fruit in the West has resulted in a number of publications that well describe the variety of fruit types but there is little information concerning ornamental characteristics of peach, and there are few breeding programs outside of China that regard the peach as a subject of ornamental interest. Therefore, the current volume by Drs. Hu and Zhang is an important contribution to the literature on the peach. It opens wide a view of peach germplasm that has been too narrow for too long in the West. This volume is the culmination of many years of careful observation and study by the authors and represents their tireless efforts. We are indeed privileged that this volume is available in English. Its general discussions that include the history, classification, and genetic and taxonomic information lay a strong and ample foundation for the detailed descriptions of a great number of ornamental genotypes. These descriptions include ornamental peaches from China, Japan, the U.S. and other areas. The authors describe in detail the characteristics of each genotype and the photographs are a feast for the eyes and display the beauty of form and color that exists in peach, revealing its potential to take a place among the premier ornamental plant species. In this volume the authors clearly describe and illustrate the existing ornamental peach germplasm and they clearly suggest to us future breeding possibilities. Much has been done but much remains to be done in ornamental peach genetic improvement. The authors have provided readers, professionals and amateurs, with the keys to unlock a genetic treasure chest. We can all look forward to the exchange of ornamental peach germplasm that can result from the publication of this work and to the development of new ornamental types that in the future will be available to adorn our gardens, parks, and cities.

Ralph Scorza Ph.D.
USDA-ARS Appalachian Fruit Research Station
2217 WIltshire Road
Kearneysville, WV 25430 USA

April, 2010

　　毋庸置疑，中华民族对桃花的热爱，无论是对其所结的果实，还是对花朵本身的美丽外形和色彩都不例外。经过多年的选择和培育，原产于中国的桃在遗传能力上不断扩展，似乎已经达到了极限。然而，这种遗传变异看来并没有尽头。树型、花色和花型的多样性最能体现桃的遗传可塑性，从而使得桃成为一个观赏植物种类。正如作者在本书中所指出的，观赏桃在中国人民喜爱的节日中成为人们庆祝活动的欢庆角色。然而，在西方国家，桃基本上只被看作是一种果树。无数的育种研究，尤其是在欧洲和北美，当然也包括世界上其他一些地区，使得最初来自中国的早期桃的果实品质得到了极大的提高。根据市场对果实口味、成熟时期、大小、色泽、紧实度等方面的需求，已经培育出了大量的果桃变异品种。现在，既有可从当地农场的桃树上直接采摘即可食用的品种，也有可以承受数周旅途从千里之外运到市场上的品种。然而很少有人在享用这些美味而有益健康的桃子时，考虑过桃花的观赏价值。实际上，也许除了在中国艺术作品中，许多西方人甚至从未见过桃花。在西方，出于对果实的关注，产生了众多详细描述果桃类型的文献，却几乎没有关于桃花观赏特性的研究。在中国之外，也很少有育种研究项目将桃花的观赏特性作为主要目标。因此，胡东燕博士和张佐双教授所著的这部书对丰富桃的研究文献做出了重要贡献，它为西方国家长久以来对于桃种质资源的狭隘认识打开了广阔的视野。这部书是作者多年仔细观察和研究的成果，体现出了他们不倦的努力。我们对能够看到这部书的英文译文深感荣幸。书中对桃花历史、类型、遗传和分类学方面的广泛讨论，为对众多观赏桃基因型的详尽描述奠定了坚实而充分的基础。这里包括对来自中国、日本、美国和其他一些地区的观赏桃品种的描述。作者对每个桃花品种的特征进行了细致描述，书中图片仿佛一场视觉盛宴，体现了各个桃花品种花型、花色之美，显示出桃花能够在重要观赏植物中占据一席之地的潜力。在这部书中，作者清晰地描绘和勾勒出了现有的观赏桃种质资源，同时也明确地提出了对今后育种可能性的建议。虽然育种工作已经做了很多，但在观赏桃遗传改良方面仍有很多工作可做。在书中作者为读者，包括专业人士和业余爱好者，提供了开启桃花遗传宝库的钥匙。我们期待着通过这部著作的出版，促进桃花种质资源的交流，培育出更多新的桃花品种类型，为我们的庭院、公园和城市增色添彩。

拉尔夫·斯科泽 博士

美国农业部 阿巴拉契亚果树研究站

2010年4月

前 言
Introduction

　　桃花是原产中国的传统名花，在中国已经有3000多年的栽培历史。作为人们生活中一种重要的果实，桃已经广为世人所知；作为温带地区一种重要的果树资源，桃已经被果树学家广泛研究。然而，关于桃在观赏园艺方面的价值始终未能得到充分的认知和重视。

　　桃在其枝型、花型、花色、叶色等方面具有丰富的形态多样性，赋予了桃极其丰富的观赏性。作为一种观赏植物，本书提出的观赏桃的概念区别于人们以往惯有的果桃的概念，而更偏重其在观赏价值方面以及在城市园林应用方面的特性。

　　对观赏桃的系统性研究始于20世纪80年代，从形态学入手，孢粉学、生物化学、分子生物学等手段相继应用于观赏桃的品种分类、品种演化及品种间亲缘关系的研究中，前人关于桃及其近缘种的研究成果对于我们观赏桃的研究是不可或缺的宝贵资料。

　　本书是作者集20年来在桃花方面的研究成果编撰而成，总结了20年来对国内外观赏桃资源及品种的调查、分类及应用研究，以及在新品种选育等方面取得的进展及成果，从观赏桃的栽培、管理、繁殖等方面系统地加以介绍，希望更多的园艺界同行们和民众能对观赏桃这一我国宝贵的观赏植物资源有一个较为全面的认识和了解。为了使那些母语为非汉语的读者们能够同样有机会领略到观赏桃的魅力，胡东燕博士将本书的主要内容重新撰写成英文缩略版本附于书后，以利于东西方在此方面的沟通和交流。

　　产生写书这一想法是在 2004年，由于时间和精力等原因一直搁置至今。现在看来这也不失为一件幸事。2004年其后的日子里，更多的来自世界各地的研究桃的专家、学者们的信息和资料，不断开阔着视野，可以肯定地说，现在呈现在大家面前的这本书，今后还会有很多的补充和修正，但至少到目前为止，这是我们尽自己的努力所能完成的资料最为全面的一部关于观赏桃的作品。

　　目前，世界范围内尚没有专门研究观赏桃的机构，更没有人对观赏桃品种资源进行系统而广泛的收集和研究。北京市植物园自1989年举办第一届桃花节以来，开展桃花资源的系统研究，目前已经在世界范围内收集到观赏桃品种60余个，栽植桃花5000余株，无论是观赏桃的品种、数量都堪称世界之首。本书中介绍的大部分品种都可以在北京市植物园觅得芳踪。

胡东燕　张佐双

2010年2月于北京

Introduction

Peaches originated in China and have a cultivation history spanning more than 3000 years. The species is well known as an important and valuable fruiting resource and has been studied intensively in pomological area. However, as an ornamental plant it has not been sufficiently recognized or raised enough attention.

The ornamental peach exhibits immense diversity of growth habit, form and color in flower and leaf enabling it to be an attractive ornamental plant for landscape use. Contrary to the common fruit peach the ornamental peach addressed in this book strengthens its ornamental characteristics and adaptability to urban application.

Systematics studies of ornamental peach were begun in 1980s. All methods from morphological, palynological, biochemical to modern molecular were applied to develop methods and understanding of classification, evolution and genetic relationships among the cultivars. Research results from studies of fruit peaches varieties and their relatives proved to be precious and indispensable resources for the studying of ornamental peach.

Our world wide investigation and research of the last 20 years this book provide the latest research achievements on ornamental peach classification, application, and breeding development. Seventy-four cultivars of ornamental peach have been illustrated and detailed descriptions provided. From cultivation and propagation to maintenance we hope to provide all horticulturalists and readers with a detailed overview of ornamental peach.

The preliminary idea of writing this book was in 2004 after completion of my dissertation. However, I feel fortunate to have deferred the final writing until 2009 as I have had a great many more opportunities to meet breeders and researchers of the ornamental peach since 2004. Their awe-inspiring efforts and new materials created have helped me on my path to understanding this plant in an as broad and in depth way as possible. This book represents the synthesis of my 20 years of passion for and pursuit of the ornamental peach. Undoubtedly there will always be room for improvement and I welcome constructive comments regarding ornamental peach performance from every reader. This English version is a much abridged version of the original Chinese text, intended as an introduction to my work for non Chinese readers. Lady Amy Victoria Loveday-Hu read through the manuscript and improved the English expression. For detailed reading or referencing the original Chinese text should be used.

For the majority of species and cultivars, I have observed, touched, sifted, collected, photographed and made copious notes. The photographs included here were chosen to represent the cultivars at their garden best.

Beijing Botanical Garden has held Ornamental Peach Festival since 1989. Under their consistent research, 61 cultivars have been collected from around the world and more than 5000 ornamental peach trees have been planted in the garden, making it the largest single collection ite in the world. Most cultivars mentioned in this book can be found in the Beijing Botanical Garden.

Dongyan Hu, Zuoshuang Zhang
Feb. 2010 in Beijing

作者简介
Introduction of author

胡东燕，教授级高级工程师。1989年毕业于北京农学院，1999年和2004年先后获得北京林业大学观赏园艺硕士和博士学位。在北京市植物园从事桃花研究20余年。曾先后对国内27个省、市、自治区的桃花资源进行过全面的调查和分析，1998年之后对日本、美国、乌克兰等桃花资源相对丰富的国家进行了系统的调查、研究与引种；曾经多次以研究学者身份赴日本农林水产省(1998年)、美国缅因大学(2002-2004年，在哈佛大学完成博士论文的分子生物学实验部分，在缅因大学完成博士论文的写作)、美国农业部(2007-2008年)进行考察访问与合作研究，先后在国内外专业刊物上发表桃花研究论文10余篇；多次参加国际园艺学大会、美国园艺学会、国际植物学大会等专业学术交流，并数次在会议上发言。先后参加北京市《桃花引种及花期控制》、《桃花新品种选育》、《主要木本观赏植物种质资源收集和种质创新研究》等课题的研究，《桃花引种及花期控制》等课题先后获得多项省部级和市级科研成果奖；并主持国家林业局《桃花新品种DUS测试指南及已知品种数据库的建立》项目。2005年荣获北京市先进工作者。

Dr. Dongyan Hu, a professor, has been working in Beijing Botanical Garden since her graduation from Beijing Agricultural College in 1989. She received her M.S. and Ph.D. from Beijing Forestry University in 1999 and 2004, respectively. In her 20 year study of ornamental peaches she has investigated ornamental peach resources in 27 provinces, cities and autonomies in China. Since 1998 she has visited Japan, the USA, and the Ukraine to investigate and explore their rich ornamental peach resources and has collaborated with the Japanese Agricultural Department (1998), University of Maine (2002-2004), and USDA (2007-2008) as a visiting research scientist. Her publications include more than 10 research papers on ornamental peaches that have been featured in national and international professional journals. Dr. Hu has regularly attended conferences sponsored by IHC, ASHS, ISHS, and IBC since 2002 and has presented her research and writings on the ornamental peach several times. She is an active part of national peach research having joined several research projects on the ornamental peaches sponsored by the Beijing Municipal Government and the National Bureau of Forestry, respectively.

　　张佐双，教授级高级工程师。享受国务院政府特殊津贴。1962年开始在北京市植物园工作。1983年担任北京市植物园副园长，1995年至2008年担任北京市植物园园长。曾任原北京市园林局副总工程师，北京市公园中心副总工程师、顾问。国际生物多样性保护计划中国国家科学委员会委员，中国植物学会植物园分会理事长，中国植物学会植物迁地保育专业委员会主任，中国花卉协会月季分会理事长，中国生物多样性基金会副理事长，中国花卉协会牡丹分会副会长，中国花卉协会梅花蜡梅分会副会长，中国花卉协会兰花分会副会长，中国环境科学学会植物园保护协会副理事长，北京植物学会副理事长，北京植物病理学会常务理事。曾经先后考察了40多个国家，60多个世界著名植物园，多次出席国际植物学会和植物园年会，并成功引种多种新优植物。先后主持《桃花引种及花期控制》、《樱桃沟自然保护区调查与保护》等10余项课题研究，先后获得国家行业及省部级科研成果奖10余项；先后出版《植物园学》、《中国月季》、《园林植物景观设计与营造》等10余部著作，论文数十篇。曾荣获建设部建设园林城市先进个人，全国绿化先进工作者。

　　Prof. Zuoshuang Zhang, is the former Director of Beijing Botanical Garden (1995-2008) and the President of the Chinese Botanical Garden Association and the Deputy President of the Chinese Biodiversity Foundation. He has visited more than 40 countries and over 60 famous botanical gardens around the world. He has presided over several research projects such as the prevention and control of *Caloptilia* sp., the test project of natural conservation in Cherry Valley, plants introduction, cultivation and conservatory display in Beijing Botanical Garden. His research into the introduction of ornamental peach cultivars has won the science and technology progress prize in Ministerial and Municipal class several times. As the editor, he has published more than 10 works including *Botanical Garden Science*, *Chinese Roses*, *Chinese Herbaceous Peonies* and *Chinese Flare Peony*. He has been awarded "special allowance" from the State Council in China for his outstanding work.

目 录
Contents

桃

这是明代王思义所编绘的《三才图会》（最早出版于1368年）中，由明代画家王圻描绘的桃花。

桃的起源、分布及传播

ORIGIN, DISTRIBUTION AND DISPERSAL OF THE PEACH

桃原产中国，分布广泛。

以观赏为主要目的的桃花，

是指在广为人知的桃的食用功能之外，

以体现桃花观赏层面功能的一个新类型，

所以要谈论桃花的起源、分布和传播，

必定是要从桃谈起的。

一、桃的起源

桃花的拉丁名*Prunus persica*经常会给人一种误解，即误以为桃花是起源于波斯的。因为种加名"*persica*"表示波斯（现伊朗一带），在欧洲国家这一名称总会给人一种想象，即这种植物的起源来自波斯。

伊朗学者札林库伯博士（1985）在论述张骞（？－前114年）访问西域时说道："这时候，苜蓿、葡萄从波斯传入中国并且开始种植。同时，杏和桃的种子传入波斯，并且种植。"由此可见，就连波斯人自己都有着明确的记载，桃的确是在约公元前1世纪～前2世纪从中国输入波斯的。

中国考古发现，在浙江河姆渡出土的新石器时代的文物中有距今6000～7000年的野生桃核（汪祖华，庄恩及 2001）。据出土文物考证，河南郑州二里岗、江苏海安青墩、浙江吴兴钱山漾、杭州水田坂、余姚河姆渡以及广西钦州独料等地，均出土6000多年前的新石器时代的桃核。尤其是河北藁城的商代遗址中出土的桃核（1.6cm × 1cm 和 2.0cm × 1.2 cm），其形状、大小、沟纹及缝合线与现代的桃核极为相似。

《山海经》中有6处记载涉及桃的分布。"不周之山"、"边春之山"、"岐山"、"灵山"、"卑山"、"夸父之山"分别位于现在的甘肃、新疆、青海、陕西、河南等地，表明中国上古时代，西部和北部广泛分布着野生桃。《尚书·武成》"归马于华山之阳，放牛于桃林之野"，是我国桃树大面积原始林最早的历史记载。《夏小正》中"柂桃，山桃也"和《尔雅·释木》中"樧，山桃"的记载也说明在3000多年前的我国黄河中下游地区仍有野生山桃的存在。

最早认为桃起源于中国的人是瑞士植物学家德坎道尔（De Candolle），他在1855年就明确表示桃原产中国（凯尔曼 1987）。Royel曾指出，喜马拉雅山南麓产野桃的地方甚多（De Candolle 1825）。美国植物学家Meyer在20世纪初发现中国陕西、甘肃、西藏等地仍有野生的桃花（Hedrick 1917）。

目前在中国还能找到桃的原生种和4个野生近缘种的成片林地，它们是青藏高原的光核桃(*Prunus mira* Koehne)，黄河上游的甘肃桃 (*Prunus kansuensis* Rehd.)，华北的山桃(*Prunus davidiana* Franch.) 及新疆的新疆

桃 (*Prunus ferganensis* Kost. et Riab.) (汪祖华，庄恩及 2001)。

无论是栽培历史的悠久，还是野生种类的分布，以及桃资源种类的多样性，都可以肯定桃的原产地和起源中心是中国，桃是中国给予这个世界的一件伟大礼物。

二、桃的分布

(一) 桃的自然分布

中国丰富的古代文献资料为桃的自然分布提供了最好佐证。

《山海经》里6次提到桃的天然野生分布，表明在上古时代，新疆、青海、甘肃、陕西、河南等我国西北部、中部地区就已经有野生桃广泛分布。

近、现代植物学家的采集活动，也为桃的自然分布提供了佐证。我国科学家（宗学普，段玉春 1987）在雅鲁藏布江、金沙江、澜沧江、怒江流域的海拔1700～4200m处，发现有大片的野生光核桃林，集中分布在海拔3000m左右，其中发现一株树龄逾千年，其干周达10m以上的光核桃古树。

(二) 桃在中国的栽培分布

从我国考古发现的桃核出土地点可以看出，最早栽培桃(殷商时期)的地区应当在黄河流域的河南（郑州二里岗殷商文化遗址）、河北（藁城市台西村商代遗址）等地。随着江苏海安青墩、浙江吴兴钱山漾、杭州水田坂、余姚河姆渡等长江流域江浙一带新石器时代和商周时期的毛桃核或桃核的发现（汪祖华，庄恩及 2001），说明桃是一种分布广泛、适应性强的树种。

中国最早的诗歌总集《诗经》中涉及桃的分布的描述有5处，表明早在先秦时代，桃在陕西（《召南·何彼秾矣》、《大雅·抑》）、山西（《魏风》）及河南（《卫风·木瓜》、《周颂·小毖》）等黄河流域中游地区已有栽培。春秋时期《韩非子》卷11《外储说左上》的记载亦可为证："子产退而为政五年，国无盗贼，道不拾遗，桃枣荫（洒满）于街者莫有援（捡取）也，椎刀遗道三日可反，三年不变，民无饥也。"即郑国的都城新郑广种桃树，说明此时桃的栽植在河南郑州一带已经相当普遍。

详细记述中国春秋时代历史的《左传》出现"桃"字12处。其中一处是鲁国地名，在今山东汶上县境。以"桃"字命名地名，可以看出当地桃已经比较普遍。宣公二年还有"赵穿攻晋公于桃园"的记载，即当时晋国已建有桃园。当时晋国的都城绛（今山西翼城县境），位于北纬35°～36°，正是适宜桃树生长的地区。

这与野生桃核出土所体现出的天然野生分布也很一致，明确体现了我国早期桃的集中栽培地区。

桃的栽培从西北地区开始，向东在河北、山东黄河流域形成第一次生中心；往南在江苏、浙江、湖北、安徽、湖南等长江流域形成第二次生中心；由此再向南，在福建、广东、广西、台湾等地，形成对低温需求量少，成熟期较早的品种类型；此外以陕、甘为中心，向西至新疆，向北至宁夏、内蒙古，向西南至四川、云南、贵州及西藏，除十分严寒的地区，我国各地几乎均有桃的栽培（韩振海1995）。

三、桃在世界范围的传播

桃走向世界是通过丝绸之路首先传到波斯，再从中亚继续传到欧洲各国，并随着欧洲殖民者的足迹遍布世界各地；另一条途径则是直接向东传到日本（图1.1）。

(一) 从中国到东亚

日本古代桃的记载始于弥生时代（前200年－300年）（星川清亲1981），毫无疑问，栽培历史肯定要早于这一记载。至今在宫崎、山口两地仍能发现野生桃（张宇和1982）。

根据前苏联学者瓦维洛夫1935年发表的《栽培植物起源中心》中的理论（赫西 1981），在一种植物野生种的原产地周边，应存在着最大可能性的变异。前苏联学者茹考夫斯基《育种的世界植物基因资源》(1974) 中有关大基因中心和小基因中心学说，更进一步说明了这一观点。一种植物往往以其初生中心为大基因中心，而在其次生分布中心存在着特有的种质资源。原产中国的桃从弥生时代传入日本，经过2000多年的繁衍、变异，逐渐形成了桃的次生分布中心，因此从日本江户时代出现记载的'帚桃'、'云龙'桃、'京更纱'等众

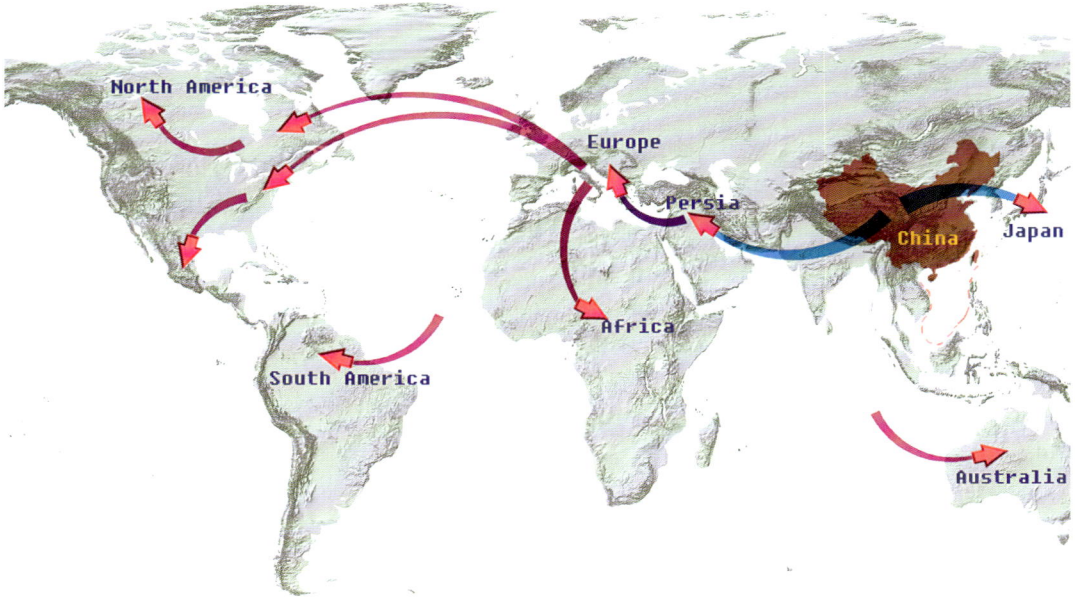

图1.1 桃在世界范围内的传播
Fig.1.1 Dispersal of the peach around the world

多观赏桃品种，很可能是桃花在其次生分布中心中特有的种质资源（伊藤 1995）。

(二) 从中国到欧洲

桃经由波斯传到欧洲最早是在公元前4世纪，首先传到希腊（Haw 1987），之后传遍欧洲，以后又随着欧洲殖民者的不断开拓，桃被带到了世界各地。

1. 希腊

桃最早出现在希腊的记载是在公元前332年，Theophrastus称其为"波斯的水果"（Hedrick 1917），这是古代历史学家经常会产生的误解，即一种植物从何而来，那个地方就会被想当然地认为是这种植物的发源地；公元64年，Dioscorides（Hedrick 1917）首次提到了桃的药用功能。

2. 意大利

在意大利的文献记载中，古罗马大政治家Virgil（前71年～前19

年）最早提到桃（Hedrick 1917）；随后，公元40年，Columella 记载这是由波斯进献给埃及的一种有毒的礼物。由此可见，桃在意大利的栽培始于纪元之前。

Pliny（1855）关于桃的记载在研究桃的栽培发展历史中具有里程碑的意义（Hedrick 1917）。Pliny在记载中多次提到桃是经由希腊和波斯引入的一种果树，并详细描述了桃6个变种的特性：'malum persicum'、'duracinus'、'Gallic'、'Asiatic'、'supernatia'、'popularia'。其中'popularia'的名字的意思是"种植得到处都是"，由此也可以看出，桃在当时的意大利已经是一种栽植相当普遍的果树。Pliny还记载了桃有益于健康的例子：果汁提取液可以做酒或醋，桃叶有阻止脑溢血的功效，桃核碾碎和酒、醋混合可以作为头痛的镇痛剂。

3. 法国

桃在法国开始栽培的时间应该和意大利大致相同（Hedrick 1917）。当时最精通桃栽培技艺的是那些修道院的修士们。法国是欧洲桃的兴发地，德国、比利时、荷兰的桃均来自于法国。桃在法国的栽培历史，乃至在世界桃的发展历史中都非常重要。

Andre Leroy在其*Dictionnaire de Pomologie*（1879）（Hedrick 1917）中对桃在法国的历史进行了总结：桃在法国的记载第一次是出现在Pliny 和Columella的文献中（Hedrick 1917），第二次出现是在530年，第三次是784年，第四次记载桃有不同种类，第五次则出现在860年。1604年Olive de Serres出版的*Theatre de Agriculture*是桃发展历史中突出的里程碑，其中记载了12种不同桃的品种，这是以往仅仅从种的角度培育桃首次发展到对变种、品种认识的开始。从此桃在法国进入了品种迅速发展的阶段。从1628年的27个，发展到1667年的39个，1768年的42个，1805年的60个，1865年的148个，1876年已经达到355个品种的记载（Hedrick 1917）。这其中出现了桃用于墙垣遮挡用途的记载，这是桃在其通常果树用途之外的又一体现。

4. 英国

英国的桃也是经由希腊传入的。13世纪，桃及其栽培技艺随着法国修道士进入英国，桃也逐渐出现在园林中，准确年代据*Chronicle of Roger of Wendover*中的记载，应该不早于1216年（Hedrick 1917）。

(三) 从欧洲到美洲

在16世纪，西班牙人最早将桃带到美洲大陆（Hedrick 1917）。

美洲桃的栽培始于里奥格兰南部地区。权威记载关于桃在墨西哥的栽培，可以在Molina1571年吕版的西班牙文书中见到（Hedrick 1917）。桃从墨西哥传播到了现在美国的新墨西哥、亚利桑那和加利福尼亚等地区，然后向东被带到密西西比河以东地区。到17世纪末，桃树已经适应了纽约地区的土壤，随后又被带到安大略湖滨。除了康涅狄格和马萨诸塞州部分地区之外，桃树在新英格兰地区很少有栽植。

桃在南美洲也得到了广泛的栽植。达尔文在他著名的旅行中曾经描述到阿根廷巴拉那河口密植的桃树（Hedrick 1917）此外，还有很多关于在太平洋海岸发现野生桃的记载（Hedrick 1917）。Oakenfull写道，桃尽管是来自国外，但巴西的气候显然很适合桃的生长（Oakenfull 1912）。Wight在1913年也有关于桃在巴西、智利、秘鲁和玻利维亚有所栽培的记载（Hedrick 1917）。

1866年美国就已经有136个桃的品种，并开始第一次用人工杂交的方法获得新的品种（Dowing 1866）。

(四) 从欧洲到非洲和大洋洲

非洲北部的晴朗天气，强光照，长生长季和足够热的气候，都十分适合桃的生长。在埃及和地中海沿岸，在栽培技术加以改善的前提下，桃的种植相当普遍。

欧洲人将桃带过了赤道，并使其在南半球成为了和在北半球一样广受喜爱的树种，应用于果园和园林中（Hedrick 1917）。

在温带大洋洲，桃已经在园艺领域占据了重要地位。在新西兰，有很大面积的桃树林（Hedrick 1917）；在澳大利亚也有桃的种植。

桃在其漫游世界的旅程中，每到一处，都会在其生长的地理区域内出现自己新的变型。这首先是白于桃本身是一种可塑性非常强的树种，可以根据各种不同的环境条件，适应性地演化成不同类型；其次，桃这种原产中国的古老树种，在其长达几千年的栽培过程中，保持了足够的遗传稳定性，使其成为果树中最为稳定的物种。正是这些因素，才使桃得以在世界范围内广泛传播和种植，成为全球温带地区重要的树种之一。

以观赏为主的桃花品种以及类型的形成和应用，也正是伴随着桃在世界范围内的传播才得到了更为广泛的发展。目前世界上观赏桃栽培分布较为广泛的，仍是那些果桃研究相对发达的国家及地区，比如：中国、日本、美国、意大利、法国等国，在近百年来均有各自育成的观赏桃品种形成。

这是由英国画家亚历山大·马歇尔Alexander Marshal
1620-1682年绘制的重瓣桃花。

2 桃花的栽培历史和发展

CULTIVATION HISTORY AND DEVELOPMENT OF THE ORNAMENTAL PEACH

相对于以果实为主要栽培目的的果桃，
桃花是人们针对以花见长的观赏桃的一种简称，
也有"碧桃"的雅称。
真正意义上出现的以花见长的观赏桃当始于唐代，
桃花的栽培历史以及品种类型的形成和应用，
也正是伴随着桃在世界范围内的传播，
随着一个国家桃的栽植程度和园艺发展水平得到了
更为广泛的发展。

一、桃花的栽培历史

(一) 中国

《诗经》（前551年 – 前478年）是中国古代第一部诗歌总集，在《诗经》中关于桃的记载有6处，其中关于桃花的记载有两处，说明早在3000年前的中国，桃花的美艳就已经开始被人们关注和赏识，并得到讴歌和赞美。当然，很难准确确定纯粹或者主要用于观赏的桃花起源于何时，桃花的栽培历史始终伴随着果桃的发展，用于观赏用途栽培的桃花的发展和演变，在中国古代文献记载中可见一斑。

《逸周书·时训解》、《礼记·月令》、《夏小正》、《吕氏春秋·仲春纪》等文献中先后出现了有关桃的花期的记载，表明人们已经开始观察到桃花的物候学特点。

晋代葛洪在《西京杂记》中记有："汉武帝修上林苑，诏群臣献奇花异果。"其中就有"秦桃、榹桃、湘核桃、金城桃、绮叶桃、紫纹桃、霜桃"等7个品种。由此可以看出，尽管汉代已经开始有出现于园林中的桃花品种的记载，但大部分仍然是以一种"异果"的形式存在。

真正出现以花见长的观赏桃是从唐代开始的，全新的花色和花形赋予了桃花明确的品种概念。五代王仁裕《开元天宝遗事》所记的仅栽培于御苑之中的'千叶'桃在中唐韩愈等人的诗作中相继出现，并涌现出'百叶'桃、'绯桃'（清·汪灏《广群芳谱》卷二十五记有："俗名'苏州桃'，花如剪绒，比诸桃开迟，而色可爱"）、'碧桃'、'绛桃'等新品种。

宋、元之际，桃花的新品种不断增加，北宋周师厚的《洛阳花木记》中记载了30个桃的品种，其中出现了'二色'桃（清·汪灏《广群芳谱》卷二十五记言"二色桃花"为"粉红，千瓣极佳"）、'合欢二色'桃（一朵二色）、'白碧'桃、'千叶绯'桃、'紫叶大'桃、'千叶碧桃'等新品种。

明代周文华在《汝南圃史》中除记载了前人的16个桃花品种之外，还增加了"瑞香桃，又名孩儿桃，矮桃，高一二尺"，即为今日的寿星桃。明代王象晋在其《群芳谱》中共记有桃品种21个，其中可供观赏的桃花品种8个，'日月'桃、'美人'桃、'鸳鸯'桃为3个新品种。此外李时珍在《本草纲目》中首次将花色引入桃的分类标准（"桃品甚多，其花有红、紫、白、千叶、二色之殊，其实有红桃、绯桃、碧桃……皆以色名者也"），成为有关桃花最早的较为科学的分类。

清代，桃花的发展达到盛期。陈淏子的《花镜》中共记有24个桃品种，其中观赏桃8个，新出现的品种有"墨桃，花色紫黑，似墨葵，亦异种难得者"，这一品种还出现在《古今图书集成》里上海县、新城县、金华县的县志记载中。该书共记有17个桃的品种，其中在《夏旦药圃同春》中对'美人'桃的记载更有了进一步的发展："美人桃，花瓣有百。其色水红，艳丽可爱，亦桃之属冠者，且喜有香。"这是首次出现有关桃花有香味的记载。同书215卷《桃部纪事》中也有："（安徽）太平县志记有桃花洞……洞口桃一株，自石中开，其花先放。香艳异常。"此外，广东高要县志中曾有'菊花'桃的记载。

近代，桃花栽培技术不断发展，桃花新品种不断出现。1949年黄岳渊、黄德邻等的《花经》以花期、花色为标准，对桃花进行了系统的介绍和初步分类，共记有11个观赏桃品种，其中"五色桃：花色大红，上洒白色"、"洒金桃：花瓣千重，瓣色红白相间，亦有淡红与正红相杂者"、"垂枝桃：枝性下垂，色有浓红、纯白、红色、淡红色，花瓣千重"，为首次见于记载的新品种。台湾杨恭毅（1984）在《杨氏园艺植物大名典》中曾有塔形桃的记载。许士捷在《养花技艺》中也曾记有绿花桃。

值得一提的是，在历代文献中，涉及到用于观赏的桃花品种几乎全部记载为"华（花）而不实"，比如"矮桃、碧桃、绯桃，皆千叶花而不实"（嘉靖《会稽志》）、"又有百叶桃，花蕊稠于别桃，鲜红，一名绛桃，不结子。"（弘治《八闽通志·泉州志》，万历《泉州府志》）。这些记载恐怕也是造成现在很多人，甚至专业书籍中都认为瓣重花密的桃花不会结实的误会的根据和来源。

(二) 日本及其他亚洲国家

1. 日本

日本最早的一部日语诗歌总集《万叶集》（竹内等 1988）收录了日本从4世纪至8世纪的著名和歌，其中有8首称颂桃花的诗歌。

　　观赏桃从奈良时代（710－794年）开始受到重视。实际上从奈良时代到平安时代（8世纪～12世纪），桃花在日本的应用主要是以庭园观赏为主的，主要原因在于此时的果实品质极差，直到19世纪后期随着从中国、美国、欧洲引进果桃品种，桃在日本才作为一种果树栽植（Pawasut *et al.* 2004）。

　　到江户时代（1603－1867年），园艺家们逐渐开始重视到花色艳丽的桃花，栽培面积有了较大的发展，大量观赏桃的园艺品种开始涌现，并逐渐使桃花成为能与樱花、梅花齐名的观赏花木。

　　1681年的《花坛纲目》（塚本 1994）中出现了8个桃花品种的记载，如其中的'矢口'（红花重瓣），至今仍为切花的主要品种。

　　1695年伊藤伊兵卫三之丞在其著名的《花坛地锦抄》中共记载了21个桃花品种。此时已有'寒白'（白色花重瓣），'源平'（粉色花和白色花共于一花或一枝），'京更纱'（单瓣，白色与粉色跳枝），'中生白'（白花重瓣），'源平垂枝'（单瓣，粉色与白色跳枝，垂枝桃），'残雪垂枝'（白花重瓣垂枝桃），'帚桃'（半重瓣花，粉色与白色跳枝，树型窄高），'菊桃'（粉色花，重瓣，花似菊花，花瓣细窄），'相模枝垂'（红花重瓣，垂枝桃），'羽衣枝垂'（半重瓣粉色花，垂枝桃）等优秀品种的记载。这其中的很多品种至今还应用于园林中。而'帚桃'更是现今世界上很多这一枝型品种的重要亲本。

　　1828年《本草图谱》（花百科周刊 1996）中记载了26个桃花品种，其中出现了寿星桃等品种。

　　1891年《日本园艺杂志》（塚本 1994）共记有21个桃花品种。

　　2. 土耳其斯坦

　　桃花出现在土耳其斯坦（Turkestan）的记载是在1876年，Schuyler（Hedrick 1917）记述到：在早春城市的边缘地带，几乎是整个Zarafshan溪谷中，大片粉色的桃花和白色的杏花种在一起，花香蔓延数里。这是当时最受人们喜爱的一种园林景观。

（三）欧洲

　　希腊是最早出现有关桃的记载的欧洲国家，意大利首次记载了桃的药用功能，法国在桃的发展历史中具有举足轻重的地位，英国和法国、比利时一起成为欧洲最早开始在园林中应用桃花的国家。桃花

作为观赏植物，在欧美园林中的应用有三百多年的历史
（Bean 1950；Everett 1967）。

1. 法国

法国是欧洲桃的兴发地，不仅在世界桃的栽培历史中
非常重要，而且对于当今观赏桃品种的形成与发展也起着
相当重要的作用。

17世纪，伴随着众多果桃品种的不断发展，用于
观赏的桃花品种在法国也开始逐渐涌现。现在仍然被广
为使用的桃花品种——淡粉色、半重瓣的'Duplex'，
1636年第一次出现于法国（Jacobson 1996）；高度重瓣，
白色花上有红色或粉色条纹的'Versicolor'1863年前已
经在法国（Krussmann 1986）有所栽植应用。

2. 英国

由于桃喜光，不喜水湿，在一定程度上限制了桃作
为果树在英国的栽植。但也许正是这些困难，反而激发
了英国人发展新品种的兴趣。英国和法国、比利时一起
成为欧洲最早开始在园林中应用桃花的国家。Thomas
Johnson在1633年（Hedrick 1917）提到一种种植在园林
中的、具有重瓣美丽花朵，而且果实极少的桃品种，称
其为'Persica flore pleno'重瓣桃花。这也是桃花以其
特殊的观赏特性用于欧美园林的最早记载（图2.1）。
随后，观赏用途的桃花品种不断出现。单瓣，白花的
'Alba'1829年首次命名于英国（Krussmann 1986）；重
瓣纯白色的'Alboplena'1849年引入英国。

3. 比利时

关于果桃，比利时并没有什么永留于史的记载，
但作为观赏桃品种发展的见证，比利时有着举足轻重
的地位。

深红色大花重瓣、具褶边的'Camelliiflora'和有着深
红色条纹、半重瓣的'Dianthfolia'1858年第一次记载于比
利时（Jacobson 1996）；紫叶粉花单瓣的'Purpurea'自从
1873年以来就已经栽植于比利时（Krussmann 1986）。

此外，红色花半重瓣的'Rubro-plena'1840年从中

图2.1 英国贺年卡中的重瓣
桃花
Fig. 2.1 Double-flowered
peach blossom by Alexander
Marshal (1620-1682)

国进口欧洲，1854年首次出现特征描述记载；垂枝桃'Pendula'首次出现在欧洲的记载是在1839年（Jacobson 1996）。

（四）美洲

桃树在美洲主要是作为一种果树广为栽植，然而美国人对于那些主要用来观花却并不结果的桃树的浓厚兴趣，早在1805年就可以从美国总统杰佛逊在其弗吉尼亚自家栽种的一株重瓣花桃这件事上有所体现（Malcolm, History of the peach tree.）。

美国农业部自1899年开始集中从中国采集桃（Werner and Okie 1998）。尽管这一举动的主要目的是为了收集果桃品种，但其中也不乏一些观赏桃。比如，玫瑰色大花重瓣、花期很晚的'Pi Tao'（很可能是我们的'碧桃'的音译）和复瓣的'Shau Thai Tao'均于1925年来自中国。花期很晚的红色重瓣垂枝桃'Red Weeping'自1931年引入后，成为以后的艳粉色大花重瓣晚花品种'Jerseypink'（Gofferda *et al.* 1992）和加拿大在国际木本植物登录机构登录的桃花品种（'Harrow Rubirose'，'Harrow Candifloss'，及'Harrow Frostipink'）（Layne 1981）的主要亲本来源。后3个品种均为相当耐寒的砧木。

用于容器栽培的矮型桃最早见于1911年（Hedrick 1917）。

苗圃业的发展也为美国引进了很多优秀的桃花品种。加州的W. B. Clarke苗圃至少引进了10个品种（Jacobson 1996）。尽管其中有些品种（如'Double Maroon'，和'Edward H. Rust'）已经近乎绝迹，但'Early Double White'，'Versicolor Weeping'则仍用于当今的苗圃生产及销售中。粉色、红色及复色的重瓣花矮生型桃花类型'Dwarf Mandarin'，也是经W. B. Clarke苗圃1940年从中国引进，从而大大提高了用于盆栽的矮生桃的观赏性。

（五）非洲和大洋洲

桃随着欧洲人又被带过了赤道，在南半球也成为和其在北半球一样广受喜爱的树种，应用于园林和果园。

在南非，桃是最为常见的水果。用于观赏的桃花品种也多近似于欧洲品种（*Illustrated Encyclopedia of Gardening in South Africa*，1984）。

据澳大利亚的记载（Harrison 1963），在阿德莱德，一些重瓣花的桃花品种，基部干径达到46cm，高6m。在西部珀斯及周边地区也有桃花种植。来自日本的粉花、白花和红花的'Carson'系列桃花品种主要包括：'Pink Cloud'和'Rose Brilliant'；而晚花品种'Alba Plena'、'Carnation Striped'、'Clara Meyer'、'Harbinger'、'Hiawatha'、'Iceberg'、'Rosea Plena'、'Windle Weeping'等大都来自于欧洲。

二、桃花品种的发展

目前世界上观赏桃品种发展程度和研究水平较高的国家，大都出现在那些果桃栽培分布广泛，研究程度相对发达的国家及地区，比如：中国、日本、美国、乌克兰等国，在近五十年来均有各具特色的观赏桃花新品种育成。

（一）中国

桃花在中国园林中的应用已经相当普遍，北京市植物园、广州石马、杭州西湖、江西庐山、上海南汇及龙华、四川成都、兰州安宁、湖南桃源及桃江，以种植大量桃花或举办一年一度的桃花节而成为中国十大桃花观赏胜地。其中庐山、成都、桃花源、龙华、杭州以及北京，就是以栽植观赏桃而闻名的，许多新品种都产生于此。北京市植物园以收集的桃花品种为基础，培育出'粉花山碧'桃和'粉红山碧'桃两个早花桃花品种（胡东燕，张秀英 2001），均为山碧桃系品种。二者均以'白花山碧'桃为父本，保留了父本高大的树体和早花的特性；同时分别继承了其母本'合欢二色'桃和'绛桃'的鲜艳花色和复瓣特点，成为能够很好衔接山桃与桃花之间花期断档空缺的不可多得的早花品种。'斑叶'桃（Hu et al. 2005）则是北京市植物园从'紫叶'桃中选育出的新品种，以其绿色叶上不规则的紫色斑点或条纹而独具特色。

全国各地开展的桃花品种资源调查工作也显示出桃花品种应用在中国呈现明显的上升趋势。例如北京从1991年的26个品种（张秀英，陈忠国 1991）增加到61个（Hu and Zhang 2008）；湖南桃花

源近年来也开始进行了广泛的观赏桃引种，目前已经拥有桃花品种
41个（顾振华等 2009）。此外，一些新的桃花观赏景区也在不断涌
现。云南昆明郊野公园近年来也逐渐开始丰富桃花品种，使得昆
明成为中国西南地区首屈一指的桃花观赏圣地（网站：http://news.
sina.com.cn/o/2006-02-26/10158305466s.shtml）；长江三峡风景名
胜区宜昌景区，也已经以桃花形成了景区的主要观赏特色（高玲
2002）；山东（臧得奎等 1998）、武汉（陈丹维等 2006）、合肥（何
浩和何晓平 2007）等地的桃花种植规模的形成，奠定了中国桃花栽
培的新格局（图2.2）。其中山东农业大学利用 γ 射线辐射‘紫叶’
桃种子得到的桃花新品种‘紫奇’（沈向等 2007），又增添了1个早花
紫叶品种。

图2.2　中国主要桃花观赏胜地
Fig. 2.2　Famous sites for the ornamental peach in China

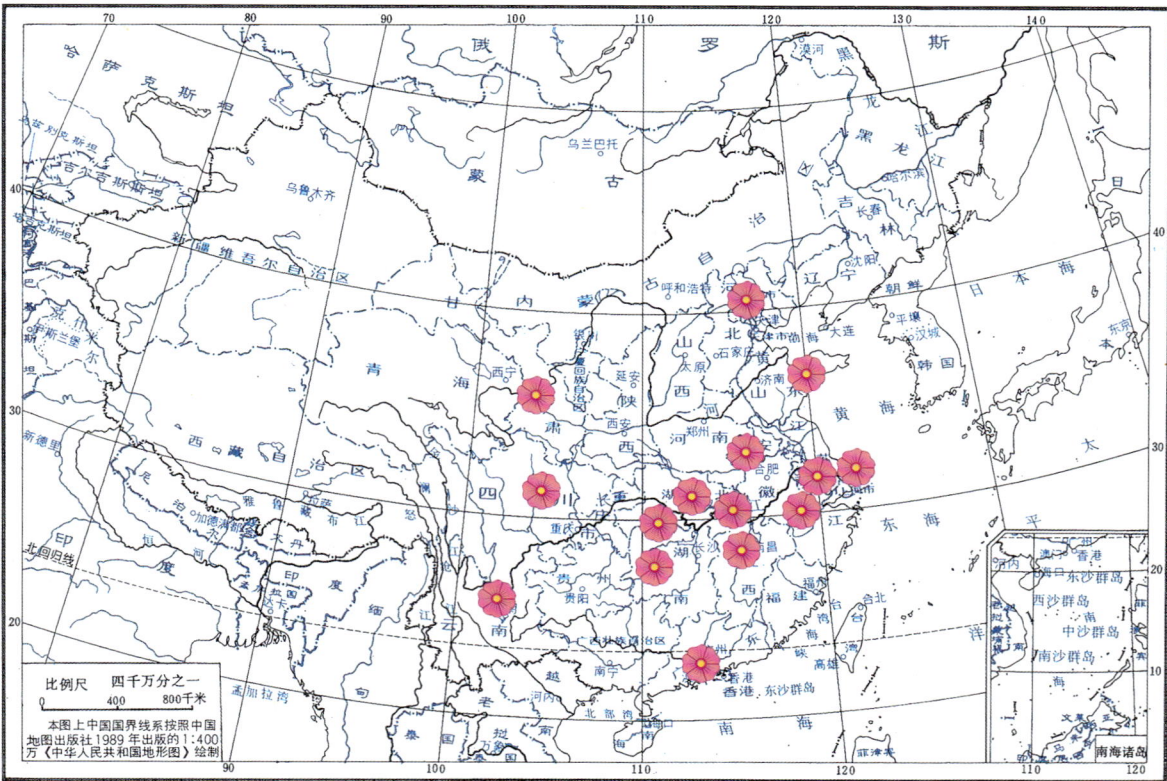

（二）日本

据《桃图鉴》（吉田雅夫 1985）记载，日本共有31个桃花品种。在过去的30年里，日本桃花品种的筛选及育种相当成功。例如，从菊花桃的芽变中筛选出的红花菊花型品种'京舞子'（Yoshida *et al.* 2000），和用菊花桃与寿星桃杂交得到的白花菊花型寿星桃品种'幸白'（前田克夫等 2005）；以及用单瓣紫叶桃和寿星桃杂交得到的紫叶寿星桃品种'赤叶寿星'（Yoshida 1974）和一系列花色各异的帚型桃品种，如'照手红'、'照手桃'、'照手白'（Yamazaki *et al.* 1987）和'照手姬'（Horikoshi *et al.* 1992）等先后育成并发表。

（三）美国

近二十年来，美国在帚型桃和紫叶垂枝桃方面的育种也取得了相当进展。'Corinthian Mauve'、'Corinthian White'、'Corinthian Rose'和'Corinthian Pink'（Werner *et al.* 1985, 2000a,b,c, 2001）为4个帚桃类品种，其中'Corinthian Rose'和'Corinthian Pink'均为紫叶帚桃品种；'Crimson Cascade'和'Pink Cascade'（Moore *et al.* 1993）为2个紫叶垂枝桃品种，这些品种的育成为丰富观赏桃花类型，延长桃花观赏期均有重要的作用。

（四）乌克兰

乌克兰尼基塔植物园在以往果桃收集的基础上，1958～1959年间，发现从中国南京植物园得到的种子播种实生苗中出现了重瓣观赏桃植株，并由此开始了观赏桃花的研究。经过三代人的努力，特别是在桃花的远缘杂交及抗病害品种筛选和培育工作效果显著。先后培育出复瓣红花的'Ogon Prometeya'、复瓣粉红色花的'Assol'等抗病性较好的桃花品种（Komar-Tyomnaya 1998）；以及桃（*Prunus persica*）与光核桃（*P. mira*）的杂交品种'Tcarevna-Lebed'（复瓣白花）、'Lyubava'（复瓣粉红色花），'Vesna'（复瓣淡粉色花）等；桃（*P. persica*）与扁桃（*P. amygdalus*）的杂交品种'Feya Zari'（复瓣粉红色花）等抗病性较好的桃花新类型（Komar-Tyomnaya 2007, Komar-Tyomnaya 1999），丰富了桃花品种种质资源。

山桃花枝

3 桃及其近缘种的形态特征

TAXONOMY AND PEACH RELATIVE SPECIES

Prunus persica (L.) Batsch.发表于1801年。

这一名称在西方国家的园艺学领域广为接受。

英国皇家园艺学会及美国园艺学会均采用该名称作为桃的学名。

鉴于该名称完全符合《国际植物命名法规》，

因此，本书也采用这个名称进行观赏桃的以下分析。

山桃、甘肃桃、新疆桃和光核桃是桃最为重要的4个近缘种。

一、桃的分类学界定

从种的级别来看，桃的命名一直问题重重。不同植物学家有着不同的分类学见解。

Amygdalus persica L.是赋予桃的第一个学名，林奈于1753年发表。米勒于1768用*Persica vulgaris*重新命名（Wu & Raven 2003）。

最初的植物学家将*Prunus*（广义的）分为*Amygdalus*，*Cerasus*，*Prunus*（狭义的）和*Padus* 4个属。后来，林奈将其减为2个属（1754）：*Amygdalus* 和 *Prunus*。德坎道尔（De Candolle 1825）则认为应该有5个属：*Amygdalus*，*Persica*，*Prunus*，*Armeniaca*和*Cerasus*，最后一个属包括*Padus*和*Laurocerasus*。

作为另外一种意见，本萨姆和虎克（Bentham & Hook 1865）把*Prunus*（广义的）看作单独的属，并将其分为7个部分：*Amygdalus*，*Armeniaca*，*Prunus*（狭义的），*Cerasus*，*Laurocerasus*，*Ceraseidos*和*Amygdalopsis*。科隆纳（Koehne 1893）则认为这些部分应该为亚属。

根据《中国植物志》（Wu & Raven 2003）的观点，*Amygdalus* [和其他5个属*Armeniaca*，*Cerasus*，*Laurocerasus*，*Padus*和*Prunus*（狭义的）一起] 经常被看作是亚属或位于*Prunus*（广义的）之下。在李亚科里，3个和樱桃相关的属（*Cerasus*，*Laurocerasus*和*Padus*），相对于其他几个属来讲关系更为密切，而大多数分类学家认为*Prunus*是一个独立的属。最近，基于分子生物学数据的系统发育学研究表明，这3个和樱桃相关的属并没有任何单独起源的证据，因而将它们分离出来，看起来并不合理（Bortiri *et al.* 2001）。至于*Amygdalus*和*Prunus*的区别，*Amygdalus*的花芽在两侧，中间为叶芽；幼叶对折在芽中。而*Prunus*有单生腋芽，幼叶包裹在芽中。二者之间并没有根本性的区别。所以采用*Prunus*（广义的）属，而不将其分成*Amygdalus* 和*Prunus*（狭义的）两个亚属显然更加合理。

Prunus persica (L.)Batsch.发表于1801年（Batsch）。这一名称在西方国家的园艺学领域广为接受（Bean 1950；Fernald 1987；Krussmann 1986；Rehder 1927）。英国皇家园艺学会（Huxley&Griffiths，1999）及美国园艺学会（Brickell & Zuk 1997）均采用该名称作为桃的学名。鉴于该名称完全符合《国际植

图3.1 山桃的冬芽无毛
Fig.3.1 Glabrous winter buds of *Prunus davidiana*

图3.2 桃的冬芽有毛
Fig.3.2 Pubescent winter buds of *Prunus persica*

物命名法规》(Geruter *et al.* 2000),因此,本书全部采用这个名称用于以下的分析。桃的近缘种也均以*Prunus*作为属名加以讨论。

二、桃的近缘种的分类检索表

以成熟时的肉质果肉不开裂为确定桃属(*Prunus*)近缘种的特征,山桃、甘肃桃、新疆桃和光核桃是桃最为重要的4个近缘种。目前在中国还能找到桃的原生种和这4个野生近缘种的成片林地,它们是青藏高原的光核桃、黄河上游的甘肃桃和山桃、华北的山桃及新疆的新疆桃,体现出桃近缘野生种的多样性。多类型的桃原生种和近缘野生种的普遍存在,也充分说明中国是桃的栽培起源中心。

表3.1 桃及其近缘种的分类检索表

(参考俞德浚1984年版《落叶果树分类学》,稍有修改)

Table3.1 Key to the peach and its related species

1. 侧脉直出至叶缘,呈弧形上升,不结合成网状;核面有
   平行沟纹,无点纹 ·· **新疆桃**
1. 侧脉在近叶缘结合呈网状
  2. 核表面平滑有浅沟,叶片椭圆披针形,锯齿圆钝 ·········· **光核桃**
  2. 核表面有深沟和孔纹
    3. 萼筒和萼片外被绒毛
      4. 冬芽外被密毛;叶片椭圆披针形或倒披针形,边缘锯齿
        较密;花柱与雄蕊等长或稍短;核表面有沟纹和点纹 ············· **桃**
      4. 冬芽无毛,叶片卵圆披针形,边缘锯齿较稀;花柱长于
        雄蕊;核表面有沟纹,无点纹 ·· **甘肃桃**
    3. 萼筒和萼片外面无毛;核面有点纹,无沟纹;核球形 ·········· **山桃**
(图3.1、图3.2)。

三、桃及近缘种的特征

(一) 桃 [*Prunus persica* (L.) Batsch.]

桃是二倍体(2n=16)植株,高度可达8m。小枝光滑无毛,

有光泽，红褐色、黄褐色或绿色。冬芽圆锥形，外被短柔毛，常2～3簇生，中间为叶芽，两侧为花芽。叶互生，椭圆披针形；叶长8～15cm，宽2～3.5cm，先端渐尖，叶基广楔形；最宽部位在叶中部偏上；叶缘有粗锯齿；叶柄长1～2cm，常具有腺点。花为子房上位周位花，先于叶开放，花径1.0～6.0cm不等。花梗较短，近基生。花萼片卵圆形或三角形卵状，外被短柔毛。果肉厚多汁，表面被柔毛，同时也有表皮光滑无毛的油桃类型和果实扁平的蟠桃类型。

桃在中国北方地区通常3月中下旬萌动，4月开花。

(二) 山桃 [*Prunus davidiana* (Carr.) Franch.]

山桃原产中国西北地区，树高可达10m以上。抗寒耐旱，较耐盐碱土壤。

山桃与桃最大的区别是在树皮和冬芽。山桃树皮光滑，冬芽无毛，而桃则树皮粗糙，冬芽有毛。山桃叶片光滑，最宽部位接近叶基部。叶基部广楔形，边缘具细锐锯齿，萼筒和萼外面无毛。果核近球形，核面有点纹，无沟纹。

山桃耐寒性好，在中国北方地区多作为桃的砧木使用，但却很少在其他地区有此类应用。

山桃花期早，3月上旬萌动，3月下旬或4月初开花，本身也可以作为观赏应用。华北地区的山桃除了常见的粉色花之外，还有白花、红花两个变种，在枝型上也有所变化，像垂枝型（图3.3）和帚型（图3.4）等变异丰富的栽培品种也已经开始用于庭院栽植。根据2009年春天的观测，山桃的单株花期达到11天，这和当年适宜的温度和湿度有密切关系。此外，山桃古铜红色的树皮（图3.5）和铜红色的秋叶（图3.6）也增添了其观赏性。

山桃对缩叶病、白粉病（Pisani and Rpselli 1983）、蚜虫均有一定抗性。山桃很容易和桃进行杂交。山桃与桃的杂交种在增强对PPV、白粉病以及缩叶病等的抗性方面都有一定效果（Moing *et al.* 2003）；对桃蚜也有一定抗性（Massonie 1979；Massonie *et al.* 1982）。

图3.3 垂枝山桃
Fig.3.3 Weeping form of *Prunus davidiana*

图3.5 山桃树皮
Fig.3.5 Beautiful bark of *Prunus davidiana*

图3.4 曲枝帚型山桃
Fig.3.4 Twisted columnar form of *Prunus davidiana*

图3.6 山桃秋色叶
Fig.3.6 Autumn leaf of *Prunus davidiana*

（三）甘肃桃 (*Prunus kansuensis* Rehd.)

甘肃桃的大部分特征和桃很接近，最明显的区别是甘肃桃冬芽光滑，果核表面具有沟纹，无点纹。

甘肃桃主要产于陕西、甘肃，多生于海拔1000～2300m的山地。高可达3～7m。小枝细长，无毛，具不明显小皮孔；冬芽卵形至长卵形，无毛。叶卵状披针形或披针形，长5～12cm，宽1.5～3.5cm，先端渐尖，基部宽楔形，叶上面无毛，叶背近基部中脉有柔毛；叶最宽处在叶片的中部；叶缘具稀疏细锯齿，叶柄长0.5～1cm，常无腺体。花单生，先叶开放，花径2～3cm，萼筒钟形，外几无毛；萼片卵形至卵状长圆形；花白色或浅粉红色，花瓣近圆形，基部渐狭成爪；花柱长于雄蕊，与花瓣近等长。果实近球形，直径约2cm；核近球形，扁平，顶端圆钝，表面具浅沟纹，无孔穴（俞德浚 1979）。

甘肃桃通常3月上中旬萌动，3月下旬至4月初开花，因此被认为抗早霜（Meader and Blake 1939）。甘肃桃是一种抗旱、耐寒、抗线虫、利用价值很高的种质资源（汪祖华和庄恩及 2001）。在北京地区表现良好，耐旱抗病，但并没有得到充分应用。

甘肃桃和桃也很容易杂交，而且生长势旺盛，具有二者的中间性状（Meader and Blake 1939; Grasselly 1974）。

（四）新疆桃 [*Prunus ferganensis* (Kost. and Rjab) Kov. and Kost.]

原产于中国西部的野生类型。有些分类学观点将新疆桃归为桃的亚种，区别在于平行叶脉直至叶缘（图3.7），果核上有较深的平行纹路，无点纹。

新疆桃在原产地高度可达8m；小枝无毛，有光泽，具多数皮孔；冬芽2～3个簇生于叶腋，被短柔毛。叶片披针形，长7～15cm，宽2～3cm，先端渐尖，基部宽楔形至圆形；叶缘锯齿顶端有小腺体，叶侧脉弧形上升，直达叶缘，但彼此并不像桃一样结合成网脉。花瓣近圆形至长圆

图3.7 叶脉示意图
新疆桃（上）叶脉直至叶缘，下为桃叶脉
Fig.3.7 Comparison of the leaf veins of *Prunus ferganensis* (U) and *Prunus persica* (D)

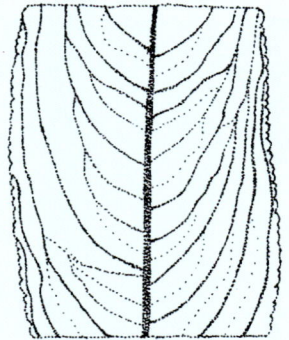

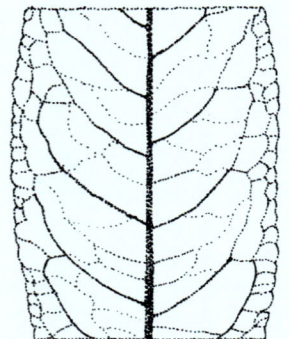

形；花淡粉色，花径3～4cm；萼片外被短柔毛；雄蕊多数，与雌蕊近等长。果实长3.5～6cm；果核扁球形，表面具沟纹，无点纹（俞德浚 1979）。花期3～4月。

新疆桃和桃的杂交种通常不育（Komar-Tyomnaya 2007；Komar-Tyomnaya 2008）。

新疆桃在北京地区表现良好，但目前仅作资源保存，并无特别品种可作观赏。

（五）光核桃（*Prunus mira* Koehne）

大片的野生光核桃林发现于20世纪80年代，在雅鲁藏布江、金沙江、澜沧江、怒江流域的海拔1700～4200m处的山坡杂木林和山谷沟边均有分布（宗学普和段玉春 1987），约有30万株，其中尤以雅鲁藏布江下游及支流尼洋河和帕龙藏布江最为集中（董国正 1991）（图3.8）。在印度一些地区作为桃的砧木（Scorza and Okie 1991）。

光核桃分布的生态条件年均温为6～13℃，绝对最低温为－15.3℃，绝对最高温为30.2℃，无霜期为160～280天，年降水量为500～700mm。冬暖夏凉为其主要气候特征。光核桃生境土壤pH值介于6.5～7.6之间（董国正 1991）。

图3.8 原生境中的光核桃（沈向）
Fig.3.8 *Prunus mira* in its natural habitat (Photograph courtesy Xiang Shen)

光核桃枝条细长，无毛，嫩枝绿色，老时灰褐色，具紫褐色小皮孔。叶披针形或卵状披针形，长5～11cm，宽1.5～4cm，先端渐尖，基部圆形；叶背沿中脉基部有绒毛；叶缘具圆钝浅锯齿，近顶端处全缘，齿端常具小腺体；叶柄无毛，常具紫红色扁平腺体4～6个。花单生，先于叶开放，粉色或白色，直径2.2～3cm；雄蕊多数，明显比花瓣短；花梗长0.1～0.3cm；萼片长圆筒形，紫褐色。果实近球形，直径约3cm；核扁卵圆形，长约2cm，表面光滑，仅具少数不明显浅沟纹（俞德浚1979）。花期3～4月。

光核桃以其核面光滑的特点与桃明显不同。表3.2列举了光核桃与桃的主要具体区别点（修改自汪祖华和庄恩及的《中国果树志·桃卷》，2001）。

光核桃根系发达，生长迅速。7年生树根水平分布3.5m，主根深1.5m，树高4～8m，胸径17.2cm，干周0.72m，冠幅大于3m；30年生树高9～10m，干周3m，冠幅4.5～5m；百年大树干周7～8m（董国正1991）；千年大树干周则高达10m以上（汪祖华和庄恩及2001）。

光核桃具有耐旱、耐寒、耐瘠薄、抗病、寿命长等特点，是培育优良抗性桃花新品种的良好原始材料。乌克兰曾经培育出光核桃的重瓣观赏品种，如'Lel'（复瓣，粉花）（Komar-Tyomnaya 2008），具有良好抗病性（Komar-Tyomnaya 1998）。

光核桃在北京地区播种能够成活，但由于夏季高温高湿的气候，导致光核桃生长不良。

表3.2 光核桃与桃的主要区别

Table 3.2   The major characteristics difference between *Prunus persica* and *Prunus mira*

	树龄	树体	叶色	叶腺	花器	果核	子叶
桃	短，15～30年	矮小，开展或垂枝，1～5m	绿色或红色，大而薄	肾形或圆形，少（2～4个）	各色，花大，花期迟	大，深沟间点纹	不出土
光核桃	长，百年以上	高大，直立7～20m	绿色，小而厚	肾形，多（4～6个）	粉白色，花小，花期早	小，光滑，浅沟，深沟型	出土

4 桃花的生物学特性

BIOLOGICAL AND PHENOLOGICAL
CHARACTERISTICS

生物学特性主要包括枝、芽、花、叶、果等的基本形态
特征，
还包括生长习性，生长发育特点，
开花、展叶、结实等各个时期的物候也是生物学特性的
主要方面。

二色桃变异

一、桃花的基本形态特征

1. 树型

桃花具有丰富的枝型，除通常的直枝类型（枝条斜出，节间较长的小乔木），还有枝条自然弯曲的龙游型；小枝下垂、伞形树冠的垂枝桃；以及节间紧密、树型低矮的寿星桃；枝条直上、树冠窄高的帚型桃。

2. 枝条

桃花的1年生枝条通常有绿色、紫红色，或绿色上带红色条纹或斑点，和花色有一定相关性。唯一的例外是'绯桃'花色鲜红，但小枝绿色。

桃成枝力较强，且分枝角度常较大，故干性弱，层性不明显，中心主干易早期自然消失。

3. 芽

桃的芽分为单芽和复合芽，单芽有单花芽（图4.1）也有单叶芽。复合芽中以3芽并生的复芽最为常见，通常中间的为叶芽，两边的为花芽（图4.2），也有或双花芽、双叶芽、或1花芽1叶芽（图4.3）等现象，桃花中还有多个花芽并生在一起的现象（图4.4，图4.5），顶芽为叶芽。桃花的芽具有明显的异质性，即芽的大小和饱满程度有明显差别。在春秋梢交界处还会出现盲节。

4. 花

桃花花色丰富，通常以果桃所具有的粉色最为原始，其次为红色或白色，红—白或红—粉或红、白、粉等不同复色类型出现得较晚；其中粉、红又有深浅不同的多种变化，杂色也有色斑、条纹之分。

桃花的花瓣数及花瓣型具有明显的差异，由原始的5瓣（单瓣），发展到10～25瓣（复瓣），更有很多品种在花萼及花丝瓣化之后达到40～50瓣，甚至70～80瓣之多。

花型有铃型、单瓣型、梅花型、月季型、菊花型、牡丹型，从平展到圆球，从疏散到紧密，各不相同，构成了丰富的多样性。

图4.1　单花芽
Fig.4.1　One flower bud

图4.2　3芽并生复合芽
Fig.4.2　Two lateral flower buds and one vegetative bud in the middle

桃花品种中雄蕊相对于花瓣的长度，以及雌蕊相对于雄蕊的高度均存在着差异；花萼颜色、雄蕊数目以及瓣化程度均可作为区分品种的重要特征。

5. 叶

桃花在叶色上也有绿色和紫色之分，更有绿叶紫斑的品种类型。桃花的叶缘有波状和平整之分，叶基、叶尖品种之间也有所差异。

6. 果、核

果实并非桃花的主要观赏对象，如果可能，应该尽量减少果实对桃花观赏特征的削弱。但作为品种特征，果实的形状、是否有毛，以及果核核面的沟纹分布情况等，还是可以作为区分品种的特征加以考虑。

7. 根的特性

根系发达，属浅根性，须根尤多；吸收根分布深度通常在土层下50～60cm之内，水平分布明显大于垂直分布。生长迅速，伤后恢复能力强，移栽易于成活。

图4.3　双芽
Fig.4.3　One flower bud is present besides the vegetative bud

图4.4　多花芽
Fig.4.4　Three flower buds together

图4.5　多花芽
Fig.4.5　Three flower buds and one vegetative bud together

二、桃花生长习性和发育特点

桃花原产中国西北高原地区，喜光，喜沙质壤土，耐瘠薄，不耐水湿，适宜栽植在光照充足、排水良好的坡地。桃花较耐寒，华北地区一般多可露地越冬。幼树抗寒力弱，苗期有可能出现冻梢。

桃花先花后叶，花芽分化大多集中在7～8月，属于夏秋分化型。花芽分化分为生理分化期、形态分化期、休眠期和性细胞形成期4个阶段。大部分桃花品种是在花芽形成柱头和子房后的夏秋即进入休眠期。甘肃桃则是在完成了性细胞分化，即形成胚珠原基后越冬（王珂2005）。

桃花花芽分化属于向心式（江雪飞2003）。花芽的形态分化期分为：花芽分化始期、萼片分化期、花瓣分化期、雄蕊分化期和雌蕊分化期（图4.6）。

桃花的叶芽萌发早于花芽，但抽梢在开花之后，1年能抽发数次副梢。叶芽萌动所需的温度要比花芽低，所以常常在大蕾期或初花期时，温度较高的年份就会出现花叶同放的现象。

桃花大多能自花结实，果实在6～9月陆续成熟。通常误认为观赏用的桃花不结果，首先是由于桃花难以形成与其花量相对应的果质优良的果实；其次，由于桃花自身的特点，在果实形成过程中会出现落果高峰（张秀英2001）；此外，桃花由于存在多柱头花，或者本身柱头发育不完全等原因，导致结实性差，出现多果（图4.7）或完全不能结实等现象。

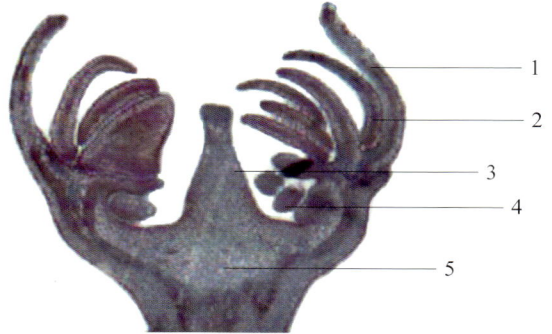

图4.6 '绛桃'的花芽分化
1.花萼；2.花瓣；3.雌蕊；4.雄蕊；5.花托
Fig.4.6　The initiation and early stages of flower bud of 'Jiang Tao'
1. calyx; 2. petal; 3. pistil; 4. stamens; 5. sepal

图4.7 桃花的多果现象
Fig.4.7　Triple fruits phenomena of the ornamental peach

桃花的自然落叶期在11月中旬。桃花在落叶后进入深度休眠，这种休眠状态必须经过一个低温阶段才能打破。需冷量是休眠期能保证花芽正常生长所需要达到的最小低温积温，Weinberger (1950) 用低于7.2℃低温的累积来表示不同品种通过这个低温阶段所需的时间。冷量不足会造成花芽、叶芽发育异常，延迟或导致断续性发叶，花蕾脱落，开花不整齐，花色不艳丽等现象（Giesberger 1972）。通常实践中也可以利用标准品种的花期来指示未知品种的需冷量，既简单又经济（Scorza and Sherman 1996）。

王力荣等（2003）按照0℃～7.2℃需冷量模式，以秋季日平均温度稳定，通过7.2℃的日期为有效低温累积的起点，以打破休眠所需0℃～7.2℃的累积低温值为品种的需冷量计算，我国重瓣桃花品种的需冷量分布范围从400h到1250h，分布范围广泛，并以900h以上的品种为主（朱更瑞等2004）。

随着品种需冷量的增加，花芽分化的各个阶段开始的时间随之推迟，萼片分化期进行的时间随之延长，雌蕊分化进行的时间随之缩短，长需冷量桃品种由于花芽分化开始得较晚，尤其是到雌蕊分化阶段，代谢物质的供应和分化的时间都将受到限制，从而影响花芽分化的质量（孙旭武 2003）。这也可能是造成晚花品种结实性差的原因之一。

大多数晚花品种，即高需冷量的品种，在冷量不足的地区无法正常开花结实的原因，一方面是由于冷量不够，但高夜温也可能是造成这种结果的原因之一（Edwards 1987；Rouse and Sherman 2002）。

三、桃花的观赏期物候

桃花最主要的观赏期即为花的观赏期。主要包括（根据张秀英，2001略作修改）：

花芽膨大期：鳞片开始松散，花芽膨大，以全树25%为准；

露萼期（图4.8）：鳞片裂开，露出内部萼片；

露瓣期（图4.9）：花萼绽开，花瓣露出，在花蕾露瓣达到80%以上时即进入"大蕾期"（图4.10），标志着最佳观赏期的开始（刘佳棽等2000）；

初花期（图4.11）：全树有5%～25%的花开放；

图4.8 露萼期'照手姬'
Fig.4.8 Calyx stage

图4.9 露瓣期'粉花山碧'桃
Fig.4.9 Pink stage

图4.10 大蕾期'绛桃'
Fig.4.10 Balloon stage

图4.11 初花期
Fig.4.11 Early bloom stage

图4.12 盛花期
Fig.4.12 Full bloom stage

图4.13 落花期
Fig.4.13 Late bloom stage

盛花期（图4.12）：在达到50%～75%的花开放时进入盛花期，为桃花的最佳观赏期；

落花期（图4.13）：全树有5%的花正常脱落花瓣为落花始期，95%的花瓣脱落为落花终期。

各个时期持续的时间会由于温度变化及不同品种之间自身的差异存在着一定差别。

桃花的观赏期一般从"大蕾期"开始，在盛花期达到最佳观赏期的高峰，但在落花终期之前均可观赏。有些品种，例如'云龙'桃，花瓣落后突显出来的旋转的雄蕊仍有很好的观赏价值（图4.14）。正常条件下，一个品种的观赏期可以达到10～12天左右，视当时、当地的温度、风力等外界因素的影响而不同。

图4.14 '云龙'桃的雄蕊
Fig.4.14 Stamens of 'Unriumomo'

粉花山碧桃

5 桃花品种分类系统研究

SYSTEMATIC STUDIES ON THE
ORNAMENTAL PEACH

桃花因其出众的观赏特性及新品种的不断涌现，已经广泛栽植于亚洲、欧洲、美洲和大洋洲。迄今有所记载的文献中有267个桃花名称。不同国家，不同地理区域，有关桃花的各种分类解释，以及种名的更迭，导致了各种不同的命名。一些品种的合理性和特征描述很值得怀疑。

桃花品种的形态特征以及准确的性状特征描述是品种分类研究的根本。规范形态特征描述，建立世界范围内统一的桃花品种特征描述项和等级标准，对于桃花品种在世界范围内的培育、研究和应用有着非常重要的意义。

桃花品种分类系统研究起源于20世纪80年代，最早的分类研究源于最基本的形态学调查和分析，随后，孢粉学、生物化学、数量分类学以及分子生物学等技术的引入，将桃花品种系统研究进一步深化。以种性为基础，以树型、花型等为主要分类依据的桃花分类系统包括直枝桃、寿星桃、帚桃、垂枝桃、曲枝桃和山碧桃等6个品种群。

一、桃花品种的形态特征及性状描述

（一）现有桃花品种性状描述及命名中存在的问题

桃花在约300年前开始应用于西方园林中（Everett 1967）之前，就已经在中国有了3000多年的栽培历史。现在，因其出众的观赏特性及新品种的不断涌现，桃花已经广泛栽植于亚洲、欧洲、美洲和大洋洲。然而，桃花的命名和在文献中的分类一直都很矛盾和混乱。迄今有所记载的文献中有267个桃花名称（胡东燕 2004）。不同国家，不同地理区域，有关桃花的各种分类解释，以及种名的更迭，导致了各种不同的命名。一些品种的合理性和特征描述很值得怀疑。其中一些桃花品种在文献中的名称相同，但却并没有相似的形态特征描述（Krussmann 1986；Moore 1993），例如，有两个品种的名称都为'Crimson Cascade'，但其中一个是绿叶，另一个则是紫叶。另一种情况，尽管一些品种有着不同的名称，但形态特征却几乎完全相同（Jacobson 1996），例如，'Cardinal'，'Magnifica'和'Russel's Red'全都具有重瓣红色花。而且不同国家对相同特征的品种也可能有着各自不同的名称，例如，具有菊花状花瓣的粉花品种在美国叫'Chrysanthemum'，在日本叫'Kikumomo'，在中国则直接称为'菊花'桃，尽管所有这些名称在各自的语言里都是"菊花"的意思，而文献记载中却出现了完全不同的3个名称。

桃花品种的形态特征以及准确的性状特征描述，是品种分类研究的根本。在不能完整描述品种特征的前提下，就无法准确确定一个品种的真实性和可靠性。在以往文献中，众多文章对品种的描述仅限于"粉色、大花"或"大花，白色"等最基本的特征，根本不能确定该品种区别于已知品种的特异性，因此规范形态特征描述，建立世界范围内统一的桃花品种特征描述项和等级标准，对于桃花品种在世界范围内的培育、研究和应用有着非常重要的意义。

（二）制定性状特征描述项和标准时应该遵循的一些原则

1. 对色彩的描述

对颜色的描述，不同人对花色会有不同的认识，各国均有各自描绘颜色的形容词，却很难准确定位一个品种公认的色度。因此在制定色彩描述标准时，应采用在国际园艺界公认的色谱来确定花色的规范代码，进行统一描述和比对，以规范颜色的标准。

2. 对大小的描述

无论是树体高度、花径，还是花瓣大小等特征，一旦涉及到大小规格，在没有任何参照的前提下并无实际意义。因此制定统一的级次标准和标准品种，对于准确定义品种特征非常重要。

3. 对于可能会因为环境因素造成影响的数量性状的描述

例如树高、叶子大小等特性，会随环境的变化而变化。因此针对不同部位，基于常年观测得出的相对稳定的数据，才可以作为判定一个品种大小的标准和依据。

4. 对于质量性状的分级标准进行图解描述

例如花蕾形状、花瓣形状、花型、树型等，作为桃花品种性状中最为关键的观赏要素，客观性的文字描述与图解结合更为直观，便于实际应用。

(三) 桃花的主要形态特征及性状描述

桃花的花型、花色、花径、花期早晚、雌蕊发育状况等和观赏有关的生殖器官特性相对于营养器官来言更为保守，在品种内更稳定，既是桃花最主要的观赏要素，也是桃花品种特征描述的重要指标和系统分类与品种识别鉴定的主要依据。

1. 花

花蕾形状：桃花的花蕾主要分为以下5类（图5.1）：长卵形、卵形、阔卵形、扁球形、圆球形。

花瓣的重瓣性：桃花中最为原始的单瓣通常为5瓣，个别花朵会出现6瓣、7瓣等不规则现象。在桃花重瓣性的划分中，以花瓣少于10

图5.1 花蕾形状
长卵形、卵形、阔卵形、扁球形、圆球形（顺序从左到右）
Fig.5.1 Shape of bud: oblong, ovate, oval, oblate-round and round (from left to right)

瓣为单瓣；花瓣10～40瓣，两层以上排列的为复瓣；40瓣以上的则称为重瓣。桃花的重瓣性主要是由花丝、雌蕊及萼片的瓣化造成的。

花型：桃花的花型变化丰富，主要分为以下6类（图5.2）：单瓣型、梅花型、月季型、菊花型、牡丹型、铃型。小花铃型花瓣内抱，雄蕊通常在大蕾期会伸出花瓣。普通桃的单瓣型花瓣平展宽大，花近盘状。同为复瓣类型的梅花型和月季型则主要区别在花瓣数和花瓣形状上，前者花瓣数在15～25瓣左右，花型平展，花瓣平整，3～5层规则排列；后者则有25～40瓣左右，外层花瓣外翻，内层花瓣内抱，多层排列，花瓣重叠。复瓣类型中还有花瓣细碎狭长的菊花型，花瓣通常在30瓣左右。重瓣类型包括花朵呈半球形或球形的牡丹型，花瓣大多在40瓣以上，多可达60～80瓣。

花瓣形状：桃花的花瓣和花型有一定的相关性，主要分为以下5种类型（图5.3）：匙形、披针形、椭圆形、阔椭圆形、近圆形。匙形花瓣为铃型小花的基本花瓣形式；卵圆形花瓣则多为单瓣型花的基本花瓣形；菊花型的花瓣一般为披针卵形，在牡丹型和月季型中的内轮花瓣中也有类似瓣形或由于花丝瓣化而来的表现得更为线形的花瓣类型；长卵形和卵圆形分别为月季型和梅花型的基本瓣形，与花瓣数量多少也有一定的相关性，通常花瓣越少，花瓣更为宽大，而相对随着花瓣的增多，花瓣也随之变窄，即与梅花型→月季型→牡丹型花瓣型的变化，相对应地呈现出花瓣从卵圆形→卵形→长卵形大小递减的趋势。

图5.2　花型类型
单瓣型、梅花型、月季型、菊花型、牡丹型、铃型（顺序从左到右）
Fig.5.2　Flower Type: single, mei-flower, rose,chrysanthemum, peony and bell-shaped (from left to right)

图5.3　花瓣形状
匙形、披针形、椭圆形、阔椭圆形、近圆形（顺序从左到右）
Fig.5.3　Shape of petal: spatulate, lanceolate, elliptic, oval, nearly round (from left to right)

图5.4 花色
白色、粉色、粉红色、粉紫色、红色
Fig.5.4 Single color white, pink, dark pink, purple pink, and red

花蕾、花瓣颜色：桃花花色丰富（图5.4），通常以果桃所具有的粉色最为原始，其次为红色或白色，红、白或红、粉或红、白、粉等不同复色类型出现得较晚；在这其中，粉、红又有深浅不同的多种变化，杂色也有色斑、条纹之分。花蕾颜色之于花朵则会相对较深。对于花色的描述以英国皇家园艺学会（RHS）的色卡为准，尽可能地减少不同国家、不同地区对颜色描述中可能出现的偏差。

花萼颜色：桃花的花萼与花色有一定的相关性，一般来说，粉色花和红色花多为红褐色萼片，白色花则为绿色萼片。花瓣为白粉、淡粉色与红色，或者白色与红色相间的品种，其花萼也表现为红褐色与绿色相间（图5.5）。

图5.5 萼片和花的相关性
'洒红' 桃萼片和花的相关性；'鸳鸯垂枝'复色花萼片二色（左）
Fig.5.5 Color relationship between flower and calyx 'Sahong Tao' 'Yuanyang Chuizhi'

图5.6　'合欢二色'桃萼片5枚
Fig.5.6　5 calyx of 'Hehuan Erse Tao'

图5.7　'人面'桃萼片2轮10枚
Fig.5.7　10 calyx of 'Renmian Tao'

花萼数目：桃花的花萼为卵状三角形，单瓣型品种萼片通常为5枚，但复瓣花中也有5枚萼片的品种，如：'合欢二色'桃（图5.6）；复瓣花和重瓣花品种大多为两轮萼片10枚，通常内轮较为窄小（图5.7）；重瓣程度高的品种会出现萼片瓣化，萼片成不规则数目（图5.8）。

雌蕊：桃花一般有4种类型的雌蕊形态，一种为雌蕊分化不完全，或完全无法形成正常的雌蕊形态，如'白花山碧'桃（图5.9）；另一类则是雌蕊明显高于雄蕊，如'五宝'桃的柱头已经伸出花蕾（图5.10）；此外桃花中多雌蕊花的现象很普遍，如'晚白'桃（图5.11），往往会出现一花多果。而桃花中大部分品种多为雌蕊与雄蕊近等高。雌蕊的有无以及相对于雄蕊或花瓣的高度对桃花的结实性有直接影响，是桃花分类中一个重要特征。

雄蕊：桃花的花药分为白色、黄色和橘红色3种；白色花药为败育类型，黄色花药为白色花，橘红色花药多为粉色花和红色花。桃花的花丝与花色有一定的相关性，白色花的花丝为白色，授粉前后无明显变化；粉色花的花丝通常为粉白色，

图5.8　'簪粉'萼片瓣化
Fig.5.8　Irregular number of calyx of 'Zan Fen'

图5.10　'五宝'桃雌蕊高于花瓣
Fig.5.10　Pistil longer than buds 'Wubao Tao'

图5.11　'晚白'桃多柱头
Fig.5.11　Polycarpous 'Wanbai Tao'

图5.9　'白花山碧'桃无雌蕊
Fig.5.9　'Baihua Shanbitao' without pistil

红色花的花丝通常为水红色，粉色花和红色花在授粉后花丝颜色变深，花心也会随之呈粉红色或深红色。花丝的长度通常分为3类，小花类品种的花丝伸出花瓣，'白花山碧'桃等品种的花丝与花瓣近等长，而大部分品种的花丝明显短于花瓣。

花径大小：桃花中除小花类的品种花朵直径在2.0cm左右外，大部分品种的花朵直径在3.5cm以上。一些重瓣花品种的直径可达5～8cm。花朵直径还与气候条件以及栽植条件有关，评价一个品种花朵大小不能仅凭1朵花、几朵花，1年或2年的极端现象来判断，需要连续数年观测，才能得出相对稳定的数据，从而作为判定该品种花径大小的标准和依据。

开花习性：桃花的花期是判断品种最为显著和直观的特征。按照花期早晚，一般分为3个级别，早花，中花和晚花。最早开花的类型是山桃与桃的杂交品种，一般会在3月下旬或4月初始花，一直持续到4月中旬。晚花品种的始花期大多在4月20日以后，花期可以一直延续到5月上旬。而开花最为繁盛，品种最多的为中花品种，即观赏桃花的主要观赏期为4月中旬至4月底之前（以上时间均以北京地区为准，其他各地相应有所变化，但各个品种开花的顺序保持不变。第六章品种介绍中除做特殊说明外，花期均指北京地区）。同样，花期的观测也需要连续进行几年，由于温度、湿度、移栽所造成的花期变化，不能够作为判定一个品种花期早晚的标准。

2. 果实和种子

桃花的结实性因品种而异。但根据雌蕊类型，不具备完好雌蕊形态的类型，结实性相当低或根本无法正常结实，是为观赏桃中非常独特的一类。对于能够结实的品种，果实形状，果实表面是否有毛，以及果核表面核纹的模式与多少，都是可以用来区别品种的主要形态特征。

果实形状：主要分为圆形、阔椭圆形、尖圆形、椭圆形、扁平（图5.12）5种类型。

果核形状：桃的果核形状主要有扁平、卵圆形、倒卵圆形和椭圆形（图5.13）4种类型。

相对而言，营养器官特征会随环境的变化而变化，但经观察，桃花品种成年植株的树型、叶色、叶形、叶缘锯齿等性状在品种内也是比较稳定的，因此，营养器官的特征也是进行品种分类的主要依据之一。

图5.12　果实形状
圆形、阔椭圆形、尖
圆形、椭圆形、扁平
（顺序从左到右）
Fig.5.12　Shape of
fruit: round, oval,
conoid, elliptic, and
oblate (from left to
right)

图5.13　果核形状
扁平、卵圆形、倒
卵圆形和椭圆形
（顺序从左到右）
Fig.5.13　Shape of
stone: oblate, oval,
obovate, and elliptic
(from left to right)

3. 枝、茎

枝型：桃花的枝型相当丰富。小枝弯曲的龙游类型是观赏桃花中独特的类型（图5.14）；枝条下垂是垂枝型桃花的特点；枝条直上或斜出是桃花中最为普遍的类型，根据分枝角度的不同有枝条直上、分枝角度狭小的帚桃类型（图5.15），和枝条斜出的直枝类型。

节间长度：是决定树体高度的直接原因，寿星桃（图5.16）是直枝类桃花类型中节间紧密、树体矮小的一个特殊类型，平均节间长度通常小于10mm，仅相当于普通直枝桃节间的1/5到1/4。作为以观赏为主要目的的桃花，单位节间长度内的着花量也可以反映节间的密度。

小枝颜色：和花色也有一定的相关性，通常白色花的小枝为绿色，粉色花或红色花的小枝为紫红色，复色花在小枝颜色上也有一定体现，绿色小枝上通常可以看到斑驳的紫红色斑点和条纹。但也有个别例外，比如'绯桃'，经过多年观测，'绯桃'的小枝色并不是与其红色花相关为紫红色，反倒是始终为绿色。

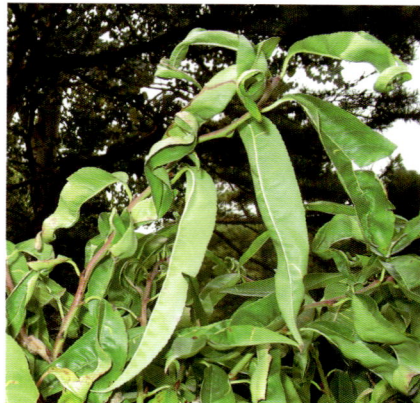

图5.14　小枝自然弯曲的'云龙'桃
Fig.5.14　'Unriumomo'
with twisted branches

图 5.15　枝条下垂的垂枝桃和枝条直上的帚型桃
Fig.5.15　Weeping and columnar (right) growth habits

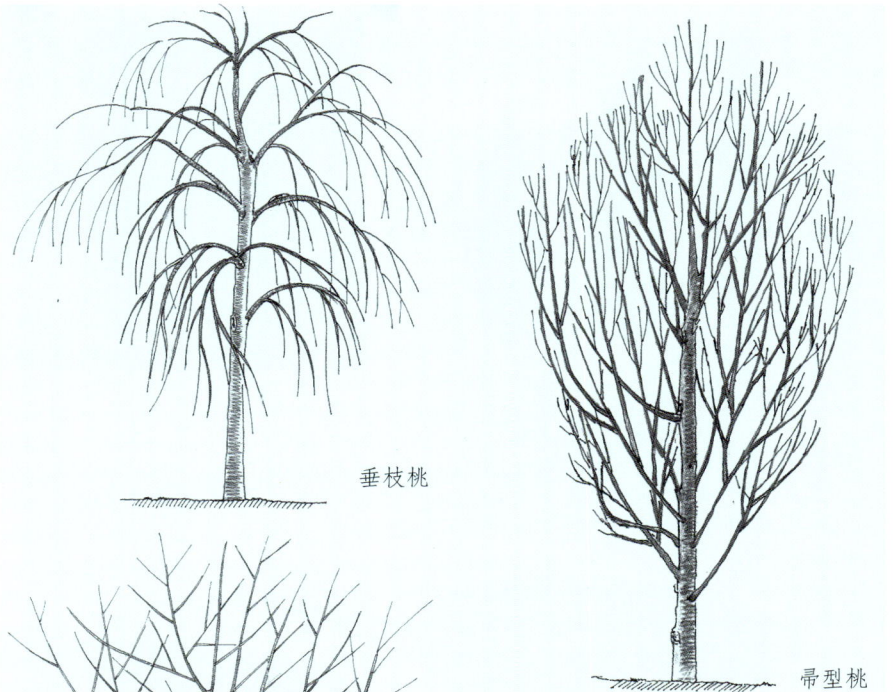

垂枝桃

帚型桃

图 5.16　枝条斜出的直枝桃及节间紧密的寿星桃
Fig.5.16　Standard and dwarf (right) growth habits

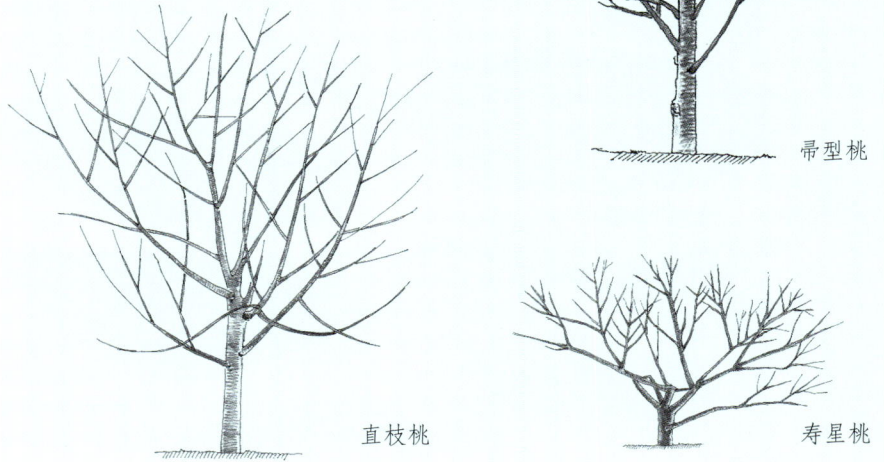

直枝桃

寿星桃

4. 叶

叶形：桃花的叶形主要有以下3个类型（图5.17）：披针形、椭圆披针形、卵圆披针形。

叶色：桃花以绿叶品种居多，也存在着紫红色叶（图5.17），此外还有绿色叶上有红色斑块或斑纹的斑叶类型（图5.18）。

叶缘：桃花的叶缘（图5.19）有细锯齿、粗锯齿和细圆齿3种类型。

叶基：桃花的叶基（图5.20）有锐尖、楔形和广楔形3个类型。

叶尖：桃花的叶尖（图5.21）有渐尖、急尖两种。

图5.17 桃花的叶形
披针形、椭圆披针形、卵圆披针形（顺序从左到右）
Fig.5.17 Leaf shape: lanceolate, elliptic-lanceolate, and ovate-lanceolate (from left to right)

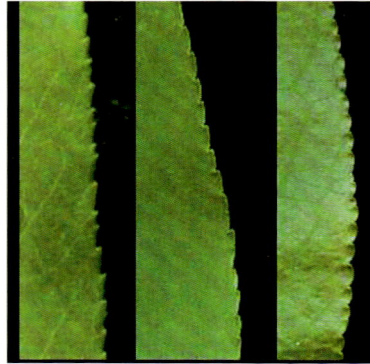

图5.19 桃花的叶缘
细锯齿、粗锯齿和细圆齿（顺序从左到右）
Fig.5.19 Leaf margin: finely serrate, coarsely serrate, and crenulate (from left to right)

图5.18 斑叶桃的叶子
Fig.5.18 Variegated leaves

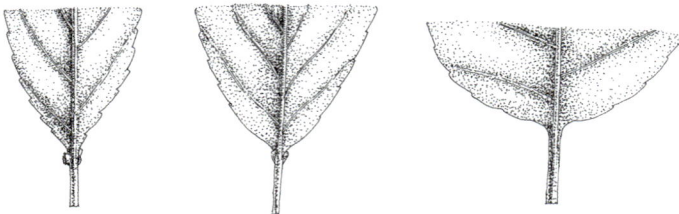

图5.20 桃花的叶基
锐尖、契形、广契形（从左到右）
Fig.5.20 Leaf base: attenuate, cuneate and broad cuneate (from left to right)

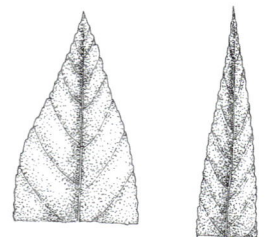

图5.21 桃花的叶尖
渐尖（左）、急尖（右）
Fig.5.21 Leaf apix: acuminate (L) or acute (R)

(四) 桃花形态特征登记表

以上详细介绍了20余个桃花主要形态性状。作为桃花品种的调查、登记，以及新品种登录所必须的基本参考数据，共制定了41个形态特征性状，如表5.1所列。这些性状的确定也参考了美国桃花品种专利登记、品种登录，以及日本桃花形态性状研究等方面的资料，分级标准则是根据实际观测而确定。

分组性状是进行桃花品种识别的主要依据。采用树型、花型、始花期、叶色、果实是否存在等5个性状作为品种分组性状。

质量性状是表现植物不连续变异状态的性状（UPOV 2002），如雌蕊有无、果实是否存在等都属于质量性状。

数量性状是能以一维的、线性等级进行描述的性状，它显示性状从一个极端到另一个极端的连续变化（UPOV 2002），如树体大小、花径大小、花瓣数目等都属于数量性状。

假质量性状的性状表达至少有部分是连续的，但其变化范围是多维的，所有单个表达状态要在性状描述范围内确定（UPOV 2002），如花蕾形状、花瓣形状、花瓣颜色等都属于假质量性状。

表5.1　桃花形态特征登记表

Table 5.1　Morphological characteristics for the ornamental peach cultivar

序号	类型	性状	性状分级
1	QL[Z]	树皮质地	光滑／粗糙
2	QN[Y]	树体	极小／小／中／大／超大
3	PQ	树型*	直立／柱状／矮型／垂枝型／曲枝型／混合型
4	PQ[X]	花枝颜色	绿色／红色／绿色带红斑或条纹
5	QN	节间长度	短／中等／长
6	PQ	花蕾形状	长卵形／卵形／卵圆形／扁球形／球形
7	PQ	花蕾颜色	粉色／红色／白色／粉、白色／粉、红色／粉、白、红色
8	PQ	花型*	单瓣型／梅花型／月季型／牡丹型／菊花型／铃型
9	QN	花径大小	极小／小／中／大／极大

（续表）

序号	类型	性 状	性状分级
10	PQ	花瓣颜色	粉色／红色／白色／粉、白色／粉、红色／粉、白、红色
11	PQ	花瓣形状	匙形／披针形／长卵形／卵形／卵圆形
12	QN	花瓣大小	极小／小／中／大／极大
13	QN	花瓣数目	单瓣／复瓣／重瓣
14	PQ	花萼颜色	绿色／红色／绿色带红斑或条纹
15	QL	花萼瓣化现象	不瓣化／瓣化
16	QL	花萼数目	5／多于5
17	QN	雄蕊相对于花瓣的长度	短于／几等长／长于
18	QN	雄蕊数目	少／中等／多
19	QN	雄蕊瓣化现象	少／中等／多
20	PQ	花药颜色	橘红／黄／白
21	QL	是否有雌蕊	无／有
22	QN	雌蕊相对于雄蕊的长度	短于／几等长／长于
23	QN	雌蕊数目	0／1／多于1
24	QN	花梗长度	短／中等／长
25	QN	始花期*	很早／早／中等／晚
26	QN	花期持续时间	短／中等／长
27	PQ	叶形	狭披针形／长椭圆披针形／卵圆披针形
28	PQ	叶面颜色*	绿色／紫红色／绿色上有紫斑
29	QN	叶长	短／中等／长
30	QN	叶宽	窄／中等／宽
31	PQ	叶面	平滑／皱缩／波状／卷曲
32	PQ	叶缘	细锯齿／粗锯齿／圆齿
33	PQ	叶基	锐尖／楔形／广楔形
34	QN	叶尖	渐尖／急尖
35	QN	叶柄长度	短／中等／长
36	QL	叶柄有无蜜腺	无／有
37	QL	果实是否存在*	无／有
38	QL	果实是否有毛	无／有
39	PQ	果实形状	扁平／圆形／卵圆形／尖圆形／椭圆形
40	PQ	果核形状	扁平形／卵圆形／倒卵圆形／椭圆形
41	QN	核纹	少／中等／多

*为品种分组性状；Z为质量性状；Y为数量性状；X为假质量性状。

二、桃花品种分类系统

(一) 桃花品种分类的原则

桃花品种分类最主要的依据是形态学方面的特征，同时在分类中遵循以下一些原则：

1. 以种为基础

桃花品种中最主要的品种均来自桃（*P. persica*）。但随着育种工作的深入，特别是桃的近缘种的引入，更加丰富了桃花品种种质资源。种源的组成是品种分类的前提性标准。

最典型的例子就是山桃（*P.davidiara*）与桃的天然杂交种：'白花山碧'桃，这一杂交种所具有的双方亲本的优势互补（与山桃相比为复瓣花，增强了观赏性；与桃花相比，花期早1周以上，提前了花期，并延长了桃花的观赏期）及远缘杂交带来的对后代育性的影响（'白花山碧'桃由于雌蕊发育不完全而不结实），对于丰富桃花品种类型和提供新型种质无疑是不可多得的重要资源。

因此在桃花品种分类中，首先应将种性作为第一要素加以考虑，根据《国际栽培植物命名法规》（ICNCP）（Brickell *et al.* 2004）的最新规定和要求，将桃与山桃的杂交种单独列出一个山碧桃组，与桃花的其他枝型类型并列作为品种群的分类单位，取消了原来分类中系或分枝的概念。

尽管已经有光核桃（*P. mira*）与桃花、甘肃桃与桃花等远缘杂交品种的育成（Komar-Tyomnaya 2000），但鉴于尚未进行系统鉴定，暂不列入分类检索表中。

2. 体现品种演化规律

桃花无论是枝型还是花型、花瓣数目，都在一定程度上体现出品种的演化规律。在桃花品种分类中也遵循了由简单到复杂、从原始到进化的原则。

枝型作为桃花最具观赏特色的特征，在分类中起着至关重要的作用。不同枝型的产生有其特定的年代，从文献史料中可知，直枝类型最为普遍，是桃花最为原始的枝型（周文华《汝南圃史》）；14世纪（中国明代）开始出现寿星桃的记载；帚型桃和垂枝桃在日本最早见于记载，是在17世纪左右（伊藤 1695年）；曲枝桃则是20世纪才出现的新枝型类型（Yoshida *et al.* 2000）。以枝型作为区分桃花品种类群的标准，既可以直观地反映出桃花的品种特性，又在一定程度上体现了品种演化过程的不同阶段。

桃花的花瓣以单瓣最为原始，随后相继出现了复瓣和重瓣的不同花型类型，在分类中以花型作为次级分类依据，也考虑到了品种之间的演化规律。

3. 便于应用

观赏桃花最主要的用途是园林栽植，所以在分类时也尽可能地满足实际应用的需求，除了直观的枝型特征之外，花、叶的颜色也是实际应用中最容易被人接受的识别品种的标准之一。因此考虑枝型、花型的同时，叶色和花色也作为进一步区分品种的标准。

（二）桃花品种分类等级

1. 种系（Species, hybrid）

根据目前所有的桃花品种类型，种间杂交的种系暂且确定仅有山桃花系（*P. persica* × *P. davidiana*），与来自桃（*P. persica*）血统的真桃花系相对应。真桃花系品种的枝、叶、芽、花均具有桃的典型特征。山桃花系品种则兼有桃和山桃的双重性状。

从现有品种和育种趋势来看，目前已经出现了桃与光核桃、甘肃桃、甚至扁桃之间新的种间杂交品种（*P. persica* × *P. mira*、*P. persica* × *P. kansuensis*、*P. persica* × *P. amygdalus*）（Komar-Tyomnaya 2007），更进一步证明了以种系作为第一级分类原则的必要性，随着对这些品系及品种的系统鉴定，未来的分类系统将会出现光核桃系、甘肃桃系，甚至扁桃系等新的种系，桃花品种也将会得到进一步的丰富和完善。

2. 品种群（Group）

最新的ICNCP第3.1条中明确规定：品种群是基于一定相似性的品种、植物个体或植物集合体的正式类级。组成并维持一个品种的标准，因不同使用者的目的而异。在一个属、种、杂交属、杂交种或其他命名等级内，凡是两个或多个特性相似的品种所应用的其他名称如：sort、type或 hybrids等，如果是与品种群含义相同的术语名称，均应以品种群取代。根据这一规定的改变，将以前各个分类阶段中（张秀英，陈忠国 1991；张秀英 1993；Hu *et al.* 2003）的

"系"取消，即将桃花品种直接分为6个品种群（Group），即直枝桃品种群、寿星桃品种群、帚型桃品种群、垂枝桃品种群、曲枝桃品种群和山碧桃品种群。每一个品种群均具有明显可以区别的特征，而品种群内各品种既有其特异性，又可互相区别，但其共同特征能够将其维持在各个品种群中。

3. 品种（Cultivar）

品种是栽培植物品种分类的基本单位。按照ICNCP的规定，栽培品种是指具有一致而稳定的明显区别特征，而且采用适当的方式繁殖（有性或无性）后，这些区别特征仍能保持下来的一个栽培植物分类单位，不管这些区别特征是形态、生理、细胞、生化方面的还是其他方面的，只要一个栽培植物群本与另一个存在着明显区别，并且这些特征是可以通过繁殖保持的，就可以作为一个独立的品种。桃花品种根据花型、花色分布在不同品种群内。

(三) 观赏桃花品种分类系统

根据最新的ICNCP的规定和要求，取消原来桃花品种分类中山桃花系和真桃花系的划分，将桃花品种直接分为6类（Group），即山碧桃类、垂枝桃类、帚桃类、直枝桃类、寿星桃类和曲枝桃类。以花型作为次级分类标准包括铃型、单瓣型、梅花型、月季型、牡丹型和菊花型。

种系，树型，及花型是桃花分类系统的3个主要的分类依据。

表5.2　桃花品种群分类检索表

Table 5.2　Key to groups of ornamental peach cultivars

1. 花期中等偏晚，树皮粗糙
  2. 小枝平直
    3. 分枝角度35°～70°，小枝斜出
      4. 分枝角度在40°～70°
        5. 节间长度正常 ················ I 直枝桃品种群 Standard Group
        5. 节间短，通常小于10mm ·············· II 寿星桃品种群 Dwarf Group
      4. 分枝角度35°～40° ··············· III 帚型桃品种群 Pillar Group
    3. 分枝角度>70°，小枝下垂 ············ IV 垂枝桃品种群 Weeping Group
  2. 小枝弯曲成之字形 ················ V 曲枝桃品种群 Twist Group
1. 花期早，树皮光滑，小枝纤细 ················ VI 山碧桃品种群 David Group

(四) 每一个品种群的主要特征

直枝桃品种群：枝条斜出，节间较长的直枝桃是桃花品种中最为常见、品种最多、变化幅度最广的一类。按照花型分为6个型，即铃型、单瓣型、梅花型、月季型、菊花型和牡丹型。

寿星桃品种群：是枝条节间很短，着生紧密的一类桃花。叶披针形，叶较大，叶缘呈不同程度的波状。

帚型桃品种群：这一类桃花最显著的特点是树体直立，枝干开张角度狭小，枝条细，丛生，树冠窄而高，形同扫帚，干性极强。

垂枝桃品种群：是桃花中枝姿最具韵味的一个类型，小枝拱形下垂，树冠伞形。花开时节，宛如花帘一泻而下，蔚为壮观。无论是孤植于庭院，还是群植，都有很好的观赏效果。

曲枝桃品种群：是桃花中枝性最为特殊的一类，小枝自然弯曲，呈"之"字状。

山碧桃品种群：是山桃与桃的杂交类型，枝、叶、芽、花具有桃和山桃的双重性状，树体高大，树皮光滑，花雌蕊发育不完全或早期萎蔫无雌蕊形态。山桃花系桃花品种的花期，明显早于真桃花系品种。

在每一个类群中，都存在着花型、花色及重瓣性的相关性（陈耀华 1997；周建涛，李惠芬 1998），即每一个类群中都应存在着不同瓣形、不同花型、不同花色，以及不同重瓣性的品种类型。有的品种已经发现或育成，而目前还没有的品种都极有可能在不久的将来出现，比如说目前帚型桃里还没有菊花型品种，但根据不同类群中花型的相关性，拥有菊花型花型的帚型桃品种也必将在不久的将来育成。这也是今后主要目标育种研究的重要依据。

三、桃花品种系统研究

桃花品种系统分类研究起源于20世纪80年代。最早的分类研究源于最基本的形态学调查和分析，随后，孢粉学、生物化学、数量分类学，以及分子生物学等技术的引入，将桃花品种系统研究进一步深化。

（一）孢粉学研究

花粉的形态是由基因控制的（Heslop-Harrison 1968；汪祖华1990），花粉具有环境影响小、形态特征稳定的特点，可以用来进行品种群的划分及品种鉴定。

经电镜扫描观察，桃花花粉均为大型花粉，具三孔沟，根据埃尔特曼（1978）NPC分类系统，桃花花粉粒形式为N3P4C5（Heslop-Harrison 1968）。花粉形状为长球形（P/E ≦ 2μm）或超长球形（P/E >2μm）（图5.22）。花粉外壁纹饰呈条纹状，条纹排列方式因品种的不同而异，条纹间有不同程度的穿孔或无孔。

张秀英等（1997）基于对22个桃花品种及3个桃花近缘种的花粉形态的电镜观察分析，桃花花粉表面纹饰分为3个类型：条纹状穿孔纹饰；条纹－穿孔纹饰；条纹状纹饰。具有直枝、矮型（寿星型）及垂枝性状的品种，分别各自具有以上3组花粉表面纹饰。这一结果表明，桃花品种亲缘关系较近的品种，具有相近的花粉形态。

2006～2009年间，对直枝类、帚桃类、寿星桃类、曲枝桃类、寿星桃类和山碧桃类共6个类型的39个桃花品种，进行了花粉电镜扫描。其中不仅增加了曲枝类（'云龙'桃1个品种）、帚桃类（'照手红'、'照手桃'、'照手姬'、'照手白'、'Corinthian Rose'5个品种）和山碧桃类（'白花山碧'桃、'粉花山碧'桃和'粉红山碧'桃）3个类型，而且增加了3个垂枝桃类品种（'源平垂枝'、'黛玉垂枝'、'单白垂枝'）、3个寿星桃类品种（'单寿白'、'单寿粉'、'瑕玉寿星'）和8个直枝桃类品种（'斑叶'桃、'京舞子'、'八重蟠桃'、'簪粉'、'北京紫'、'晚白'桃、'洒红'桃、'五宝'桃），从而更为全面地反映出桃花品种花粉外壁模式的整体图画。

观察分析结果同样显示上述穿孔状、条纹穿孔状和条纹状3种花粉外壁模式，在条纹穿孔状类型中又有规则条纹和不规则条纹之分（图5.23）。

穿孔状纹饰表面平滑，穿孔较多，条纹分支短，不整齐，沟脊均不明显。

条纹穿孔状纹饰是桃花花粉外壁纹饰最为普遍的类型，条纹和穿孔同时存在，根据条纹分布状况，又分为条纹排列整齐、分布均匀的规则条纹穿孔状；以及条纹明显弯曲呈螺旋状的不规则条纹穿孔状。

条纹状纹饰以条纹为主，仅有少量穿孔或穿孔不明显，条纹脊宽，脊间距较窄。

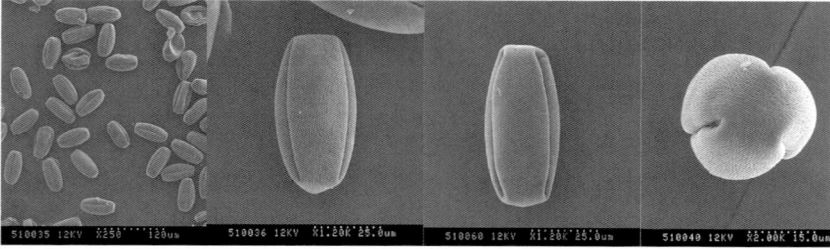

图5.22 桃花品种花粉外壁纹饰
从左至右顺序为：'绿萼垂枝'群体 SEM（×250）；
'绿萼垂枝'赤道面观，SEM（×1200）：长球形；
'京舞子'赤道面观，SEM（×1200）：超长球形；
'云龙'桃极面观，SEM（×2000）
Fig.5.22　Pollen morphology of the ornamental peach (from left to right)
'Lü E Chuizhi' Group. [SEM(×250)]
'Lü E Chuizhi' Equatorial view: prolate. [SEM(×1200)]
'Kyoumaiko' Equatorial view: perprolate.[SEM(×1200)]
'Unriumomo' Polar view. [SEM(×2000)]

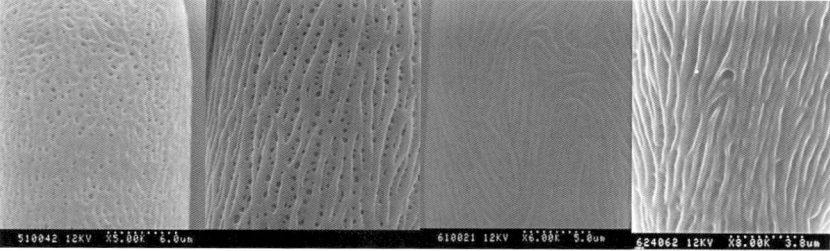

图5.23 桃花品种花粉外壁纹饰类型
从左至右为穿孔状类型、规则式条纹穿孔状类型、不规则式条纹穿孔状类型、
条纹状类型。
Fig.5.23　Types of the ornamental peach pollen exine sculpture: perforate; regularly striate and perforate; irregularly striate and perforate, and striate (from left to right)

图5.24 '山碧'桃和'云龙'桃的花粉外壁纹饰
从左至右顺序为'白花山碧'桃、'粉花山碧'桃、'粉红山碧'桃、'云龙'桃
Fig.5.24　Pollen exine sculpture of Shanbitao and Unriumomo: 'Baihua Shanbitao', 'Fenhua Shanbitao', 'Fenhong Shanbibtao', 'Unriumomo' (from left to right)

　　埃尔特曼（1978）认为被子植物的花粉外壁纹饰演化是由没有结构层（光滑）向穿孔（穴状）发展，再由穿孔继续演化成条纹类型。6个不同类型的桃花品种的花粉外壁模式，在一定程度上体现出了这一观点。每个品种群内都有进化程度不同的品种存在。

　　山碧桃类3个品种均为条纹穿孔状类型，同以'白花山碧'桃为父本的'粉花山碧'桃和'粉红山碧'桃的花粉外壁纹饰与'白花山碧'桃形态接近（图5.24），体现了其品种间较近的亲缘关系。

　　作为曲枝类型唯一的1个品种'云龙'桃，其花粉外壁纹饰为穿孔状类型（图5.24），进化程度不是很高，单瓣花型的'云龙'桃属于较为原始的桃花品种。鉴于该类型仅有这1个品种，尚无法体现这一枝型的特点。

　　垂枝类的6个品种的花粉外壁纹饰均属于条纹穿孔状（图5.25），除'黛玉垂枝'为规则条纹穿孔状类型外，'绿萼垂枝'、'源平垂枝'、'单白垂枝'、'朱粉垂枝'和'红花紫叶垂枝'均属于不规则条纹穿孔状类型。垂枝类型同样是在17世纪～19世纪出现（伊藤1695），进化程度较高。

　　帚桃类的5个品种具有条纹穿孔状和条纹状两种类型，'照手白'、'照手姬'、'照手红'属于规则条纹穿孔状，'Corinthian Rose'属于不规则条纹穿孔类型，'照手桃'则属于条纹状（图5.26），均为较进化的外壁纹饰类型。帚桃类型是在17世纪以后才出现的新型枝型（伊藤 1695），在桃花品种演化上属于比较进化的类型，与花粉外壁纹饰的演化程度相吻合。

　　寿星类的6个品种均属于条纹穿孔状，除'单寿白'为规则条纹穿孔状外，其余'单寿粉'、'寿白'、'寿粉'（图5.26）、'寿红'和'瑕玉寿星'均为不规则条纹穿孔。寿星桃类型的出现是在明代，晚于普通直枝类性，相对也较为进化。

　　直枝类的18个品种，其花粉外壁纹饰呈现出形态多样化，具备上述各种纹饰类型（图5.27）。'绛桃'、'碧桃'属于穿孔状；'单白'、'单红'、'北京紫'、'二色'桃为规则条纹穿孔状，'红碧'桃、'洒红'、'簪粉'、'绯桃'、'晚白'桃、'五宝'桃、'白碧'桃、'八重蟠桃'、'京舞子'、'紫叶'桃、'斑叶'桃为不规则条纹穿孔状；'菊花'桃为条纹状，体现出不同的进化程度。直枝桃是桃花最为基本的类型，栽培历史悠久，在长期栽培应用中品种不断演化，出现了不同进化程度的丰富品种。

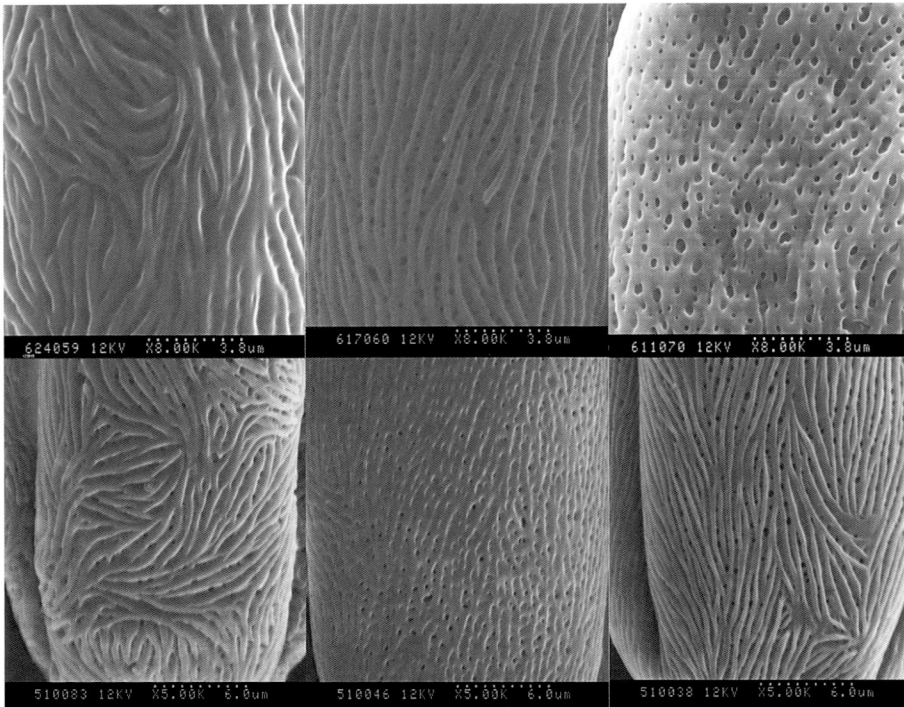

图5.25 垂枝桃花粉外壁纹饰
第一排从左至右：'源平垂枝'、'朱粉垂枝'、'红花紫叶垂枝'
第二排从左至右：'单白垂枝'、'黛玉垂枝'、'绿萼垂枝'
Fig.5.25 Pollen exine sculpture of weeping ornamental peach cultivars
Up: 'Yuanping Chuizhi'，'Zhufen Chuizhi'，'Crimson Cascade'
Down: 'Danbai Chuizhi'，'Daiyu Chuizhi'，'Lü E Chuizhi'

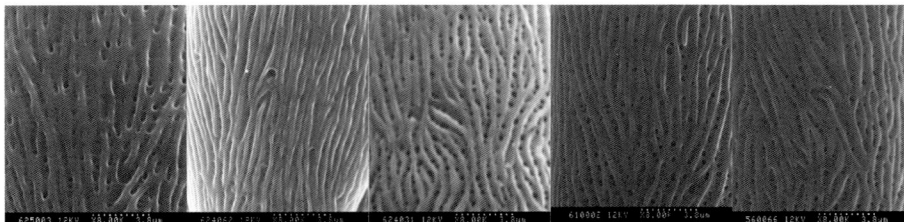

图5.26 帚桃和寿星桃花粉外壁纹饰
从左至右为'照手姬'、'照手桃'、'Corinthian Rose'、'寿白'、'寿粉'
Fig.5.26 Pollen exine sculpture of columnar and dwarf ornamental peach cultivars
    'Terutehime'，'Terutemomo'，'Corinthian Rose'，'Shoubai'，'Shoufen' (from left to right).

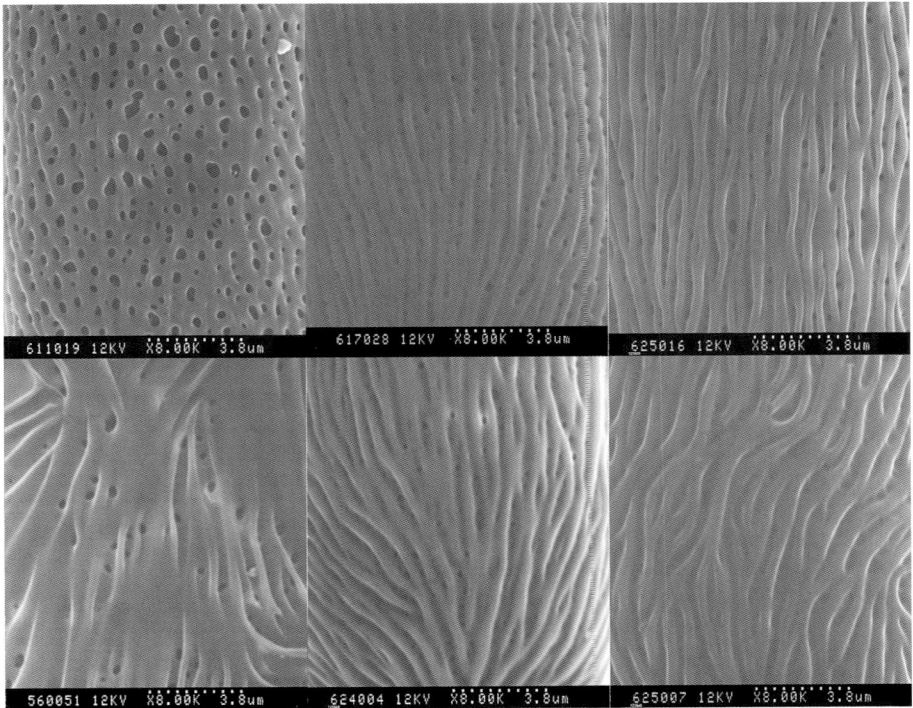

图5.27   直枝桃花粉外壁纹饰
第一排从左至右：'绛桃'、'晚白'桃、'红碧'桃
第二排从左至右：'绯桃'、'北京紫'、'菊花'桃
Fig.5.27   Pollen exine culpture of standard ornamental peach cultivars
Up: 'Jiangtao', 'Wanbaitao', 'Hongbitao'
Down: 'Feitao', 'Beijing Zi', 'Kikoumomo'

(二) 数量分类研究

1998年，张春英通过对27个桃花品科及4个近缘种的花、叶、果、核等23个性状特征的采集及数量分类分析，Q型聚类结果显示甘肃桃与桃的亲缘关系较远，而新疆桃与桃的亲缘关系较近，'白花山碧'桃介于桃与山桃之间，而在桃花品种之间也体现出与形态分类相一致的结果，支持枝型作为高一级分类标准（张春英 1999; 胡东燕 1999）。

鲁振华等（2009）的最新研究表明，用Permut Matrix软件对观赏桃杂交组合（1987-7-1-矮丽红）及其后代群体部分性状的调查，首次采用Permut Matrix软件对观赏桃花瓣数、萼片形状，以及花朵直径的遗传方式进行数值聚类分析，结果显示：花瓣、萼片遗传距离较近,和花径的大小也有一定的遗传关系,这一结论可以为观赏桃的育种和遗传特征提供一定的参考依据。尽管这种方法目前仅仅应用了一个桃花品种的测定,但这种方法可以借鉴应用到更多桃花品种中的更多性状上,实现使数值向质量性状的转变,可以对多种数值进行质变分析。

(三) 同工酶分析

同工酶分析主要是通过同工酶酶谱谱带分析,研究品种间的亲缘关系。

同工酶即酶的多种分子形态,是基因的产物。它可以反映不同品种间的基因差异 (汪祖华等 1991)。对35个桃花品种的过氧化物同工酶的研究 (胡东燕 1999) 结果支持枝型应该作为较高一级的分类标准。所有寿星类桃花品种均具有一条共有带,说明寿星桃类的品种,相对于其他枝型的品种,具有较近的亲缘关系。所有垂枝类型的品种除具有一条共有特征谱带外,均拥有比其他枝型品种数目要多的谱带。同时,所有起源于 *P. persica* 的品种均有3条共有特征谱带,极有可能是该种的特征谱带。来自山桃花系的一个品种——'白花山碧'桃分别具有桃花及山桃两个种的谱带,说明'白花山碧'桃是一个多起源的杂交品种,在其形成过程中,桃花及山桃的亲本均极有可能参与其中。

但是同工酶只能产生数量有限的谱带 (Messeguer *et al.* 1987),限制了大量桃花品种之间的遗传关系研究及分析。

(四) 分子生物学研究

分子生物学研究近年来在观赏植物研究领域应用得越来越广泛。在桃花品种系统研究中,先后应用RAPD、ISSR、AFLP等多种分子标记技术,对桃花品种进行遗传特性和品种间的差异,以及相互之间的演化及亲缘关系等研究;并应用AFLP标记绘制了DNA指纹图谱,以分子数据作为鉴别桃花品种的依据;并利用SSR技术确定了垂枝性状(pl)的相对精准的遗传距离。

1. 分子标记

分子标记是指在生物系统与进化研究中,每个能反映遗传变异的、能提供系统学信息的多态位点称为一个分子标记(邹喻萍等 2001)。在遗传育种研究中,每个与特殊性状或目的基因连锁的多态位点也称为一个分子标记。能提供分子标记的分子生物学技术称为分子标记技术。

　　和常规的形态学特征或生理学性状相比，分子标记可以检测到植物体DNA水平上的差异（Debener 2002），而且不会因生长阶段和环境因素的改变而改变。分子标记可以在品种鉴定上提供直接的帮助，从而最终导致相对完善的分类系统；许多分子技术还可以阐明品种间的等级关系（Culman & Grant 1999）。

　　(1) RAPD　RAPD是首先应用于桃花品种研究的分子标记。张春英等（1998）应用RAPD进行桃花品种分类研究，实验结果表明，寿星桃类的品种之间有着较近的亲缘关系，垂枝桃类和复色品种均有类似情况。关于'白花山碧'桃，RAPD结果表明，相对于山桃（*P. davidiana*），它与桃花（*P. persica*）的亲缘关系更为接近。程中平等（2002, 2003）采用RAPD技术探讨碧桃类群内及寿星桃种质、红叶桃种质的系统发育，运用特殊谱带建立了红叶桃、寿星桃的分子检索表，提出重点保存的碧桃种质、寿星桃种质和红叶桃种质。

　　(2) AFLP　胡东燕等应用AFLP（Hu *et al.* 2005a）分子标记技术对桃花品种进行了遗传演化关系的研究。

　　应用6对EcoRI/MseI引物组合，在75到500碱基对间，供试的51个桃花分类单位（表5.3）共生成275条有效标记。其中，265条谱带为多态性标记。相对每个分类单位的谱带变化范围为90到140，平均每个分类单位的谱带数目为120，显示出了桃花品种AFLP遗传多态性（图5.28）。

　　在这275条AFLP有效标记中，遗传距离的变化范围在0.044到0.404。从遗传学上看，种间的遗传距离应当高于种内各品种间的遗传距离。从PAUP软件得到的UPGMA聚类分析树状图中可以看到两个明显的分支，一支为山桃分支［（Davidiana (CD)）］，另一支为桃花分支［（Persia (CP)）］（图5.29）。

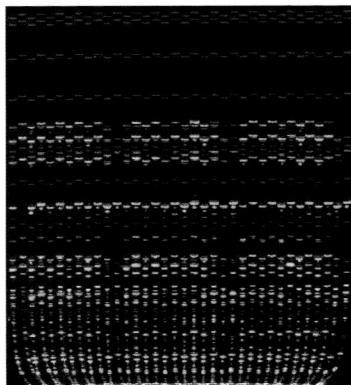

图5.28　供试桃花样品AFLP荧光染色成像图
E-ACTM-CTC(蓝色)，E-AGGM-CTC(绿色)，
E-ACCM-CTC(黄色)，红色为内置标尺
Fig. 5.28　AFLP gel image of ornamental peach taxa with primer-pair combination of, and size standard (red).

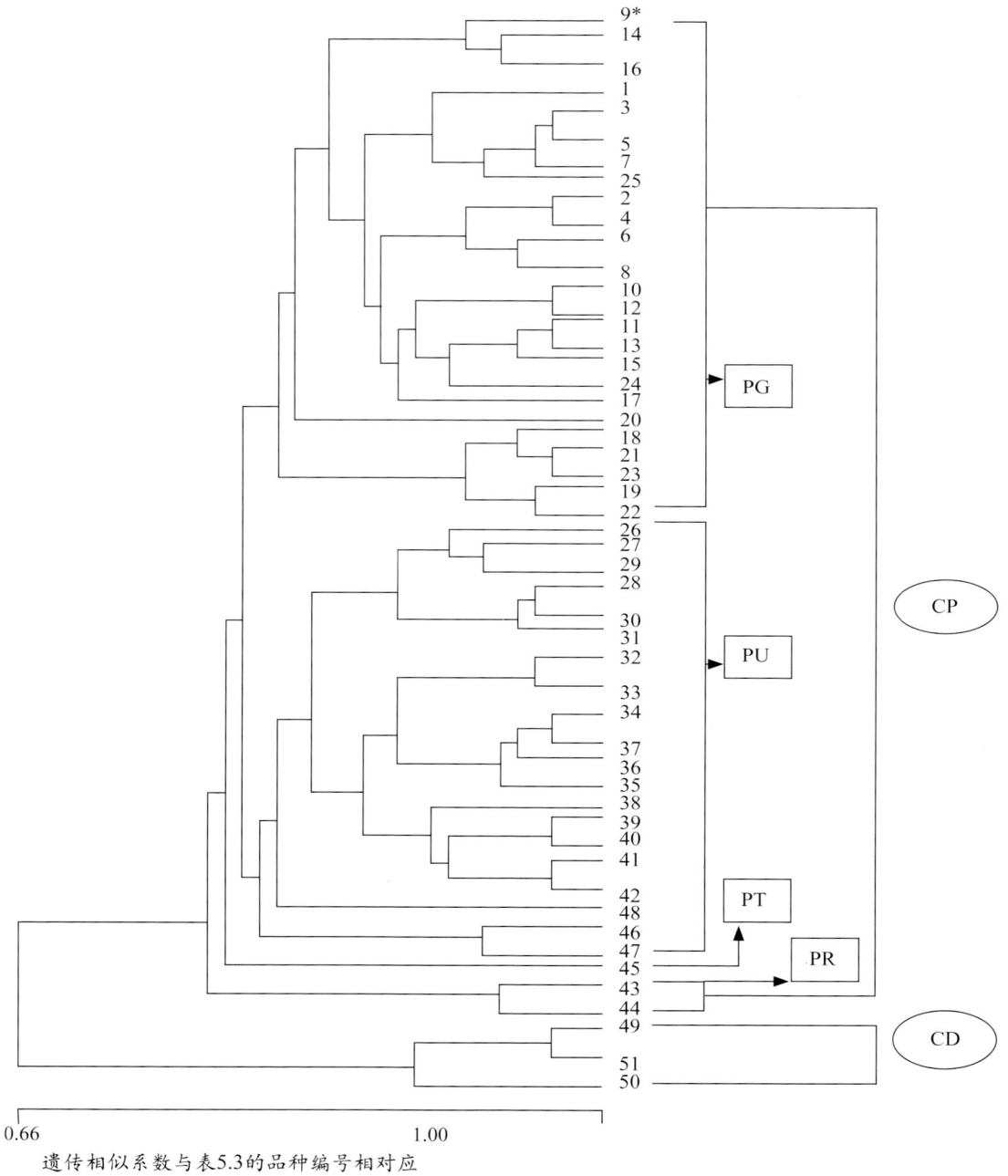

遗传相似系数与表5.3的品种编号相对应

图5.29 基于51个供试桃花样品AFLP数据的UPGMA聚类分析树状图
Fig. 5.29 UPGMA tree generated from PAUP for 51 ornamental peach
cultivars based on AFLP markers.

表5.3 AFLP供试桃花品种

Table 5.3 DNA samples of ornamental peach *taxa* for AFLP

编号 No.	品种名 Name	拉丁名 Latin name	编号 No.	品种名 Name	拉丁名 Latin name
1	'照手桃'	*Prunus persica* 'Terutemomo'	26	'赤叶寿星'	*P. persica* 'Red Dwarf'
2	'Corinthain Mauve'	*P. persica* 'Corinthain Mauve'	27	'Bonfire Patio'	*P. persica* 'Bonfire Patio'
3	'照手白'	*P. persica* 'Teruteshiro'	28	'单粉'	*P. persica* 'Danfen'
4	'Corinthain Mauve'	*P. persica* 'Corinthian White'	29	'北京紫'	*P. persica* 'Beijing Zi'
5	'Corinthian Pink'	*P. persica* 'Corinthian Rose'	30	'单红'	*P. persica* 'Danhong'
6	'Corinthian Rose'	*P. persica* 'Corinthian Pink'	31	'单白'	*P. persica* 'Danbai'
7	'帚桃'	*P. persica* 'Houkimomo'	32	'寒红'桃	*P. persica* 'Hanhong Tao'
8	'单白垂枝'	*P. persica* 'Shiroshidare'	33	'二色'桃	*P. persica* 'Erse Tao'
9	'绿萼垂枝'	*P. persica* 'Lü E Chuizhi'	34	'簪粉'	*P. persica* 'Zan Fen'
10	'赤垂枝'	*P. persica* 'Akashidare'	35	'碧桃'	*P. persica* 'Bi Tao'
11	'红雨垂枝'	*P. persica* 'Hongyu Chuizhi'	36	'红碧'桃	*P. persica* 'Hong Bitao'
12	'Clarisse'	*P. persica* 'Clarisse'	37	'绯桃'	*P. persica* 'Fei Tao'
13	'朱粉垂枝'	*P. persica* 'Zhufen Chuizhi'	38	'五宝'桃	*P. persica* 'Wubao Tao'
14	'黛玉垂枝'	*P. persica* 'Daiyu Chuizhi'	39	'菊花'桃	*P. persica* 'Juhua Tao'
15	'源平垂枝'	*P. persica* 'Genpaishidare'	40	'菊桃'	*P. persica* 'Kyoumomo'
16	'鸳鸯垂枝'	*P. persica* 'Yuanyang Chuizhi'	41	'白碧'桃	*P. persica* 'Bai Bitao'
17	'五宝垂枝'	*P. persica* 'Wubao Chuizhi'	42	'晚白'桃	*P. persica* 'Wanbai Tao'
18	'单瓣寿粉'	*P. persica* 'Danban Shoufen'	43	'紫叶'桃	*P. persica* 'Zi Ye Tao'
19	'寿粉'	*P. persica* 'Shoufen'	44	'斑叶'桃	*P. persica* 'BBG2'
20	'亮粉寿星'	*P. persica* 'Liangfen Shouxing'	45	'云龙'桃	*P. persica* 'Unriumomo'
21	'寿白'	*P. persica* 'Shoubai'	46	'白花山碧'桃	*P. persica* 'Baihua Shanbitao'
22	'单瓣寿红'	*P. persica* 'Danban Shouhong'	47	'粉花山碧'桃	*P. persica* 'Fenhua Shanbitao'
23	'寿红'	*P. persica* 'Shouhong'	48	'粉红山碧'桃	*P. persica* 'Fenhong Shanbitao'
24	'NCSU Dwarf Double Red'	*P. persica* 'NCSU Dwarf Double Red'	49	粉花山桃	*P. davidiana* var. *rubra*
25	'瑕玉寿星'	*P. persica* 'Xiayu Shouxing'	50	曲枝山桃	*P. davidiana* 'BBG1'
			51	白花山桃	*P. davidiana* var. *alba*

山桃分支 [Davidiana (CD)] 中的3个分类单位：白花山桃 (*P. davidiana* var. *alba*)，粉花山桃 (*P. davidiana* var. *rubra*) 和'曲枝山'桃山桃有100% 聚类鉴别效率支持。这3个分类单位与其他所有桃花分类单位的平均遗传距离是0.339。从种的水平来看，相对于和其他桃花品种，这3个分类单位之间的遗传距离很近。

桃花分支 [(Persica (CP)] 包含4个亚组，红叶亚组 (Clad PR)，曲枝亚组 (Clad PT)，直枝亚组 (Clad PU) 和混合枝型亚组 (Clad PG)。

红叶亚组包括两个红叶品种：'紫叶'桃和'斑叶'桃，有100% 聚类鉴别效率支持。'斑叶'桃 (Hu *et al.* 2005b) 是一般'紫叶'桃的芽变。2001年从北京植物园选出，'斑叶'桃有着独特的紫色与绿色的斑叶效果，它与'紫叶'桃的遗传距离为0.12。形态特征和分子数据都支持'斑叶'桃为一个新品种。

曲枝亚组 (Clad PT) 仅有'云龙'桃1个品种。这是一个来自于油桃的芽变品种 (Yoshida *et al.* 2000)，在UPGMA聚类分析树状图中与其他桃花品种的平均距离为0.237。极有可能是独立起源，是今后桃花育种重要的种质资源。'云龙'桃是唯一一个具有曲枝枝型的桃花品种，尽管花并不很出众，但奇异的枝型赋予了该品种独特的观赏性，也为观赏桃花增添了一种新枝型的育种趋势。正如其他枝型的品种大部分均各自聚为一组，曲枝枝型极有可能也会产生具有不同花色、花型的一系列新品种。AFLP的这一结果也证明了枝型在桃花分类系统中应当占据显要的等级地位。

共有20个桃花品种聚在直枝亚组 (Clad PU) 中。从形态特征上看，全部具有直出小枝。包括18个直枝类型的品种及2个寿星桃品种。不同花色及不同花型的品种交迭出现在各个聚类组中。

在混合枝型亚组 (Clad PG) 内，又分成寿星桃类型小组 (Clad PGD) 和混合类型小组 (Clad PGM)，即帚型、垂枝型和寿星桃类型3种枝型混在一起。所有供试的7个帚型桃品种和10个垂枝桃类型均聚在PGM组。

在AFLP分析中，并没有明显的同一花色或花型的品种聚合在一起，这说明花色和花型可能不足以成为桃花分类系统中的一个等级标准。

对51个桃花分类单位进行的AFLP分析结果显示，桃花的枝型应该被置于分类系统中高级别的地位。花瓣的数目也可以在分类中加以考虑。AFLP分子标记所产生的大量扩增片段揭示了桃花品种间丰富、详尽的遗传关系，表明AFLP分子标记可以用于检测桃花品种起源，分析桃花品种间的遗传关系。

(3) ISSR　胡东燕等应用ISSR（Hu *et al.* 2006）分子标记技术对桃花品种进行了遗传演化关系的研究。

通过对16个桃花分类单位（表5.4）的ISSR分析，在300到1400碱基对之间，16个供试桃花品种共有132条谱带从10个引物中产生，62%的谱带为多态性谱带。每个供试样品平均产生80条谱带。每个ISSR引物扩增出4～12条谱带，平均值为8条。图5.30显示的是以引

表5.4　应用于ISSR研究的桃花DNA样本
Table 5.4　DNA samples of ornamental peach for ISSR

编号 No.	品种名 Name	拉丁名 Latin name
1	'Corinthian Pink'	*Prunus persica* 'Corinthian Pink'
2	'帚桃'	*P. persica* 'Houki Momo'
3	'绿萼垂枝'	*P. persica* 'Lü E Chuizhi'
4	'Clarisse'	*P. persica* 'Clarisse'
5	'瑕玉寿星'	*P. persica* 'Xiayu Shouxing'
6	'赤叶寿星'	*P. persica* 'Red Dwarf'
7	'Bonfire Patio'	*P. persica* 'Bonfire Patio'
8	'北京紫'	*P. persica* 'Beijing Zi'
9	'碧桃'	*P. persica* 'Bi Tao'
10	'京舞子'	*P. persica* 'Kyoumaiko'
11	'白碧'桃	*P. persica* 'Bai Bitao'
12	'紫叶'桃	*P. persica* 'Zi Ye Tao'
13	'云龙'桃	*P. persica* 'Unriumomo'
14	'白花山碧'桃	*P. persica* 'Baihua Shanbitao'
15	'粉花山碧'桃	*P. persica* 'Fenhua Shanbitao'
16	粉花山桃	*P. davidiana* var. *rubra*

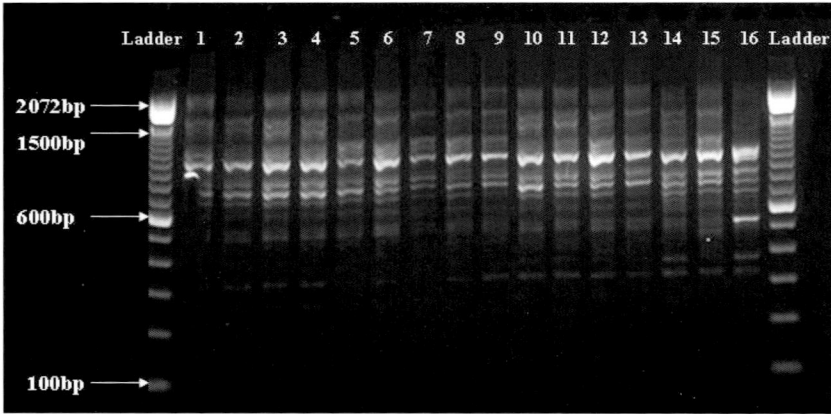

图5.30 用UBC868引物生成的16个供试桃花样品的琼脂凝胶电泳结果
Fig. 5.30 Agarose gel imagine generated from primer UBC868 for 16 ornamental peach taxa

物UBC868为例的琼脂凝胶电泳结果。

在132条ISSR有用谱带的基础上，遗传距离变化范围从0.030到0.402。桃是天然授粉自花受精的树种，因此遗传变异本身就很低（Hesse 1975；Scorza *et al.* 1985；Scorza&Okie 1990），自身的遗传基础相当狭窄（Scorza *et al.* 1985）。AFLP分析也证实了桃的遗传多样性水平很低这一观点（Aranzana *et al.* 2001）。ISSR分析得出的结论再次证实了桃花品种的遗传多样性相当狭窄。

ISSR数据分析证明，山桃在一定程度上参与的桃花品种的形成，具有山桃血统的品种与来自桃的品种在遗传距离上有着明显的差距。ISSR分子标记技术可以提供桃花品种间的遗传关系，特别是在说明桃花品种群体组别关系上更为有效，对于今后研究桃花与其他近缘种的杂交后代之间的亲缘关系上有极大应用潜力。

2. 指纹图谱

DNA标记可以区分不同品种甚至个体差异，通常称做"DNA指纹图谱"。这一技术包括产生出一系列可以以大小区分的谱带，即DNA片段，这些谱带也许并不能完全具有代表性，但可以详细说明某些特殊DNA序列的存在与否（Grant&Culham 1997）。分子数据通过提供有效的DNA水平的证据，可以极大地帮助桃花品种鉴别和育种起源的鉴定。

以6对AFLP引物组合建立的桃花品种DNA指纹图谱库，可以用来鉴别或支持桃花品种的描述及登录，证实桃花品种的育种起源，为今后新品种的保护、鉴定和登录奠定了坚实的基础 (胡东燕 2004)。

'粉花山碧'桃和'粉红山碧'桃都是以'白花山碧'桃为父本杂交而成的。从形态上看，这3个品种都具有高大的树体和细长的小枝，而且花期都比一般桃花品种要早。在6个引物组合产生的条带中，3个品种共有谱带为61条。'白花山碧'桃有5条特异谱（M-CAT/E-ACT110, M-CTC/E-ACT158, M-CTC/E-AGG110,143,265）。'粉花山碧'桃（M-CTC/E-AGG84）和'粉红山碧'桃（M-CAT/E-AGG234）都存在各自的特异条带。AFLP结果支持这3个品种共同属于一个杂种品种群。

'云龙'桃是唯一一个具有曲枝枝型的桃花品种。独特的枝型同样可以从来自3个引物组合的16条特异谱带中得到反映（M-CAT/E-ACC62, 77, 133, 187, M-CAT/E-ACT68, 75, 96, 120, 190, 219, 330, 360, M-CAT/E-AGG62, 77, 133, 187）。

和其他寿星桃品种比较，'瑕玉寿星'有1条特异谱带（M-CAT/E-ACT270-274-277），AFLP数据证实这是一个合理的品种。

'五宝垂枝'是有着淡粉色与粉红色相间的重瓣花的垂枝桃。M-CTC/E-AGG368这一独特的谱带可以视为该品种合理性的分子依据。

分子标记不仅可以用来区分基因型，还可以提供基因型之间的遗传关系。

'赤叶寿星'（'Red Dwarf'）是一个单瓣粉花、紫叶寿星桃品种，由吉田雅夫育成，于1974年发表 (吉田&清家 1974)。该品种是一个单瓣粉花、紫叶直枝桃品种（'Akame'，日本品种）与绿叶寿星桃品种的F_2代。'赤叶寿星'有一段特异谱带模式（M-CTC/E-AGG-336-36-447）。而'Bonfire Patio'则是美国人摩尔（Moore *et al.* 1993）通过一个单瓣粉花紫叶直枝桃品种天然授粉得到实生苗。AFLP的实验结果证明了这2个紫叶桃花品种的杂交起源。显而易见，AFLP分子标记可以成为检测桃花品种起源及遗传来源的有利工具。

AFLP及ISSR等分子标记的引用，对于进一步从分子水平揭示桃花品种之间的亲缘关系及育种亲本，提供了可靠的依据和技术保证，也为今后进一步开展分子辅助育种奠定了基础。

然而，分子数据并不可能完全代替形态特征。一些形态特征对于品种鉴别仍然十分重要，有些时候往往很难仅仅凭分子特征在品种间建立必然的紧密联系。

3. 遗传距离

Dirlewanger和Bodo (1994) 通过RAPD标记发现了垂枝性状 (pl) 的遗传距离为11.4cM和17.2cM。李亚蒙通过SSR分子标记技术，发现位于G1染色体52.6cM和65.1cM处的微卫星标记BPPCT020、CPPCT029与垂枝基因的连锁关系密切，其中标记CPPCT029与pl基因共分离，为以后进一步精确定位奠定了基础 (李亚蒙 2006)。这一结果也使垂枝桃的分子生物学研究得到了进一步的深入。

京舞子

6 桃花品种介绍

CULTIVARS ILLUSTRATION

按照直枝桃、寿星桃、帚桃、垂枝桃、曲枝桃和山碧桃等6个不同品种群，

详细介绍了75个桃花品种的起源、形态和花期；

同时也对在欧美国家较为常见的20个桃花品种做了简要介绍。

一、直枝型品种群

直枝桃枝条斜出，一般分枝角度35°～50°，节间较长，是桃花品种中最为常见、品种最多、变化幅度最广的一类。花型变异丰富，有铃型、单瓣型、梅花型、月季型、菊花型和牡丹型，从平展到圆球，从疏散到紧密，各不相同；花径也有明显差异，有1～2cm的小花型，也有3～4cm的一般花型，5～6cm以上的大花型也屡见不鲜。

根据花型、花色、叶色和抗性，直枝桃品种群的品种分类检索表如表6.1。

表6.1　直枝桃品种群品种分类检索表

Table 6.1　Key to ornamental peach cultivars of Upright group

1. 枝条斜出，花单瓣，叶绿色或紫色
　2. 花径小，2cm左右，铃型花，叶绿色或紫色
　　3. 花肉粉色，叶绿色·······················'瑞光''Rui Guang'
　　3. 花深粉紫色，叶紫色···················'哈露红''Ha Lu Hong'
　2. 花径大于3.5cm，花盘状或碟状，叶绿色或紫色
　　4. 花白色，叶绿色···························'单白''Alba'
　　4. 花粉色、红色或复色，叶绿色或紫色
　　　5. 花粉色，叶绿色或紫色
　　　　6. 花粉色，叶绿色
　　　　　7. 花粉可育······················'单粉''Dan Fen'
　　　　　7. 花粉败育······················'玫粉''Mei Fen'
　　　　6. 花粉色，叶紫色······················'北京紫''Beijing Zi'
　　　5. 花红色或复色
　　　　　8. 花红色························'单红''Dan Hong'
　　　　　8. 花粉色与白色复色·············'京更纱''Kyosarasa'
1. 枝条斜出，花复瓣或重瓣，叶绿色、紫色或斑叶
　　　　9. 花复瓣，15～40瓣范围内
　　　　　10. 花瓣平整，花瓣3～5层规则排列，型似梅花
　　　　　　11. 花白色，叶绿色·······'白碧'桃'Bai Bitao'
　　　　　　11. 花粉色、红色或复色，叶绿色、紫色或斑叶
　　　　　　　12. 花粉色，叶绿色或紫色
　　　　　　　　13. 花粉色，叶紫色··········'紫奇''Zi Qi'
　　　　　　　　13. 花粉色，叶绿色
　　　　　　　　　14. 对细菌性穿孔病有一定抗性

15. 耐寒，对褐腐病、白粉病有一定抗性

  16. 粉色55C，无蜜腺 ………………………………… 'Harrow Frostipink'

  16. 粉色52C，有蜜腺 ………………………………… 'Harrow Candifloss'

15. 耐寒性不明显，对桃树腐烂病有一定抗性 ……………… 'Jerseypink'

14. 无明显抗病性，果实圆形或扁平

  17. 果实圆形，花期早，萼片5 …… **'合欢二色'桃** 'Hehuan Erse Tao'

  17. 果实扁平，花期中等，萼片10 …… **'八重蟠桃'** 'Yaezaki bantou'

12. 花红色或复色，叶绿色、紫色或斑叶

  18. 花红色，绿叶、紫叶或斑叶

    19. 花红色，叶绿色

      20. 花药败育，果实扁平 …………… **'赤花蟠桃'** 'Akabana bantou'

      20. 花药可育

        21. 抗病性不明显 ………………………………… **'绛桃'** 'Jiang Tao'

        21. 耐寒，对褐腐病、白粉病有一定抗性 ….. 'Harrow Rubirose'

    19. 花红色，紫叶或斑叶

      22. 叶紫色 ……………………………………… **'紫叶'桃** 'Zi Ye Tao'

      22. 斑叶，绿叶上有紫色斑点 ………… **'斑叶'桃** 'Zuoshuang'

  18. 花复色，叶绿色或紫色

    23. 花白色、粉色和红色复色，叶绿色 …… 'Peppermint Stick'

    23. 花粉色与红色复色跳枝, 叶紫色 …… **'凝霞紫叶'** 'Ningxia Zi Ye'

10. 花瓣翻卷或狭长，花瓣数在30~40瓣左右

    24. 花瓣外翻，内卷，瓣卵圆形，型似月季

      25. 花粉色或复色

        26. 花淡粉色 …………………………… **'人面'桃** 'Renmian Tao'

        26. 花粉色也粉红色跳枝 ………………… **'二色'桃** 'Erse Tou'

      25. 花红色，花期早 ………… **'寒红'桃** 'Hanhong Tao'

    24. 花瓣细长，型似菊花

      27. 花粉色，花瓣数30左右 …… **'菊花'桃** 'Kikoumomo'

      27. 花红色，花瓣数36左右 …… **'京舞子'** 'Kyoumaiko'

9. 花重瓣，花瓣数在40以上，花朵呈半球形或球形，型似牡丹

      28. 花白色 ………………… **'晚白'桃** 'Wanbai Tao'

      28. 花粉色、红色或复色

        29. 花红色

        30. 花粉可育，小枝绿色或红褐色

          31. 花红色，小枝绿色 ……… **'绯桃'** 'Fei Tao'

          31. 花亮红色，小枝红褐色……………………

            …………………………… **'红碧'桃** 'Hong Bitao'

        30. 花粉败育 ………………………… **'玫紫'** 'Mei Zi'

      29. 花粉色或复色

        32. 花粉色

1. '瑞光'('Rui Guang')

起源：不详。

形态：树体较大，枝条直立，小枝紫褐色。着花密，花蕾椭圆形，花梗短；萼片5枚，紫色带绿晕，边缘有白毛；小花铃型，碗状，花肉粉色，花径2cm；花瓣5~10，匙型，稍皱卷；子房发育，果可食。叶卵状披针形。着花中等。花期4月中旬。

2. '哈露红'('Ha Lu Hong')

起源：不详。

形态：小枝紫褐色。花蕾长卵形，花瓣卵形，5枚，边缘稍外翻，小花铃型，深粉紫色（67D），发黄；花丝粉红色；花径1.5cm；花梗极短；萼片紫色；着花密；子房发育。叶紫色，卵状披针形。花期4月中旬。

'瑞光'('Rui Guang')　　　　　　　'哈露红'('Ha Lu Hong')

3. '单白'（'Alba'）

起源：1829年首次命名于英国（Krussmarn 1986）。

形态：枝条直立，干皮灰褐色，小枝绿色。花白色（155D）；花蕾卵圆形；花瓣卵圆形，长2.4cm，花径5.1cm，单瓣型，花瓣数5；雄蕊数平均51.3，花丝长1.43cm，白色；花药黄色；雌蕊1枚，雌蕊低于雄蕊；着花中密(着花率1.22/cm)，花梗长0.57cm；花萼绿色，5枚。叶绿色，椭圆披针形，长11.2cm，宽3.05cm，叶长与叶宽比（L/W）为3.67，叶缘细圆齿，叶柄长0.78cm，肾形蜜腺1~2个。果实绿色，长3.3cm，宽3.2cm，圆形；果核长2.57cm，宽1.7cm，椭圆形，核面平滑。花期4月中旬。

'单白'　（'Alba'）

4. '单粉'（'Dan Fen'）

起源：不详。

形态：干皮灰褐色，小枝灰黄色。花粉色(65C)，花蕾长圆形；花瓣卵圆形，长2.37cm，花径5.1cm；花单瓣型，5枚，偶有6~9枚；雄蕊数平均38.7，花丝长1.5cm，粉白色；花药黄色；雌蕊1枚，雌蕊略高于雄蕊；着花中等（着花率0.92/cm），花梗长0.57cm；萼片5枚，红褐色，边缘有毛；子房发达。叶绿色，椭圆披针形，长12.22cm，宽3.74cm，叶长与叶宽比（L/W）为3.27，叶缘细锯齿，叶柄长0.8cm，肾形蜜腺2个。果实绿色，长4.7cm，宽4.74cm，圆形；果核长2.9cm，宽2.1cm，椭圆形，核面平滑。花期4月中旬。

'单粉'
（'Dan Fen'）

'玫粉'
('Mei Fen')

5. '玫粉'（'Mei Fen'）

起源：北京植物园发现的花粉败育变异品种。

形态：干皮灰褐色，小枝绿色。花粉色（69A），花蕾长卵形；花瓣卵圆形，长2.13cm，花径4.57cm；花单瓣型，5枚；雄蕊数平均40，花丝长1.37cm，白色；花药白色；雌蕊1枚，雌蕊与雄蕊近等长；着花中密（着花率0.8/cm）；萼片红褐色，5枚，边缘有毛；花梗长0.57cm。叶绿色，椭圆披针形，长14.02cm，宽3.65cm，叶长与叶宽比（L/W）为3.84，叶缘细圆齿，叶柄长0.73cm，肾形蜜腺2个。经多年观察，未见果实。花期4月中旬。

6. '北京紫'（'Beijing Zi'）

起源：不详。

形态：干皮灰褐色，小枝紫色。花粉色（62D），花蕾长卵形；花瓣卵形，长1.17cm，花径3.63cm；花单瓣型，5枚；雄蕊数平均39.3，花丝长1.1cm，花丝粉白色；花药黄色；雌蕊1枚，雌蕊与雄蕊近等长；着花中密（着花率1.2/cm），花梗长0.57cm；萼片红褐色，5枚，边缘有毛。新叶紫红色，夏季略减弱，偏绿；椭圆披针形，长13.02cm，宽

'北京紫'
('Beijing Zi')

'单红'（'Dan Hong'）　　　　　'京更纱'（'Kyosarasa'）

3.14cm，叶长与叶宽比（L/W）为4.15，叶缘细圆齿，叶柄长1.06cm，肾形蜜腺2～3个。果实带紫晕，长3.98m，宽3.78cm，圆形；果核长2.23cm，宽1.76cm，卵圆形，核面光滑。花期4月中旬。

7. '单红'（'Dan Hong'）

起源：不详。

形态：干皮灰黑色，小枝紫色。花红色（59D）；花蕾长卵形；花瓣卵形，长1.8cm，花径3.8cm，单瓣型，花瓣5枚；雄蕊数平均40，花丝长1.77cm，花药黄色；雌蕊与雄蕊近等长；着花中密（着花率0.98/cm）；萼片红褐色，5枚；花梗长0.5cm。叶绿色，披针形，长11.82cm，宽2.14cm，叶长与叶宽比（L/W）为5.52，叶缘细锯齿，叶柄长0.54cm，蜜腺1～2个。果实绿色，长4.7cm，宽4cm，卵圆形；果核长2.3cm，宽1.9cm，卵圆形；核面平滑。花期4月中旬。

8. '京更纱'（'Kyosarasa'）

起源：日本江户时代就有记载（伊藤 1695）。

形态：干皮灰黑色，小枝绿色。花白色与粉色相杂，单瓣，碗状，花瓣长1.6cm，宽1.3cm，花径3.6cm，花粉活力高达95.7%。叶绿色，椭圆披针形，长12cm，宽3.7，叶长与叶宽比（L/W）为3.3，叶柄长1.1cm（Pawasut *et al.* 2004）。花期4月中旬。

9. '白碧'桃（'Bai Bitao'）

起源：中国古老品种，最早出现于宋代（960－1279）。有些书上的f. *alba-plena*，也被称作'千瓣白'桃。

形态：干皮灰色，小枝绿色。花白色（155D），花蕾球形；花瓣长卵形，长2.4cm，花径5.1cm，复瓣，梅花型，花瓣数27.7（22～28）枚，雄蕊数平均59.7，花丝长1.03cm，白色，平展，花药黄色；雌蕊高于雄蕊；着花中等（着花率0.88/cm）；花萼绿色，两轮；花丝和萼片均有瓣化现象；花梗长0.48cm。叶绿色，椭圆披针形，长11.4cm，宽3.4cm，叶长与叶宽比（L/W）为3.35，叶缘细圆齿，叶柄长0.96cm，肾形蜜腺2～3个。果实绿色，长3.82cm，宽3.58cm，圆形；果核长2.44cm，宽1.48cm，倒卵圆形；核面平滑。花期4月中旬。

'白碧'桃

（'Bai Bitao'）

'紫奇'（'Zi Qi'）

10. '紫奇'（'Zi Qi'）

起源：山东农业大学沈向等2000年通过对'紫叶'桃的开放授粉种子进行辐射育成（沈向等2007）。经过6年的区域试验，综合观赏性状优良、稳定，与其母本紫叶桃比较，花期提前8～12天，属于早花品种，花色亮丽、明快，花径大，叶色紫红。2006年通过国家林业局组织的现场验收，获得新品种权。

形态：树势强健，树姿半开张，萌芽力、成枝力均强。干皮灰褐色，有大量皮孔。6年生树高4.5m，干周44.6cm。1年生枝无毛，有光泽，阳面紫红色。花亮粉红色（62B），花瓣匙型；花径3.9～4.4cm；梅花型，花瓣数12～18枚；雄蕊29～36枚，花丝、花药均为红色，基部雄蕊略瓣化；着花繁密，1年生枝上着花密度为6.36朵/cm（±1.58）；花萼紫褐色，1轮，5枚，外被白毛，长卵形；子房1枚，能发育成果实；花梗长0.5cm。叶色春季为深紫红色，夏季略有减退；叶卵圆披针形，长9.1cm（±0.59），宽3.2cm（±0.45），叶长与叶宽比（L/W）为2.84；叶面平展，叶尖渐尖，叶基楔形，叶缘锯齿状，叶柄粗壮，长1cm，肾形蜜腺2～4个。果实绿色，圆球形，果顶尖圆，纵径3.09cm（0.55），横径3.06cm（0.43），密被绒毛；花期早且长，在其育成地山东泰安3月下旬开花，花期18～21天。

可栽植范围在Zone5～9。注意夏季修剪，适当开张控旺，缓和树势，促成花芽。无特殊敏感性逆境伤害和病虫害。

11. 'Harrow Frostipink'

起源：加拿大安大略省 Harrow 研究站 1981 年培育的耐寒品种 (Layne 1981)，来源于杂交组合（'Harrow Blood'× NY555036) 的开放授粉实生苗。

形态：树势强，花大，浅粉 55C，颜色从白到浅粉 56D 到 53C 全部开放，着花密度 1.25/cm（25/20cm）。叶绿色（147A），椭圆披针形，长 9.5cm，叶宽 3.0cm，叶长与叶宽比（L/W）为 3.16；叶缘细圆齿；无蜜腺；叶柄长 0.5～1cm。耐寒品种，可以在 8a、7a、6b 安全越冬，在 6a 需要保护越冬；对褐腐病、白粉病和细菌性穿孔病有一定抗性。

12. 'Harrow Candifloss'

起源：加拿大安大略省Harrow研究站1981年培育的耐寒品种 (Layne 1981)，来源于杂交组合（'Harrow Blood'× NY555036) 的开放授粉实生苗。

形态：树势强，花大，中等粉色52C，颜色从白到浅粉56D到53C全部开放，着花中等1/cm（20/20cm）。叶绿色（147A），椭圆披针形，长9.0cm，叶宽2.5cm，叶长与叶宽比（L/W）为3.6；叶缘细圆齿；有蜜腺。耐寒品种，可以在8a，7a，6b安全越冬，在6a需要保护越冬；对褐腐病、白粉病和细菌性穿孔病有一定抗性。

13. 'Jerseypink'

起源：美国新泽西州Rutgers university于1990年发表的品种 (Goffreda *et al.* 1992)。

形态：树势强。花多，花径5.5～6cm，深粉，重瓣，花药深黄，肾形蜜腺，晚花。对细菌性穿孔病和桃树腐烂病有一定抗性。

14. '合欢二色'桃（'Hehuan Erse Tao'）（'矢口'、'迎春'）

起源：中国古老品种，最早出现于宋代（960－1279）。

形态：树势强健，干皮深灰色，小枝褐色带绿晕；花从初开的粉色（65B）到后期全树整体颜色变为（62B）；花蕾卵圆形；花瓣卵圆形，长2.2cm，宽1.4cm，花径4.6cm，复瓣，梅花型，花瓣数21.3枚（19～25）；雄蕊数平均84，花丝长1.1cm，白色，花丝有瓣化现象；花药橘红色；雌蕊明显高于雄蕊；着花密（着花率1.2/cm）；花萼5枚，红褐色偏绿；花梗长0.63cm。叶绿色，披针形，长14.5cm，宽3.7cm，叶长与叶宽比（L/W）为3.9；叶缘细圆齿，叶柄长1.0cm。果实绿色，长3.98cm，宽3.74cm，圆形；果核长2.2cm，宽1.64cm，椭圆形；核面平滑。花期4月10日左右，为真桃花系中粉花品种开花最早的一个。

'矢口'（'Yaguchi'）最早的记载出现于日本的江户时代（17世纪～19世纪）（伊藤1695），性状描述与上述'合欢二色'桃甚为相近，根据《国际栽培植物命名法规》（ICNCP）（Brickell *et al.* 2004）的规定和要求，以较早出现的品种名为准，故此将'矢口'归入'合欢二色'桃。补充性状为花粉活力85.3%（Pawasut *et al.* 2004）。

同理，甘肃农科院果树所从实生桃中选出的'迎春'（汪祖华和庄恩及2001）也归入'合欢二色'桃。补充性状为需冷量低，为600h。

'合欢二色'桃
（'Hehuan Erse Tao'）

15. '八重蟠桃'（'Yaezaki bantou'）

起源：不详。日本品种。

形态：花粉色（62B）；花蕾扁球形；花瓣卵圆形，长1.96cm，花径4.65cm，复瓣，梅花型，花瓣数26.5枚（23～29），雄蕊数平均68.6，花丝长1.24cm，花药橘红色；雌蕊略高于雄蕊；着花中密（着花率0.93/cm）；萼片红褐色，两轮，花丝有瓣化现象；花梗长0.73cm。叶绿色，椭圆披针形，长12.42cm，宽3.65cm，叶长与叶宽比（L/W）为3.7，叶缘细圆齿，叶柄长0.98cm；肾形蜜腺2～3个。果实绿色，扁平，长5.24cm，宽3.02cm；果核长1.16cm，宽1.76cm，扁平；核面粗糙。花期4月中旬。

16. '赤花蟠桃'（'Akabana bantou'）

起源：选自'八重蟠桃'（粉花重瓣，扁桃）的天然授粉后代（Yoshida *et al*. 2000）。

形态：树势中等；小枝红色，细长；花红色（65D）；花蕾扁球形；花瓣卵圆形，长1.97cm，宽1.78cm，花径4.3cm，复瓣，梅花型，花瓣数19枚（18～20），雄蕊数平均70.3，花丝长0.97cm，花药白色；雌蕊高于雄蕊；着花稀（着花率0.53/cm）；萼片红褐色，两轮；花

'八重蟠桃'
（'Yaezaki bantou'）

'赤花蟠桃'
（'Akabana bantou'）

丝和花萼均有瓣化现象；花梗长0.53cm。叶绿色，披针形，长13cm，宽2.98cm，叶长与叶宽比（L/W）为4.36；叶缘细圆齿，叶柄长1.08cm；肾形蜜腺2个。果实绿色，扁平，长4.33cm，宽2.33cm；果核长1.16cm，宽1.6cm，扁平；核面粗糙。花期4月中旬。

17. '绛桃'（'Jiang Tao'）

起源：中国古老品种，最早出现于五代（907－979）。

形态：干皮灰褐色，小枝紫色。花红色（65D）；花蕾卵圆形；花瓣卵形，长2.2cm，花径4.73cm，复瓣，梅花型，花型规则，花瓣数14.7枚（13～18）；雄蕊数平均39.7，花丝长1.23cm，红色，平展，有瓣化现象；花心白色；花药橘红色；雌蕊与雄蕊近等长；着花中等（着花率0.75/cm）；花萼红褐色，两轮；花梗长0.9cm。叶绿色，椭圆披针形，长13.3cm，宽3.7cm，叶长与叶宽比（L/W）为3.59；叶缘细圆齿，叶柄长1.0cm；肾形蜜腺2～4个。果实绿色，长3.46cm，宽3.26cm，圆形；果核长2.88cm，宽1.96cm，卵圆形；核面粗糙。花期4月中旬。

'绛桃'
（'Jiang Tao'）

18 . 'Harrow Rubirose'

起源：加拿大安大略省Harrow研究站1981年培育的耐寒品种（Layne 1981），来源于杂交组合（'Harrow Blood' × NY555036）的开放授粉实生苗。

形态：树势强，花大，红色52A，后期52B；复瓣，花瓣数15～20；着花密1.75/cm（35/20cm）。叶绿色（147A），叶长11cm，叶宽3.5cm，披针形；叶缘细圆齿；有蜜腺。

'Harrow'系列的3个品种的共性：树势强，叶绿色（147A），叶披针形，细圆齿。耐寒品种，可以在8a,7a,6b安全越冬，在6a需要保护越冬，对褐腐病、白粉病和细菌性穿孔病有一定抗性。

19. '紫叶'桃（'Zi Ye Tao'）

起源：中国古老品种，最早出现于宋代（960－1279）。

形态：干皮灰黑色，小枝紫褐色。花红色（65D）；花蕾卵圆形；花瓣卵形，长1.93cm，花径4.73cm，复瓣，梅花型，花型规则，花瓣数27枚（25～30）；雄蕊数平均46，花丝长1.07cm，花药橘红色；雌蕊1～2个，高于雄蕊；着花中密（着花率0.85/cm）；萼片红褐色，两轮，花丝和花萼均有瓣化现象；花梗长0.53cm。叶紫色，椭圆披针形，长13.8cm，宽3.8cm，叶长与叶宽比（L/W）为3.65；叶缘细圆齿，叶柄长1.08cm。果实紫红色，长4.4cm，宽3.93cm，卵圆形；果核长2.92cm，宽1.98cm，倒卵圆形；核面粗糙。花期4月中旬。

'紫叶'桃
（'Zi Ye Tao'）

'斑叶'桃（'Zuoshuang'）

'Peppermint Stick'

20. '斑叶'桃（'Zuoshuang'）

起源：北京植物园2005年从紫叶桃中发现的芽变（Hu *et al.* 2005a）。

形态：小枝紫色。红色（65D）花为主，个别红色花上有粉色（68C）花瓣；花蕾卵圆形；花瓣卵形，长1.7cm，花径4.2cm，复瓣，梅花型，花型规则，花瓣数25.3枚（24～27）；雄蕊数平均47，花丝长1.09cm，花药橘红色；雌蕊高于雄蕊；着花中密（着花率0.83/cm）；萼片红褐色，两轮，花丝和花萼均有瓣化现象；花梗长0.43cm。叶紫色、绿色（137C）或绿叶上有不规则紫色斑点或斑块（187A），叶背脉红；椭圆披针形，长13.8cm，宽3.68cm，叶长与叶宽比（L/W）为3.75；叶缘细圆齿，叶柄长0.75cm；肾形蜜腺2～4个。果实绿色，长4.34cm，宽3.72cm，卵圆形；果核长2.92cm，宽1.9cm，椭圆形，核面上短沟多且深。花期4月中旬。

21. 'Peppermint Stick'

起源：1933年由加州的W. B. Clarke苗圃（Jacobson 1996）引入美国。

形态：树势强，树体大。白色花瓣上有粉色或红色条纹，复瓣，梅花型，花瓣数24.4枚（20～32）；花丝有瓣化现象，花药黄色，雌蕊高于雄蕊，雌蕊1。

22. '凝霞紫叶'（'Ningxia Zi Ye'）

起源：在紫叶桃中发现的芽变。发表于1993年（张秀英1993）。

形态：小枝紫色。红色（65D）花为主，个别红色花上有粉色（68C）花瓣；花蕾卵圆形；花瓣卵形，长1.9cm，花径4.5cm，复瓣，梅花型，花型规则，花瓣数26枚（24～30），

'凝霞紫叶'（'Ningxia Zi Ye'）

雄蕊数平均47，花丝长1.1cm，花药橘黄色；雌蕊2～6个，高于雄蕊；着花中密（着花率0.95/cm）；萼片红褐色，两轮，花丝和花萼均有瓣化现象；花梗长0.52cm。叶紫色，椭圆披针形，长13.6cm，宽3.5cm，叶长与叶宽比（L/W）为3.89；叶缘细圆齿，叶柄长1.01cm。果实绿色带紫晕，长3.88cm，宽3.70cm，卵圆形；果核长2.80cm，宽1.78cm，倒卵圆形；核面粗糙。花期4月中旬。

23. '人面'桃（'Renmian Tao'）

起源：中国古老品种，最早出现于五代（907－979）。

形态：干皮深灰色，小枝黄褐色。花粉色（65C），无跳枝现象，授粉后变深；花蕾球形；花瓣卵形，长2.19cm，花径4.88cm，复瓣，月季型，花瓣数36枚（33～39）；雄蕊数平均56，花丝长1.22cm，白色；花药橘红色；雌蕊与雄蕊近等长，雌蕊2～5个；着花中密（着

'人面'桃（'Renmian Tao'）

花率1/cm）；萼片红褐色，两轮，花丝和花萼均有瓣化现象；花梗长0.75cm。叶绿色，椭圆披针形，长15.24cm，宽4.19cm，叶长与叶宽比（L/W）为3.64；叶缘细锯齿，叶柄长0.73cm；肾形蜜腺2～3个。果实绿色，长4.83cm，宽4.03cm，卵圆形；果核长3.1cm，宽1.98cm，倒卵圆形；核面粗糙。花期4月中下旬。

24. '二色'桃（'Erse Tao'）

起源：中国古老品种，最早出现于宋代（960－1279）。

形态：一枝上的花有粉、红二色。淡粉色（62D）上有粉红色（64D）跳枝花；花蕾球形；花瓣卵形，长2.23cm，花径5cm，复瓣，月季型，花瓣数37.3枚（35～42）；雄蕊数平均53，花丝长1.27cm，白色；花药橘红色；多心皮雌蕊2～6个，雌蕊与雄蕊近等长；着花中密（着花率1.1/cm）；萼片红褐色，两轮，花丝和花萼均有瓣化现象；花梗长0.8cm。叶绿色，椭圆披针形，长15.34cm，宽4.3cm，叶长与叶宽比（L/W）为3.57；叶缘细锯齿，叶柄长0.73cm；肾形蜜腺2～3个。果实绿色，长5.1cm，宽4.1cm，卵圆形；果核长3.3cm，宽2.0cm，倒卵圆形；核面粗糙。花期4月中旬。

'二色'桃（'Erse Tao'）

'寒红'桃
('Hanhong Tao')

25. '寒红'桃 ('Hanhong Tao')

起源：发表于1993年（张秀英1993）。

形态：干皮灰褐色，小枝黄褐色，光滑。花亮红色 (54D)；花蕾卵圆形；花瓣卵形，长1.87cm，花径4.37cm，复瓣，月季型，花高1.7cm；花瓣数44.3枚 (36～48)；雄蕊数平均43.3，花丝长1cm，花心白色，花丝水红；花药橘红色；雌蕊1～5个不等，雌蕊与雄蕊近等长；着花较稀 (着花率0.6/cm)；萼片红褐色，两轮，边缘有毛，花丝和花萼均有瓣化现象；花梗长0.73cm。叶绿色，椭圆披针形，长11.84cm，宽3.62cm，叶长与叶宽比 (L/W) 为3.27；叶缘细圆齿，叶柄长1cm；肾形蜜腺2个。果实绿色，长4.62cm，宽4.08cm，圆形；果核长3.16cm，宽1.9cm，椭圆形；核面粗糙。

'寒红'桃在北京地区的花期为红色桃花中最早的品种，仅次于山碧桃系列品种，4月上旬即可见花，比常见的'绛桃'花期还要早。

26. '菊花'桃 ('Kikoumomo')

起源：中国清代（1644－1911）即有记载。日本江户时代（17世纪—19世纪）也有记载（伊藤1695）。

形态：树势中等；干皮深灰色，小枝细长柔弱，黄褐色，节间较长，2.3cm。花粉色（65A）；花蕾卵形；花瓣披针卵形，不规则扭曲，边缘呈不规则的波状，长 2.13cm，宽 0.6cm；花径 4.53cm，复瓣，菊花型，

'菊花'桃（'Kikoumomo'）

花高 2cm；花瓣数 29 枚（22～32）；雄蕊数平均 32.7，花丝长 1cm，花丝卷曲，有瓣化现象，花药黄色，花粉活力 88.2%（Pawasut *et al.* 2004）；雌蕊略高于雄蕊；着花中等（着花率 0.75/cm）；花萼红褐色偏绿，两轮；花梗长 0.43cm。叶绿色略显灰，边缘略卷，椭圆披针形，长 10.2cm，宽 2.82cm，叶长与叶宽比（L/W）为 3.62；叶缘细锯齿，叶柄长 0.92cm；肾形蜜腺 2～3 个。果实绿色. 尖圆形，长 4.02cm，宽 3.05cm；果核长 2.92cm，宽 1.56cm，椭圆形；核面粗糙。花期 4 月中旬。

'京舞子'（'Kyoumaiko'）

27. '京舞子'（'Kyoumaiko'）

起源：是'菊花桃'的芽变（Yoshida *et al*. 2000）。

形态：干皮灰黑色，小枝黄褐色。花红色（52B）；花蕾卵形；花瓣披针卵形，长 1.7cm，宽 0.5cm，花径 4.4cm，复瓣，菊花型，花高 2.1cm；花瓣数 36 枚（33～40）；雄蕊数平均 38，花丝长 0.67cm，花丝白色，授粉后变成红色；花丝有瓣化现象；花药橘红色，花粉活力 95.2%（Pawasut *et al*. 2004）；雌蕊 2～3 个，雌蕊低于雄蕊；着花中密（着花率 0.83/cm）；花萼红褐色，两轮，花丝和萼片均有瓣化现象；花梗长 0.73cm。叶绿色，椭圆披针形，长 8.1cm，宽 2.2cm，叶长与叶宽比（L/W）为 3.68；叶缘细锯齿，叶柄长 1.2cm；肾形蜜腺 1～2 个。果实绿色，尖圆形，长 4.76cm，宽 3.56cm；果核长 3.12cm，宽 1.6cm，椭圆形；核面平滑。花期 4 月中下旬。

28. '晚白'桃（'Wanbai Tao'）

起源：发表于1991年（张秀英，陈忠国）。

形态：树体中等，干皮光滑，灰色，小枝黄褐色。花白色（155D），花蕾扁球形；花瓣长卵形，长 2.2cm，花径 5.46cm，重瓣，牡丹型，半球形，花高 2.23cm；花瓣数 63（55～68）枚；雄蕊数平均 50.7，花丝长 1.17cm，白色，花药黄色；多心皮雌蕊 2～5 个；雌蕊高于雄蕊；着花中等（着花率 0.7/cm）；花萼绿色，两轮；花丝和萼片均有瓣化现象；花梗长 0.9cm。叶绿色，椭圆披针形，长 13.9cm，宽 4.06cm，

'晚白'桃
('Wanbai Tao')

'绯桃'
('Fei Tao')

叶长与叶宽比（L/W）为 3.42，叶缘细圆齿，叶柄长 0.95cm，肾形蜜腺 2～3 个。果实绿色，长 3.62cm，宽 3.55cm，圆形；果核长 2.52cm，宽 1.54cm，倒卵圆形；核面粗糙。花期较晚，在 4 月中下旬。

除了花期明显晚于'白碧'桃外，花瓣数多的牡丹型花型也是与'白碧'桃的明显区别。

29. '绯桃'（'Fei Tao'）

起源：中国古老品种，最早出现于五代（907－979）。

形态：干皮灰黑色，小枝绿色。花绯红色（59D），花蕾扁球形；花瓣长卵形，长 2.43cm，花径 5.63cm，重瓣，牡丹型，花呈球形，花高 2.48cm；花瓣数 68.2（58～83）枚；重瓣度高，曾经发现个别植株达到 120 枚；雄蕊数平均 44，花丝长 1.22cm，花丝白色，授粉后变成红色；花药黄色；多心皮雌蕊 2～5 个；雌蕊高于雄蕊；着花较稀（着花率 0.55/cm）；萼片红褐色，两轮；花丝和花萼均高度瓣化；花梗长 0.65cm。叶绿色，卵圆披针形，长 12.86cm，宽 4.26cm，叶长与叶宽比（L/W）为3.02；叶缘细圆齿，叶柄长0.97cm；肾形蜜腺3～4个。果实绿色，长5.44cm，宽4.33cm，卵圆形；果核长3.8cm，宽1.7cm，椭圆形；核面粗糙。花期4月下旬，有些植株可延续到5月初。

独有的小枝绿色红花品种，花期时可以作为'绯桃'明显的识别特点。

30. '红碧'桃（'Hong Bitao'）

起源：中国古老品种。

形态：干皮深灰色，小枝灰黄色。花亮红色（66D），花蕾球形；花瓣长卵形，长2.43cm，花径5.63cm，重瓣，牡丹型，花呈近球形，较松散，花高2.33cm；花瓣数55.4（45～64）枚；雄蕊数平均32.8，花丝长1.33cm，花丝白色，授粉后变成红色；花丝有瓣化现象；花药橘红色；多心皮雌蕊2～3个；雌蕊高于雄蕊；着花中等（着花率0.65/cm）；萼片红褐色，两轮；花丝和花萼均有明显瓣化现象；花梗长0.73cm。叶绿色，椭圆披针形，长14.14cm，宽3.76cm，叶长与叶宽比（L/W）为3.76；叶缘细圆齿，叶柄长1.28cm；肾形蜜腺1～2个。果实绿色，长4.3cm，宽3.78cm，卵圆形；果核长2.7cm，宽1.75cm，倒卵圆形；核面粗糙。花期4月中下旬。

'红碧'桃与'绯桃'最明显的区别在于花色和花型，前者亮红，后者的红色则较深；'红碧'桃由于花瓣数较少而花型整体比'绯桃'松散。

31. '玫紫'（'Mei Zi'）

起源：北京植物园1998年发现的芽变品种（Hu and Zhang 1999）。

形态：干皮深灰色，小枝灰黑色。花玫红色（73B），花蕾球形；花瓣长卵形，长2.17cm，花径4.93cm，重瓣，牡丹型，半球形，花高1.7cm；花瓣数60.7（54～65）枚；雄蕊数平均86.67，花丝长1.07cm，白色；大蕾期雄蕊伸出花蕾；花药白色；多心皮雌蕊2～5个；雌蕊与雄蕊近等长；着花中等（着花率0.75/cm）；萼片红褐色，两轮；花丝和花萼均有明显瓣化现象；花梗长1.15cm。叶绿色，卵圆披针形，长12.74cm，宽4.02cm，叶长与叶宽比（L/W）为3.17；叶缘细圆齿，叶柄长0.8cm；肾形蜜腺2个。花期4月中下旬。

独特的花色和白色花药是'玫紫'最主要的识别特征。

'红碧'桃
('Hong Bitao')

'玫紫'
('Mei Zi')

32.'碧桃'（'Bi Tao'）

起源：中国古老品种，最早出现于五代（907－979）。

形态：干皮灰褐色，小枝紫色带绿晕。花肉粉色（62B），花蕾卵圆形；花瓣卵形，长2.87cm，花径6.25cm，重瓣，牡丹型，近球形，花高2.10cm；花瓣数50.8（42～63）枚；雄蕊数平均32.67，花丝长1.53cm，白色；花药黄色；多心皮雌蕊2～3个；雌蕊与雄蕊近等长；着花中等（着花率0.8/cm）；萼片绿色，两轮；花丝和花萼均有明显瓣化现象；花梗长0.75cm。叶绿色，椭圆披针形，长13cm，宽3.5cm，叶长与叶宽比（L/W）为3.72；叶缘细锯齿，叶柄长1.22cm；肾形蜜腺2～4个。果实绿色，长5.1cm，宽4.1cm，卵圆形；果核长3.55cm，宽1.9cm，倒卵圆形；核面粗糙。花期4月中下旬。

　　纯色肉粉色的球形花朵是'碧桃'最显著的识别特征。

'碧桃'
（'Bi Tao'）

'簪粉'
('Zan Fen')

33. '簪粉'（'Zan Fen'）

起源：发表于1993年（张秀英）。

形态：枝条开展，干皮灰色，小枝灰褐色。花粉紫色（73C），远观略带紫色晕；花蕾扁球形；花瓣长卵形，长 2.47cm，花径 5.63cm，重瓣，牡丹型，半球形，花高 2.03cm；花瓣数 62.67（53～68）枚；雄蕊数平均 60.67，花丝长 1.2cm，花丝有瓣化现象；花药橘红色；多心皮雌蕊 3～5 个；雌蕊高于雄蕊；着花中等（着花率 0.65/cm）；萼片红褐色，两轮；花丝和花萼均有明显瓣化现象；花梗长 0.86cm。叶绿色，椭圆披针形，长 13.68cm，宽 3.8cm，叶长与叶宽比（L/W）为 3.6；叶缘细锯齿，叶柄长 1.06cm；肾形蜜腺 1～3 个。果实绿色，长 4.65cm，宽 4.48cm，圆形；果核长 3cm，宽 1.9cm，倒卵圆形；核面粗糙。花期 4 月中下旬。

独特的花色是识别'簪粉'最主要的特点。

'洒红'桃（'Sahong Tao'）

34. '洒红'桃（'Sahong Tao'）

起源：中国古老品种。

形态：干皮灰褐色，小枝绿色或上有紫褐色斑点或条纹。花粉（66D）、红（59D）、白（155D）复色，以白色花为主，远观为白色上有星星点点的粉色或红色；花蕾扁球形；花瓣长卵形，长 3.02cm，花径 6.08cm，重瓣，牡丹型，花呈近球形，花高 2.1cm；花瓣数 63.3（58～66）枚；雄蕊数平均 35.3，花丝长 1.17cm，花药为黄色；多心皮雌蕊 2～4 个；雌蕊高于雄蕊；着花中等（着花率 0.7/cm）；花萼颜色与花瓣色相关，白色花为绿色花萼，白粉复色花则为红褐色与绿色相间，两轮，花萼及花丝均有明显瓣化现象；花梗长 0.86cm。叶绿色，椭圆披针形，长 15.5cm，宽 3.95cm，叶长与叶宽比（L/W）为 3.92；叶缘细圆齿，叶柄长 1.22cm；肾形蜜腺 1～4 个不等。果实绿色，长 4.86cm，宽 4.12cm，尖圆形；果核长 3.084cm，宽 1.84cm，椭圆形；核面粗糙。花期 4 月中下旬。

以白色为主体的花色，但鲜有纯白色花，总有粉色或红色斑点和条纹。

35. '五宝'桃（'Wubao Tao'）

起源：中国古老品种。

形态：干皮灰黑色，小枝黄褐色。一花二色，粉色花（65D）上洒有水粉（67D）或水红（66C）色条纹；花蕾扁球形；花瓣长卵形，长 2.57cm，花径 5.33cm，重瓣，牡丹型，花瓣数 59.2（50～74）枚；近球形，花高 2.1cm；雄蕊数平均 46.5，花丝长 1.11cm；花药橘红色；多心皮雌蕊 4～6 个；雌蕊与雄蕊近等长；着花中等（着花率 0.65/cm）；

'五宝'桃（'Wubao Tao'）

萼片红褐色，两轮，花丝和花萼均有明显瓣化现象；花梗长0.57cm。叶粗厚，纸质，绿色深于其他品种；叶卵圆披针形，长11.23cm，宽3.78cm，叶长与叶宽比（L/W）为2.98；叶缘粗锯齿，叶柄长1cm；肾形蜜腺2～3个不等。果实绿色，长5.44cm，宽4.22cm，卵圆形；果核长3.02cm，宽1.62cm，倒卵圆形；核面粗糙。花期4月下旬，有些植株可延续到5月初。

36. '二乔'（'Er Qiao'）

起源：1997年在北京植物园发现的'红碧'桃的芽变。

形态：干皮灰黑色，小枝黄褐色。一花二色，红色花（63A）上有粉色（65B）花瓣；花蕾扁球形；花瓣长卵形，长2.47cm，花径4.17cm，重瓣，牡丹型，半球形，花高1.67cm；花瓣数53.67（50～59）枚；雄蕊数平均40.33，花丝长1.2cm，花丝有瓣化现象；花药橘红色；多心皮雌蕊2～3个；雌蕊高于雄蕊；着花中等（着花率0.68/cm）；萼片红褐色，两轮；花梗长0.9cm。叶绿色，椭圆披针形，长15cm，宽3.67cm，叶长与叶宽比（L/W）为4.0；叶缘细圆齿，叶柄长1.04cm；肾形蜜腺2～4个不等。果实绿色，长5.32cm，宽4.11cm，卵圆形；果核长3cm，宽1.68cm，倒卵圆形；核面粗糙。花期4月下旬，有些植株可延续到5月初。

'二乔'（'Er Qiao'）

二、寿星桃品种群

寿星桃是枝条节间很短（一般小于10mm），着生紧密的一类低矮桃花的统称。叶通常为椭圆状披针形，叶缘呈不同程度的波状。

该品种群中的'单瓣寿粉'、'单瓣寿红'、'单瓣寿白'、'油寿白'、

'寿粉'、'寿红'、'寿白'、'瑕玉寿星'、'二乔寿星'、'幸白'等为绿叶品种；'赤叶寿星'（'Red Dwarf'）（Yoshida 1974）和'Bonfire'（重瓣白色花）（Moore *er al.* 1993）为紫叶品种。

寿星桃出现在中国的明代（1368－1644），欧美各国的寿星桃几乎均来自中国。美国最早应用寿星桃作为盆栽是在1911年（Hedrick 1917）。粉色、红色及复色的重瓣花矮生型桃花类型'Dwarf Mandarin'是经W.B. Clarke 苗圃1940年从中国引进（Jacobson 1996）。所以寿星桃品种名称以中国品种名称为准，以往有些书上使用的f. *densa*——Dense Group，亦为此类，在此均归入相应的品种异名中。

根据花型和花色，寿星桃品种群的品种分类检索表如表6.2：

表6.2 寿星桃品种群品种分类检索表
Table 6.2 Key to ornamental peach cultivars of Dwarf group

1. 枝条节间短，花单瓣，叶绿色或紫色
  2. 花白色，叶绿色
    3. 果实有毛 ·······························'单瓣寿白''Danban Shoubai'
    3. 果实无毛 ································'油寿白''You Shoubai'
  2. 花粉色或红色，叶绿色或紫色
    4. 花粉或红色，叶绿色
      5. 花粉色 ···························'单瓣寿粉''Danban Shoufen'
      5. 花红色 ·························' 单瓣寿红''Danban Shouhong'
    4. 花粉色，叶紫色
      6. 不具备对特殊病害的抗性 ··············'赤叶寿星''Red Dwarf'
      6. 对细菌性穿孔病有良好抗性 ····················'Bonfire'
1. 枝条节间短，花复瓣，叶绿色
    7. 花白色
      8. 花瓣卵形，花型似梅花 ··················'寿白''Shoubai'
      8. 花瓣披针卵形，花型似菊花 ···'幸白''Saiwai Howaito'
    7. 花粉色、红色或复色
      9. 花粉色 ····························'寿粉''Shoufen'
      9. 花红色或复色
        10. 花红色 ·····················'寿红''Shouhong'
        10. 花复色
          11. 花粉白复色 ············'瑕玉寿星''Xiayu Shouxing'
          11. 花粉红复色 ·········'二乔寿星''Erqiao Shouxing'

37. '单瓣寿白'（'Danban Shoubai'）

起源：不详。

形态：干皮灰色，小枝绿色。花白色（155D），花蕾卵圆形；花瓣卵圆形，长1.5cm，花径3.4cm，单瓣型，花瓣数5～6枚，雄蕊数平均32，花丝长1.2cm，白色；花药黄色；雌蕊高于雄蕊；着花密（着花率1.7/cm）；花萼绿色，5枚；花梗长1.4cm。叶绿色，披针形，长14.16cm，宽2.66cm，叶长与叶宽比（L/W）为5.32；叶缘皱波状细

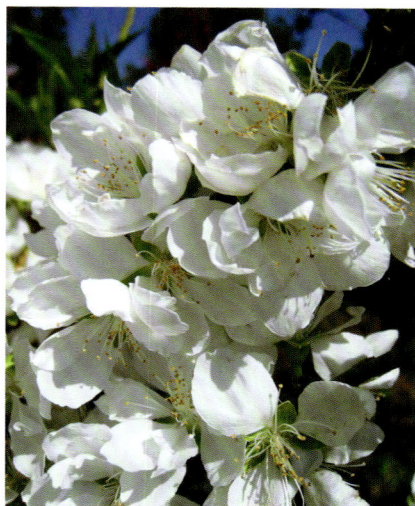

'单瓣寿白'（'Danban Shoubai'）

锯齿，叶柄长0.88cm，蜜腺肾形，2～3个。果实有毛，绿色，长4.42cm，宽3.7cm，卵圆形；果核长2.88cm，宽1.88cm，椭圆形，核面粗糙。花期4月中旬。

38. '油寿白'（'You Shoubai'）

起源：1997年在北京植物园发现的芽变。发表于1997年（张秀英，戴思兰）。

形态：'油寿白'与'单瓣寿白'的最大区别就在于果实无毛，果实长3.88cm，宽3.44cm，小于'单瓣寿白'；果核核面以点为主，区别于'单瓣寿白'的沟点并存。其他生状与'单瓣寿白'区别不大。

'油寿白'（'You Shoubai'）

'单瓣寿粉'（'Danban Shoufen'）

39. '单瓣寿粉'（'Danban Shoufen'）

起源：不详。

形态：干皮灰色，小枝红褐色。花粉色（65B），花蕾卵圆形；花瓣卵圆形，长 2.23cm，花径 4.57cm，单瓣型，花瓣数 5～6 枚；雄蕊数平均 41.3，花丝长 1.47cm，粉白色，花药橘红色；雌蕊与雄蕊近等长；着花中密（着花率 1.8/cm）；花萼红褐色，5 枚；花梗长 0.43cm。叶绿色，披针形，长 15.88cm，宽 2.9cm，叶长与叶宽比（L/W）为 5.47；叶缘皱波状细锯齿，叶柄长 0.92cm。果实绿色，长 4.2cm，宽 4.14cm，圆形；果核长 2.5cm，宽 1.74cm，椭圆形，核面光滑。花期 4 月中旬。

40. '单瓣寿红'（'Danban Shouhong'）

起源：不详。

形态：干皮灰色，小枝红褐色。花红色（59D），花蕾卵圆形；花瓣卵形，长 2.1cm，花径 4.23cm，单瓣型，花瓣数 5～6 枚；雄蕊数平均 32.7，花丝长 1.17cm，白色到水红色，花心白色；花药橘红色；雌蕊与雄蕊近等长；着花密（着花率 1.6/cm）；花萼红褐色，5 枚；花梗长 0.4cm。叶绿色，披针形，长 14.2cm，宽 2.8cm，叶长与叶宽比（L/W）为 5.14；叶缘皱波状细锯齿，叶柄长 0.88cm，蜜腺肾形，2～3 个。果实绿色。花期 4 月中旬。

'单瓣寿红'
('Danban Shouhong')

41. '赤叶寿星'（'Red Dwarf'）

起源：选自'赤芽'（单瓣，粉色花，紫叶）和另一寿星桃的F$_2$代，由吉田雅夫于1974年育成并发表（Yoshida & Seike 1974）。

形态：干皮灰色，小枝红褐色。节间0.53cm。花粉色（69C），花蕾长卵形；花瓣长卵形；长2cm，花径4.13cm，单瓣型，花瓣数5枚；雄蕊数平均40.7，花丝长1.17cm，花药橘红色；雌蕊高于雄蕊；着花密（着花率1.9/cm）；花萼红褐色，5枚；花梗长0.33cm。叶暗红色，披针形；叶长19.8cm，宽3.7cm，叶长与叶宽比（L/W）为5.35；叶缘皱波状细锯齿，叶柄长0.84cm。果实绿色略带紫晕，长4.2cm，宽3.6cm，卵圆形；果核长2.8cm，宽2.1cm，卵圆形，核面粗糙。花期在其育成地为4月中旬。

'赤叶寿星'
('Red Dwarf')

42. 'Bonfire'：Tom Thumb^{TM}

起源：'Bonfire'（紫叶寿星，单瓣粉色花）（Moore *et al*. 1993）是来自日本的 Tsukuba No.2 天然授粉的 F_2 代实生苗。

形态：干皮光滑，皮孔密。节间长度平均 8mm。花大，粉红色（红－紫系列 62D），单瓣型，5 瓣，花径 3.5cm。叶色紫红（187A），披针形，长 18.45cm，宽 3.36cm，叶长与叶宽比（L/W）为 5.49；叶腺 2～4 个，肾形，叶柄长度 1.22cm。果实小，紫红色，有毛，椭圆形。可以耐－23℃。

花期：在品种育成地阿肯色州 3 月中始花，3 月中 50%，4 月 1 日左右盛花。

'Bonfire' 的果实颜色偏紫，对细菌性穿孔病有良好的抗性，是与 '赤叶寿星' 的主要区别。

'Bonfire'

'寿白'
（'Shoubai'）

43.'**寿白**'（'Shoubai'）

起源：不详。

形态：干皮粗糙，深灰色，小枝绿色。花白色（155D），花蕾球形；花瓣卵圆形，长2.17 cm，宽1.8cm，花径4.6 cm，复瓣，梅花型，花瓣数25.2枚（25～29）；雄蕊数平均51.8，花丝长1.05cm，白色，花丝有瓣化现象，花药黄色；雌蕊高于雄蕊；着花密（着花率1.91/cm）；花萼绿色，两轮；花丝和萼片均有瓣化现象；花梗长0.77cm。叶绿色，披针形，长14.93cm，宽2.9cm，叶长与叶宽比（L/W）为5.2；叶缘皱波状，细锯齿；叶柄长1.43cm；蜜腺肾形，3～4个。果实绿色，长3.84cm，宽3.68cm，圆形，果核长2.7cm，宽1.94cm，椭圆形；核面粗糙。花期4月中下旬。

'幸白'
（'Saiwai Howaito'）

44.'**幸白**'（'Saiwai Howaito'）

起源：用寿星桃和'菊桃'杂交育成（前田克夫等 2003）。育种人：前田克夫、荻原勋、志村勋、箱田直紀及石川駿二。

形态：树姿开张，节间长度极短，树形矮小，树势中等，枝条粗，小枝绿色。花瓣白色，花瓣长椭圆形，瓣缘无波纹，花复瓣，菊花型，花蕊瓣化，花萼内壁黄色。花期在育成地日本东京为4月上旬。

'寿粉'（'Shoufen'）

45. '寿粉'（'Shoufen'）

起源：不详。

形态：树冠扁球形，干皮深灰色，小枝绿色带紫晕。花粉色（65B），花蕾卵球形；花瓣卵圆形，长1.98cm，宽1.67cm，花径4.5cm，复瓣，梅花型，花瓣数23.8枚（22～25）；雄蕊数平均43.8，花丝长1.18cm，白色，花丝有瓣化现象，花药橘红色；雌蕊高于雄蕊；着花密（着花率1.47/cm）；花萼红褐色偏绿，两轮；花丝和萼片均有瓣化现象；花梗长0.57cm。叶绿色，披针形，长15.6cm，宽2.69cm，叶长与叶宽比（L/W）为5.2；叶缘皱波状，细锯齿；叶柄长1.1cm；蜜腺肾形，2个。果实绿色，长4.15cm，宽4.15cm，圆形；果核长2.42cm，宽1.47cm，卵圆形；核面粗糙。花期4月中旬。

46. '寿红'（'Shouhong'）

起源：不详。

形态：树冠圆球形，干皮粗糙、灰黑色，小枝黄褐色。花红色（59D），花蕾球形；花瓣卵圆形，长2.07cm，宽1.87cm，花径4.33cm，复瓣，梅花型，花瓣数18.3枚（16～29）；雄蕊数平均32.7，花丝长1.3cm，水红色，花心白色；花丝有瓣化现象，花药黄色；雌蕊与雄蕊近等长；着花密（着花率1.57/cm）；花萼红褐色，两轮；花丝和萼片均有瓣化现象；花梗长0.27cm。叶绿色，披针形，长16.5cm，宽3.1cm，叶长与叶宽比（L/W）为5.3；叶缘皱波状明显，细锯齿；叶柄长0.98cm；蜜腺肾形，1～3个。果实绿色，长4.22cm，宽3.92cm，卵圆形；果核长2.74cm，宽1.94cm，椭圆形；核面粗糙。花期4月中下旬。

'寿红'
（'Shouhong'）

47. '瑕玉寿星'（'Xiayu Shouxing'）

起源：发表于1993年（张秀英）。

形态：树型不甚规整，干皮深灰色，小枝绿色或黄褐色。白色花（155C）上有粉色（62B）条纹或斑点；花蕾卵形；花瓣卵形，长2.22cm，花径4.65cm，复瓣，梅花型，花瓣数23.5枚（23～24）；雄蕊数平均54，花丝长1.53cm，花药颜色与花瓣色相关，白色花的花药为黄色，纯粉色花的花药为橘黄色，白色与粉色复色的花其花药为黄色，花丝有瓣化现象；雌蕊与雄蕊近等长；着花密（着花率1.8/cm）；花萼颜色与花瓣色相关，白色花为绿色花萼，白粉复色花则为红褐色与绿色相间，两轮，花萼及花丝均有瓣化现象；花梗长0.3cm。叶绿色，披针形，长15.8cm，宽2.5cm，叶长与叶宽比（L/W）为6.3；叶缘细锯齿，叶柄长1.28cm；肾形蜜腺2～3个。果实绿色，长4.78cm，宽4.16cm，卵圆形；果核长2.84cm，宽1.86cm，椭圆形；核面粗糙。花期4月中下旬。

'瑕玉寿星'（'Xiayu Shouxing'）

'二乔寿星'（'Erqiao Shouxing'）

48.'二乔寿星'（'Erqiao Shouxing'）

起源：1997年发现在北京植物园的芽变。

形态：圆球形，干皮粗糙、灰黑色，小枝绿色或黄褐色。花粉色63D 与红色（59D）跳枝或一朵花上同时有这两种颜色，花蕾长卵圆形；花瓣 长卵圆形，长2.12cm，宽1.8cm，花径4.13cm，复瓣，梅花型，花瓣数16.2 枚（12～19），雄蕊数平均28，花丝长1.6cm，花丝有瓣化现象，花药黄 色；雌蕊高于雄蕊；着花密（着花率1.8/cm）；花萼红褐色，两轮；花 梗长0.77cm。叶绿色，披针形，长14.83cm，宽2.57cm，叶长与叶宽 比（L/W）为5.77；叶缘皱波状，细圆齿；叶柄长0.92cm；蜜腺肾 形，4个。花期4月中下旬。

三、帚型桃品种群

帚型类桃花是桃花众多品种中极为独特的一个类型，也被称作"塔型桃"、"柱型桃"，最早起源于日本的江户时代（17世纪～19世纪）。1695年伊藤伊兵卫三之丞在其著名的《花坛地锦抄》中就已经有关于帚桃的记载。帚型桃是日本给予世界的一大贡献，故此沿用这一古老名称作为该品种群的名称。

帚型桃类最显著的特点就是：树体高大直立，枝干开张角度狭小，树冠窄高，干性极强，极适合作行道树和庭园栽植。

该品种群有'帚桃'、'照手桃'、'照手姬'、'照手红'、'照手白'、'Corinthian Mauve'（Werner *et al.* 2000c）、'Corinthian White'（Werner *et al.* 2000a） 为绿叶品种。另外两个，'Corinthian Rose'（Werner *et al.* 2000b） 和 'Corinthian Pink'（Werner *et al.* 2001） 为紫叶品种。

根据花型和花色，帚型桃品种群的品种分类检索表如表6.3：

表6.3　帚型桃品种群品种分类检索表
Table 6.3　Key to ornamental peach cultivars of Pillar group

1. 小枝直立，开张角度狭小，花复色·······················'**帚桃**''Houkimomo'
1. 小枝直立，开张角度狭小，花单色
　2. 花白色，叶绿色 ·································'**照手白**''Teruteshiro'
　2. 花粉色或红色，叶绿或紫色
　　3. 花粉色或红色，叶绿色
　　　4. 花粉色
　　　　5. 花淡粉色（62C），花瓣数 25 ～ 32 ·······'**照手姬**''Terutehime'
　　　　5. 花粉红色（73B），花瓣数 18 ～ 24 ·······'**照手桃**''Terutemomo'
　　　4. 花红色 ·······························'**照手红**' 'Terutebeni'
　　3. 花粉色，叶紫色
　　　6. 花淡粉色，叶紫色 ····························'Corinthian Pink'
　　　6. 花深粉色，叶深紫色 ·························'Corinthian Rose'

49. '帚桃'（'Houkimomo'、'Pillar'、'Pyramid'、'Fastigiata'）

起源：日本江户时代就有帚桃的记载（伊藤1695）。

形态：干皮黄褐色，小枝灰黄色、直立，分枝角度小。白色花（155D）与粉色花（65A）跳枝，以白色花为主；花蕾卵形；花瓣卵圆形，长2.07cm，宽1.6cm，花径4.5cm。复瓣，梅花型，花瓣数29枚（28～30）；雄蕊数平均35.7，花丝长1.1cm，花药颜色与花瓣色相关，纯粉色花的花药为橘红色，白色花及复色花的花药为黄色，花丝有瓣化现象；雌蕊与雄蕊近等长；着花中等（着花率0.8/cm）；花萼颜色与花瓣色相关，红色花为红褐色花萼，白色花为绿色花萼，复色花则为红褐色与绿色相间，两轮，花萼及花丝均有瓣化现象；花梗长0.83cm。叶绿色，披针形，长12.6cm，宽3.5cm，叶长与叶宽比（L/W）为3.6；叶缘细圆齿，叶柄长0.9cm。果实绿色，果核椭圆形，核面平滑。花期4月中旬。

帚桃起源于日本，尽管不同国家对白色花与粉色花跳枝的帚桃有着不同的叫法，但所有现在帚桃类型的品种最早均来源于此。在此用'帚桃'作为品种名，原来日本的假名Houkimomo也作为统一的品种名雅名，取代以往的Pillar，Columnar，Pyramid和Fastigiata等名称。

'帚桃'
（'Houkimomo'）

50. '照手桃'（'Terutemomo'、'Corinthian Mauve'）

起源：选自'帚桃'和'赤枝垂'的F$_2$代（山崎等 1987）。在美国使用的粉花重瓣品种可选用'Corinthian Mauve'（Werner *et al.* 2000c）。

形态：干皮黄褐色，小枝灰黄色、直立，分枝角度小。花粉红色(73B)，花蕾圆形；花瓣长卵形，长 2cm，花径 4.7cm，复瓣，梅花型，花瓣数 22.1 枚(18～24)；雄蕊数平均 43，花丝长 1cm，白色，花丝有瓣化现象，花药橘红色；雌蕊与雄蕊近等长；着花中密(着花率 1/cm)；花萼红褐色，两轮；花梗长 0.67cm。叶绿色，椭圆披针形，长 12.1cm，宽 3.5cm，叶长与叶宽比(L/W)为 3.5；叶缘细圆齿，蜜腺肾形，一般 3 个，2～5 个；叶柄长 0.9cm。果实绿色，长 3.62cm，宽 3.58cm，圆形；果核长 2.36cm，宽 1.64cm，椭圆形，核面平滑。花期 4 月中旬。

鉴于花色、花瓣数等形态特征，以及AFLP分析结果（Hu *et al.* 2005）显示，'Corinthian Mauve'（Werner *et al.* 2001c）与'照手桃'并无显著差异，所以在此未将其作为独立的品种列出，根据《国际栽培植物命名法规》（ICNCP）（Brickell *et al.* 2004）的规定，以首先发表的品种名为准，将其置于相应品种之后。在北美地区应用与'照手桃'有相同效果。

'照手桃'
（'Terutemomo'）

'照手桃'（'Terutemomo'）

'照手姬'（'Terutechime'）

51.‘照手姬’（‘Terutehime’）

起源：‘帚桃’的天然授粉后代（堀越等 1992）。

形态：干皮黄褐色，小枝灰黄色、直立，分枝角度小。花淡粉色（62C），花蕾卵圆形；花瓣长卵形，长 2.1cm，宽 1.4cm，花径 4.7cm，复瓣，梅花型，花瓣数 28.6 枚（25～32）；雄蕊数平均 49.7，花丝长 1.17cm，白色，花药橘红色；雌蕊与雄蕊近等长；着花中密（着花率 0.8/cm）；花萼红褐色偏绿，两轮；花丝和萼片均有瓣化现象；花梗长 0.63cm。叶绿色，椭圆披针形，长 12.3cm，宽 3.4cm，叶长与叶宽比（L/W）为 3.6；叶缘细锯齿近细圆齿，叶柄长 0.9cm。果实绿色，长 3.92cm，宽 3.78cm，卵圆形；果核长 2.54cm，宽 1.66cm，椭圆形；核面粗糙。花期 4 月中旬。

‘照手姬’
（‘Terutehime’）

'照手红'
('Terutebeni')

52.'照手红'('Terutebeni')

起源：选自'帚桃'和'赤枝垂'的F₂代（山崎等 1987b）。

形态：干皮灰黑色，小枝灰黄色、直立，分枝角度小。花红色（59D），花蕾卵形；花瓣卵形，长1.9cm，宽1.7cm，花径4.6cm，复瓣，梅花型，花瓣数21.5枚（17～24），雄蕊数平均43.5，花丝长1.17cm，红色，花药橘红色；雌蕊与雄蕊近等长；着花中等（着花率0.65/cm）；花萼红褐色，两轮，花萼及花丝均有瓣化现象；花梗长0.75cm。叶绿色，椭圆披针形，长12.3cm，宽3.4cm，叶长与叶宽比（L/W）为3.6；叶缘细圆齿，叶柄长0.9cm。果实绿色，长3.76cm，宽3.66cm，圆形；果核长2.52cm，宽1.76cm，椭圆形；核面粗糙。花期4月中旬。

53. '照手白'（'Teruteshiro'、'Corinthian White'）

起源：选自'帚桃'和'残雪垂枝'的F$_2$代（山崎等 1987b）。在美国使用的粉花重瓣品种可选用'Corinthian White'（Werner *et al.* 2000a）。

形态：干皮灰黑色，小枝灰黄色、直立，分枝角度小。花白色

（155D），花蕾卵形；花瓣卵圆形，长1.8cm，宽1.6cm，花径4.5cm，复瓣，梅花型，花瓣数34枚（31～37）；雄蕊数平均45，花丝长1.1cm，白色，花丝有瓣化现象，花药黄色；雌蕊高于雄蕊；着花中等（着花率0.6/cm）；花萼绿色，两轮；花丝和萼片均有瓣化现象；花梗长0.67cm。叶绿色，椭圆披针形，长13.8cm，宽3.7cm，叶长与叶宽比（L/W）为3.7；叶缘细圆齿，叶柄长0.84cm。果实绿色，长4.22cm，宽3.47cm，卵圆形；果核长2.42cm，宽1.48cm，倒卵圆形；核面平滑。花期4月中旬。

鉴于花色、花瓣数等形态特征，以及AFLP分析结果（Hu *et al.* 2005b）显示，'Corinthian White'（Werner *et al.* 2001a）与'照手白'并无显著区别，所以在此未将其作为独立的品种列出，根据《国际栽培植物命名法规》（ICNCP）（Brickell *et al.* 2004）的规定，以首先发表的品种名为准，将其置于相应品种之后。在北美地区应用与'照手白'有相同效果。

54. 'Corinthian Pink'

起源：选自油桃'NC174RL'和来自日本的'Pillar'（帚型桃）的F$_2$代实生苗（Werner *et al.* 2001）。专利号：PP11902。

'Corinthian Pink'

形态：树体大。枝条分枝角度5°～20°，树势强，生长快，年生长量为90～120cm，4年生苗达到4.27m高，1.37m冠幅。树皮粗糙程度中等。小枝受光面红色。花色淡粉，初期（56D）与成熟（56A）略有差别；重瓣，平均为31.4瓣，花径平均3.8cm；雄蕊多数，花药亮黄色，雌蕊高于雄蕊；花萼淡棕色183A，花梗长度1.4cm。叶色紫色，从幼叶（183A）到成熟叶（178B）颜色有所变化；叶片大，卵圆披针形，叶长14.1cm，叶宽4.8cm，叶长与叶宽比（L/W）为2.93；叶缘细圆齿，叶尖急尖，蜜腺肾形，3个，叶柄长度1.19cm；果实有毛，小。对细菌性叶斑病有一定抗性，易染桃树蛀干天牛；需冷量为950h（4C），子房可育，但坐果率低，少于1%的花可以结果。果实紫色，长4.05 cm，宽3.75cm，椭圆形；果核长2.13cm，宽1.78cm，倒卵圆形；核面粗糙。

花期在其育成地北卡罗来纳州为3月（3月25日到4月10日），花期可延续14天。

55. 'Corinthian Rose'

起源：选自油桃 'NC174RL' 和来自日本的 'Pillar'（帚型桃）的F$_2$代实生苗（Werner *et al.* 2000b）。专利号：PP11564。

形态：树势强健，树体中等，树冠窄高，生长迅速，4年生树4.3～4.88m高；树皮光滑，灰绿色（197A）；对细菌性叶斑病有一定抗性。年生长量为90～120cm，4年生的树可以长到4.3m高，1.3m冠幅。树冠窄高，分枝角度在5°～20°。树皮粗糙程度中等。小枝绿色，向阳面有紫色晕，1年生枝条颜色灰偏红（178B），2年生休眠枝条为灰棕色（199A）。花蕾大，卵形；初开花朵深粉色（56C），成熟后为红色68C；花大，重瓣，花瓣数27.4，花径3.8cm；花萼颜色183A；雄蕊多数，雌蕊大多1个，花药亮黄色，花梗长1.5cm。叶大，深紫红色，叶长11.1cm，宽3.8cm，叶长与叶宽的比值为2.92，卵圆披针形；肾形蜜腺平均3个。果实绿色略带紫晕，果径小于3cm，椭圆形；毛多。

在其育成地北卡罗来纳州初花期在3月15日～30日，盛花期在3月25日到4月10日。花期可持续10～14天，个体花期可达7～10天。需冷量为950h。

与 'Corinthian Pink' 的区别在于花色偏深，叶色偏紫，果实偏绿。对细菌性叶斑病有一定抗性。

'Corinthian Rose'

四、垂枝桃品种群

　　垂枝桃是桃花中枝姿最具韵味的一个类型，小枝拱形下垂，树冠伞形。花开时节，宛如花帘一泻而下，蔚为壮观。无论是孤植于庭院，还是群植，都有很好的观赏效果。

　　根据花型和花色，垂枝桃品种群的品种分类检索表如表6.4：

表6.4 垂枝桃品种群品种分类检索表

Table 6.4 Key to ornamental peach cultivars of Weeping group

1. 小枝下垂，单瓣或复瓣，叶绿色
   2. 花单瓣
      3. 花白色
         4. 果实有毛·························· '白枝垂''Shiroshidare'
         4. 果实无毛（油桃）················ '白色荣光''White Glory'
      3. 花粉色，粉红色或复色
         5. 花粉色·························· '单粉垂枝''Danfen Chuizhi'
         5. 花粉红色、红色或复色
            6. 花红色························ '赤枝垂''Akashidare'
         6. 花复色
            7. 白粉跳枝···················· '源平垂枝''Genpeishidare'
            7. 白色与粉色条纹，无纯色花······ '幸枝垂''Sachi-shidare'
   2. 花复瓣
      8. 花白色
         9. 花瓣在25瓣以内 ················ '绿萼垂枝''Lü E Chuizhi'
         9. 花瓣28～42瓣················ '残雪垂枝''Zansetsushidare'
      8. 花红色、粉色或复色
         10. 花红色
            11. 梅花型，花瓣在15枚以内····················
                ···················· '红雨垂枝''Hongyu Chuizhi'
            11. 梅花型，花瓣在18～30瓣····················
                ···················· '相模枝垂''Sagamishidare'
         10. 花粉色或复色
            12. 花复色，白色与粉色跳枝····················
                ···················· '鸳鸯垂枝''Yuanyang Chuizhi'
            12. 花纯粉色或不同深浅的粉色跳枝
               13. 整株淡粉色花色均匀，几乎无跳枝现象·········
                   ···················· '黛玉垂枝''Daiyu Chuizhi'
               13. 花粉色跳枝
                  14. 花以淡粉为主，有粉色跳枝 ················
                      ···················· '五宝垂枝''Wubao Chuizhi'
                  14. 花粉红色为主，偶有淡粉跳枝················
                      ···················· '羽衣枝垂''Hagoromo-shidare'
1. 小枝下垂，花复瓣，叶紫色
            15. 花粉色 ······ '粉花紫叶垂枝''Pink Cascade'
            15. 花红色 ······ '红花紫叶垂枝''Crimson Cascade'

56. '白枝垂'（'Shiroshidare'、'单白垂枝'）

起源：是从'赤枝垂'（'Akashidare'）的天然授粉实生苗中筛选出的品种（Yoshida *et al.* 2000）。

形态：小枝下垂，绿色，树势中等。花白色（155D），花蕾长卵形；花瓣卵形，长1.97cm，花径3.6cm，单瓣型，花瓣数7.5（5～10）；雄蕊数平均34，花丝长1.63cm，花药黄色；雌蕊1枚，雌蕊与雄蕊近等长；着花中等（着花率0.78/cm），花梗长0.57cm；花萼绿色，5枚。叶绿色，椭圆披针形，长13.7cm，宽3.6cm，叶长与叶宽比（L/W）为3.8，叶缘细圆齿，叶柄长0.9cm。果实绿色，长3.3cm，宽3.02cm，圆形；果核长2.28cm，宽1.68cm，椭圆形，核面平滑。花期4月中旬。

'白枝垂'
（'Shiroshidare'
'单白垂枝'）

57. '白色荣光'（'White Glory'）

起源：1985年由美国北卡罗来纳州立大学Werner育成发表（Werner *et al.* 1985），已在专门负责那些并无专门登录机构的木本观赏植物进行登录的登录机构["Cultivar Registration of Unassigned Woody Ornamentals"（CRUWO）]登录。

形态：小枝下垂，绿色，树势强健。纯白色花，花大，单瓣型，5

瓣居多，5%为半重瓣，花瓣大小为1.5～1.9cm长，1.2～1.8cm宽；花丝白色，1.0～1.8cm长，雄蕊数41～45枚，1/3的雄蕊瓣化，花药黄色；花柱长度1.4～1.6cm，20%～30%的花不形成雌蕊；花萼两轮，绿色，花梗长度1cm。叶绿色，椭圆披针形到卵圆披针形，叶长17.2cm（13～20cm），叶宽4.3cm（3.5～5m），叶长与叶宽比（L/W）为

'白色荣光'（'White Glory'）

4，蜜腺肾形，2个，叶柄长1～1.5cm。果实无毛，半球形，3～4cm。在－18℃可以存活，不受伤害（Zone 7）。需冷量约900h。

58. '单粉垂枝'（'Danfen Chuizhi'）

起源：不详。

形态：小枝下垂，红褐色，树势中等。花粉色（65D），单瓣，花瓣数5；花蕾长卵形；花瓣卵圆形，花径3.7cm，花丝白色，花药橘红色，着花中等，萼片红褐色，5枚。叶绿色，椭圆披针形，长11.8cm，宽3.0cm，叶长与叶宽比（L/W）为3.9；叶缘细锯齿；叶柄长1.1cm。花期4月中旬。

'单粉垂枝'（'Danfen Chuizhi'）

59. '赤枝垂'（'Akashidare'、'单红垂枝'）

起源：不详。

形态：小枝下垂，红褐色，树势中等。花红色（68B），花蕾长卵形；花瓣卵形，长1.53cm，花径3.9cm，单瓣型，花瓣数5；雄蕊数平均38，花丝长1.27cm，花药红褐色；雌蕊1枚，雌蕊与雄蕊近等长；着花密（着花率1.5/cm），花梗长0.35cm；花萼红褐色，5枚。叶绿色，椭圆披针形，长11.3cm，宽3.1cm，叶长与叶宽比（L/W）为3.6，叶缘细圆齿，叶柄长0.8cm。果实长3.2cm，宽3.2cm，圆形；果核长2.32cm，宽1.64cm，椭圆形；核面粗糙。花期4月中旬。

60. '源平垂枝'（'Genpeishidare'）

起源：日本江户时代就有的古老品种（伊藤　1695）。

形态：小枝下垂，绿色，树势中等。花为白色与粉色相间（155D/65B）；花蕾长卵形；花瓣长卵形，长1.9cm，花径3.5cm，单瓣型，花瓣数6.5枚（5~8）；雄蕊数平均35，花丝长1.27cm，花药黄色；雌蕊1枚，雌蕊与雄蕊近等长；着花中等（着花率0.7/cm），花梗长0.33cm；花萼绿色，5枚，会随着花瓣颜色有紫红色斑纹或条块；花丝和萼片均有瓣化现象。叶绿色，椭圆披针形，长13.6cm，宽3.5cm，叶长与叶宽比（L/W）为3.8；叶缘细锯齿，叶柄长0.9cm。果实绿色。花期4月中旬。

'赤枝垂'（'Akashidare'）

'源平垂枝'（'Genpeishidare'）

‘幸垂枝’（‘Sachishidare’）

‘绿萼垂枝’（‘Lü E chuizhi’）

61. ‘幸垂枝’（‘Sachishidare’）

起源：不详，日本品种（Yoshida *et al.* 2000）。

形态：小枝下垂，绿色，树势中等。单瓣，复色花，白花上有粉色点或条纹，几乎无纯色白花。

62. ‘绿萼垂枝’（‘Lü E chuizhi’）

起源：不详。

形态：树冠伞形，小枝绿色，拱形下垂，树势中等。花白色（155D），花蕾卵圆形；花瓣卵形，长2.7cm，花径4.5cm，复瓣，梅花型，花瓣数23枚（21～24）；雄蕊数平均39，花丝长1.3cm，花丝有瓣化现象，花药黄色；雌蕊2，雌蕊高于雄蕊；着花中等（着花率0.65/cm），花梗长0.56cm；花萼绿色，两轮。叶绿色，椭圆披针形，长13.2cm，宽3.7cm，叶长与叶宽比（L/W）为3.6；叶缘细锯齿，叶柄长1.0cm。果实绿色，果长4.3cm，宽3.5cm，卵圆形；果核长2.75cm，宽1.83cm，椭圆形，核面粗糙。花期在4月中旬。花开之前其长卵形的绿色萼片包被着白色花蕾，满树晶莹剔透，初花期白绿相间，别有一番韵味。

63. ‘残雪枝垂’（‘Zansetshidare’）

起源：是日本江户时代就有的古老品种（伊藤1695）。

形态：与‘绿萼垂枝’的主要区别在于花瓣多，叶小。花径4.3cm，花瓣长2.0cm，宽1.4cm，花瓣数28～42，花粉活力高达98.7%；叶卵圆披针形（Pawasut *et al.* 2004）。

64. '红雨垂枝'（'Hongyu Chuizhi'）

起源：不详。

形态：小枝拱形下垂，紫褐色，树势中等；花粉红色（63B），花蕾卵圆形；花瓣卵圆形，长1.53cm，花径4.3cm，复瓣，梅花型，花瓣数14枚（12～16）；雄蕊数平均40，花丝长1.1cm，花药橘红色；雌蕊高于雄蕊；着花中等（着花率0.65/cm）；花萼红褐色，两轮，花萼及花丝均有瓣化现象；花梗长0.57cm。叶绿色，椭圆披针形，长12.8cm，宽3.6cm，叶长与叶宽比（L/W）为3.5；叶缘细圆齿，叶柄长0.8cm。果实绿色，长3.68cm，宽3.44cm，圆形；果核长2.42cm，宽1.47cm，椭圆形；核面平滑。始花期在4月8日左右。

'红雨垂枝'
（'Hongyu
Chuizhi'）

65. '相模枝垂'（'Sagamishidare'）

起源：日本江户时代就有记载的古老品种（伊藤1695）。

形态：与'红雨垂枝'的主要区别在于花大，花瓣多。花径4.7cm，花瓣数18～30枚，花瓣长2.1cm，宽1.6cm，花粉活力高达97.7%。叶卵圆披针形（Pawasut *et al*. 2004）

66. '鸳鸯垂枝'（'Yuanyang Chuizhi'）

起源：发表于1991年（张秀英，陈忠国1991）。

形态：干皮灰色，小枝拱形下垂，黄褐色；粉色花（69B）与白色花（155C）共于一枝；花蕾卵圆形；花瓣卵形，长1.8cm，花径4.2cm，复瓣，梅花型，花瓣数20枚（19～22）；雄蕊数平均35，花丝长1.53cm，花药颜色与花瓣色相关，纯粉色花的花药为橘红色，白色花及复色花的花药为黄色，花丝有瓣化现象；雌蕊与雄蕊近等长；着花中等（着花率0.75/cm）；花萼颜色与花瓣色相关，红色花为红褐色花萼，白色花为绿色花萼，复色花则为红褐色与绿色相间，两轮，花萼及花丝均有瓣化现象；花梗长0.6cm。叶绿色，椭圆披针形，长12.1cm，宽3.2cm，叶长与叶宽比（L/W）为3.8；叶缘细锯齿，叶柄长0.98cm。果实绿色，长3.68cm，宽3.04cm，椭圆形；果核长2.54cm，宽1.64cm，椭圆形；核面粗糙。花期在4月下旬。

'鸳鸯垂枝'（'Yuanyang Chuizhi'）

67. '黛玉垂枝'（'Daiyu Chuizhi'）

起源：1995年在北京植物园发现。发表于1997年（张秀英，戴思兰1997）。

形态：小枝下垂，绿色带紫晕；树势中等；花淡粉色（69C），近白，无跳枝现象；花蕾卵圆形；花瓣卵形，长1.7cm，花径4.1cm，复瓣，梅花型，花瓣数21枚；雄蕊数平均35，花丝长1.23cm，花丝均有瓣化现象，花药黄色；雌蕊高于雄蕊；着花中等（着花率0.65/cm）；花萼绿色，萼萼两轮；花梗长0.63cm。叶绿色，椭圆披针形，长13.9cm，宽3.4cm，叶长与叶宽比（L/W）为4.1；叶缘细锯齿，叶柄长0.9cm，果实绿色，长3.46cm，宽3.14cm，卵圆形；果核长2.4cm，宽1.48cm，椭圆形；核面粗糙。花期4月中下旬，略早于其他垂枝桃。

'黛玉垂枝'
（'Daiyu Chuizhi'）

68. '五宝垂枝'（'Wubao Chuizhi'）

起源：不详。

形态：小枝下垂，小枝绿色；树势中等；花浅粉色（65D）与肉粉色（65B）相间共于一枝，还会出现粉晕（62B）；花蕾球形；花瓣卵形，长1.9cm，花径4.2cm，复瓣，梅花型，花瓣数21枚（18～23）；雄蕊数平均33，花丝长1.4cm；花药颜色与花瓣色相关，浅粉色花的花药为黄色，肉粉色及带粉晕的花朵花药为橘红色；雌蕊高于雄蕊；着花中等（着花率0.65/cm）；花萼红褐色与绿色相间，两轮，花萼及花丝均有瓣化现象；花梗长0.68cm。叶绿色，椭圆披针形，长14.5cm，宽3.6cm，叶长与叶宽比（L/W）为4.1；叶缘细圆齿，叶柄长0.97cm。果实绿色，长3.96cm，宽3.36cm，椭圆形；果核长2.72cm，宽1.72cm，椭圆形；核面粗糙。花期4月中下旬。

69. '羽衣枝垂'['Hagoromoshidare'、'朱粉垂枝'（'Zhufen Chuizhi'）]

起源：日本江户时代就有所记载（伊藤1695）。

形态：花径4.7cm，花瓣长2cm，宽2cm。叶长13.1cm，叶宽4.2，叶长与叶宽比（L/W）为3.1，叶柄长1.2cm（Pawasut *et al.* 2004）。花粉活力高达97.9%。

中国1991年发表的品种'朱粉垂枝'的主要性状为干皮灰黑色，小枝拱形下垂，绿色带紫晕；花以粉红色（64D）花为主，偶有淡粉色（62D）跳枝；花蕾卵圆形；花瓣卵形，长2cm，花径4.0cm，复瓣，梅花型，花瓣数26枚（22～29）；雄蕊数平均35，花丝长1.27cm，花丝有瓣化现象；花药橘红色；雌蕊高于雄蕊；着花中等

（着花率0.75/cm）；花萼红褐色偏绿，两轮，外翻；花梗长1.53cm。叶绿色，椭圆披针形，长13.6cm，宽3.7cm，叶长与叶宽比（L/W）为3.6；叶缘细锯齿近细圆齿，叶柄长1.2cm。果实绿色，长4.2cm，宽3.56cm，卵圆形；果核长2.73cm，宽1.78cm，椭圆形；核面粗糙。花期4月中下旬。

二者除了在叶子大小上略有差别，花的性状基本相似，因'朱粉垂枝'发表较晚，根据《国际栽培植物命名法规》的规定，以出现较早的名称为准，因此采用'羽衣枝垂'作为该品种今后统一的名称。

'羽衣枝垂'［'Hagoromoshidare'、'朱粉垂枝'（'Zhufen Chuizhi'）］

'粉花紫叶垂枝'（'Pink Cascade'）：花粉色，复瓣。叶紫色。

'红花紫叶垂枝'（'Crimson Cascade'）：花红色，复瓣。叶紫色。

70.'粉花紫叶垂枝'（'Pink Cascade'）

起源：来自于不知名的New Jersey紫叶系列母株（Moore *et al.* 1993）。

形态：树势强，小枝拱形下垂，粉色花，复瓣。叶紫色。

71.'红花紫叶垂枝'（'Crimson Cascade'）

起源：来自于不知名的New Jersey紫叶系列母株（Moore *et al.* 1993）。

形态：树势强，生长快，4 m高，2.5m冠幅（5年生苗），成熟可达 8 m。小枝拱形下垂。花红色（65B），花蕾卵圆形；花瓣卵圆形，长1.63cm，花径4.0cm，复瓣，梅花型，花瓣数26枚（22～29）；雄蕊数平均35，花丝长1.27cm，花药橘红色，花丝有瓣化现象，雌蕊高于雄蕊；着花中等（着花率0.55/cm）；花萼红褐色，花梗长0.63cm。叶紫色，椭圆披针形，长12.2cm，宽3.4cm，叶长与叶宽比（L/W）为3.6；叶缘细锯齿，叶柄长0.85cm。果小，绿色有紫晕，长3.27cm，宽3.23cm，圆形；果核长2.33cm，宽1.73cm，椭圆形；核面平滑，核纹以浅沟纹为主，很少有穴点。

五、曲枝桃品种群

曲枝桃类是桃花中枝性最为特殊的一类，小枝自然弯曲，呈"之"字状。目前该品种群只有1个品种。

72. '云龙'桃（'Unriumomo'）

起源：来自于油桃的芽变品种（Yoshida *et al.* 2000）。

形态：树型直立，树体中等。树皮灰褐，较光滑，小枝绿色；树势中等。花色淡粉（65D），花蕾长卵形，花蕾色56A；花瓣卵圆

'云龙'桃
（'Unriumomo'）

形，长1.9cm，花径4.0 cm，单瓣型，花瓣5枚；雄蕊数平均42.3，花丝长1.33cm，花药橘红色；雌蕊低于雄蕊；着花中等（着花率0.65/cm），花梗长0.57cm；花萼红褐色，5枚。叶绿色，扭曲，椭圆披针形，叶长10.6cm，叶宽2.8cm，叶长与叶宽比（L/W）为3.8；叶缘细锯齿近细圆齿，叶柄长1.1cm。果实无毛，绿色，受光面为紫红色，卵圆形，长2.9cm，宽2.6cm；果核长2.5cm，宽1.7cm，椭圆形，核面光滑。'云龙'桃尽管花并不很出众，但奇异的枝型赋予了该品种独特的观赏性，也为观赏桃增添了一种新枝型的育种趋势。

六、山桃花品种群

这一品种群是桃花与山桃（*Prunus davidiana*）的杂交种，枝、叶、芽、花具有山桃和桃的双重特征，树体高大，树皮光滑，小枝细长、无毛。雌蕊早期萎蔫或无雌蕊。花期明显早于真桃花系品种，在4月上旬即可达到盛花。

目前共有3个品种：'白花山碧'桃、'粉花山碧'桃、'粉红山碧'桃。

品种检索表如表6.5：

山碧桃3个品种的比较
上：'白花山碧'桃
中：'粉红山碧'桃
下：'粉花山碧'桃

表6.5　山碧桃品种群品种分类检索表

Table 6.5　Key to cultivars in David group

1. 花色白，小枝绿色，无雌蕊 ………… '白花山碧'桃 'Baihua Shanbitao'
1. 花色粉或粉红色，小枝紫红色，有雌蕊形态
　　2. 花淡粉色 ………………………………… '粉花山碧'桃 'Fenhua Shanbitao'
　　2. 花粉红色 ………………………………… '粉红山碧'桃 'Fenhong Shanbitao'

73. '白花山碧'桃（'Baihua Shanbitao'）

起源：通常认为是山桃与桃的天然杂交品种。

形态：树体高大，枝型开展。树皮光滑，深灰色或暗红褐色。小枝细长，黄褐色。花白色（155D），花蕾卵形；花瓣卵形，长1.83cm，花径4.3cm，复瓣，梅花型，花瓣数18枚（16～23）；雄蕊数平均73.5，花丝长1.83cm，雄蕊与花瓣近等长，花药黄色；无

'白花山碧'桃
（'Baihua Shanbitao'）

雌蕊；着花密（着花率0.83/cm）；花梗长0.53cm；萼片绿色，两轮，卵状；花丝和萼片均有瓣化现象。叶绿色，椭圆披针形，长12.8cm，宽3.2cm，叶长与叶宽比（L/W）为4；叶缘细锯齿，叶柄长1.5cm。花期在所有桃花品种中最早，在4月上旬即可盛花。

74. '粉花山碧'桃 ['Fenhua Shanbitao'、'探春'（'Tanchun'）]

起源：父本为'白花山碧'桃，母本为'合欢二色'桃。

形态：树型明显具有父本'白花山碧'桃的高大，树皮灰褐，较光滑，小枝绿色；树势中等。花色淡粉（69A），花蕾卵形；花瓣卵形，长1.97cm，花径4.1cm，复瓣，梅花型，花瓣数20.3枚（17～23）；雄蕊数平均53，花丝长1.43cm，花药橘红色；雌蕊明显低于雄蕊；着花中等（着花率0.83/cm）；花梗长1.07cm；花萼红褐色，两轮，花丝和萼片均有瓣化现象。叶绿色，椭圆披针形，长12.8cm，宽3.2cm，叶长与叶宽比（L/W）为4；叶缘细锯齿，叶柄长1.4cm。果实绿色，长3.4cm，宽3.2cm，圆形；果核长2.55cm，宽

'粉花山碧'桃
（'Fenhua Shanbitao'）

1.82cm，椭圆形；核面平滑。4年观测期内平均始花期在4月6日。

　　'探春'（方伟超等 2008）是以'白花山碧'桃为父本，以'迎春'为母本，和'粉花山碧'桃的亲本相同，形态特性上也无大异。因为'粉花山碧'桃发表在先，根据《国际栽培植物命名法规》（ICNCP）（Brickell *et al.* 2004）的规定，以首先发表的品种名为准，将其置于相应品种之后。在此归入'粉花山碧'桃。补充性状：需冷量400h。

75. '粉红山碧'桃（'Fenhong Shanbitao'）

起源：父本为'白花山碧'桃，母本为'绛桃'。

形态：树型有父本'白花山碧'桃的典型特征，高大、开展，树皮灰褐色，较为光滑，小枝红色；树势中等。花粉红色（55B），花

'粉红山碧'桃
（'Fenhong Shanbitao'）

蕾卵形；花瓣卵圆形，长2.1cm，花径4.27cm，复瓣，梅花型，花瓣数28枚（22～33），雄蕊数平均53，花丝长1.43cm，花丝有瓣化现象，花药橘红色；雌蕊低于雄蕊；着花中等（着花率0.78/cm）；花萼紫红色，花梗长0.93cm。叶绿色，椭圆披针形，长13.2cm，宽3.3cm，叶长与叶宽比（L/W）为4；叶缘细锯齿，叶柄长1.5cm。果实绿色，长4.24cm，宽3.84cm，卵圆形；果核长2.48cm，宽1.65cm，卵圆形，核面光滑。始花期在4月8日左右。

'粉花山碧'桃及'粉红山碧'桃为北京植物园1999年育成的2个早花品种（胡东燕，张秀英 2001）。此类桃花兼具山桃树体的高大、良好抗性和真桃花系的复瓣及花期较长的优点，而其最大的应用价值还在于花期的适时性，完全可以弥补北京地区4月上旬春花的断档期，很好地衔接上早春的连翘、大山樱、辽梅、山杏与仲春的榆叶梅、大部分品种的桃花以及樱花等，并具有花色鲜艳、花期相对持久的特性。在北京乃至全国桃花栽植区都不失为难得的新优品种。

经过多年观测，'粉花山碧'桃及'粉红山碧'桃的结实性相当低，分别为7.5%和6.5%。表现出干净整洁的叶面和良好的观赏性，是进一步培育完全不育桃花品种的良好亲本。

关于本章所列举品种取舍的原则：

以上所收录的品种均为在以往文献中出现过，有着相对明确的性状特征描述。对于其中一些根据现有描述暂且可以归入相应品种体系中的品种，如果以后经过系统鉴定，证明其确实具有独特性状特征者，今后可以考虑列为独立的品种。

而那些由于本身品种性状描述模糊不清，或描述过于简单者暂无法归入现有分类体系，暂不在本章检索表中加以考虑，在此将欧美国家此类常见的20个品种提及如下，仅供参考。

1.'Alboplena'（重瓣，白花）：1849年引入英国（Jacobson 1996）。花瓣数不清，无法判定与前述品种的异同。

2.'Alboplena Pendula'（垂枝，复瓣，白花）（Krussmann 1986）；花瓣数不清，无法判定与前述品种的异同。

3.'Aurora'（深粉色，重瓣，花径3cm）：1937年介绍到商业市场，专利号为PP1245，已在"无专门登录机构木本观赏植物"

（CRUWO）登录机构登录（Jacobson 1996），品种地位毋庸置疑，但因花色和花瓣数目不清，无法列入本章检索表，特在此列出。

4. 'Atropurpurea'（紫叶，单瓣，粉花）(Huxley 1999)：根据性状简单描述，与'北京紫'极为相近。

5. 'Cardinal'（深粉红，复瓣）(Krussmann 1986)：花色和花瓣数目不清，仅从性状描述上和'绛桃'很相近。

6. 'Camelliaeflora'（大花，重瓣，粉红色）(Notcutt 1926)：1845年之前就已在法国有所记载（Krussmann 1986）。花色和花瓣数目不清，无法判定与前述品种的异同。

7. 'Chrysanthemum'（大花，花瓣似菊花）：据记载起源于南加州（Jacobson 1996），但很难确定其是否来源于东方。从起源年代可以看出明显晚于该特征品种在日本和中国的时期。

8. 'Dianthifolia'（花大，复瓣，深红色，花瓣窄）：1845年从日本引入（Krussmann 1986）欧洲。花色和花瓣数均无特定描述，无法判定与前述品种的异同。

9. 'Duplex'（花小，淡粉色，复瓣）：1636年最早在法国出现（Krussmann 1986）。花色和花瓣数均无特定描述，无法判定与前述品种的异同。

10. 'Foliis Rubris'（紫叶，单瓣，淡粉色）：19世纪60年代起源于密西西比河流域，自欧洲人引入欧洲后，在1873年很快被更名为'Royal Redleaf'（Jacobson 1996）。和前述'Atropurpurea'性状极为相近。

11. 'Helen Borchers'（复瓣，粉色，花径6.23cm，晚花）(Dirr 1998)：花色和花瓣数均无特定描述，无法判定与前述品种的异同。

12. 'Iceberg'（复瓣，白花）：1938年由美国加州的W.B.Clarke苗圃引入（Jacobson, 1996）。花色和花瓣数均无特定描述，无法判定与前述品种的异同。

13. 'Klara Mayer'（亮粉色，重瓣花，花径约4cm宽，花期晚）(Jacobson 1996)：1890年左右由德国的Spath苗圃引入美国（Jacobson 1996）。因花色和花瓣数均无明确描述，无法判定该品种与前述品种的异同，仅在此列出。

14. 'Magnifica'（半重瓣，深红色花）：由英国的Veitch苗圃引进美国（Jacobson 1996）。因其红色和花瓣数目均无明确描述，无法判定与前述品种的异同，暂且在此列出。

15. 'Rubro-plena'（半重瓣或重瓣，红色）是在19世纪40年代从中国出口到欧洲的品种（Jacobson，1996）。没有更多的形态描述。

16. 'Versicolor'（重瓣，白色花瓣上有粉色或红色条纹）：1863年前出现在法国（Krussmann 1986）。从花色的描述上看和前述'Peppermint Stick'性状相近，几乎完全一致。

17. 'Windle Weeping'（垂枝，复瓣，粉花）（Krussmann 1986）：从花色上看和前述'羽衣枝垂'（'朱粉垂枝'）的性状相近。

18. 'Pink Peachy'，'Red Peachy'和'White Peachy'：等3个寿星桃品种颜色各异，分别为分散、红色和白色，半重瓣。株高1~1.5m。没有更多关于花色和花瓣数的记述。

粉红山碧桃

7 桃花育种

BREEDING OF THE ORNAMENTAL PEACH

桃花的育种由来已久，由于桃花非常容易发生天然芽变，人们在常年的观察中发现了很多优秀的品种，例如从古代就有的'碧桃'、'绯桃'、'人面'桃等古老品种。近年来，美国、日本、中国、乌克兰等国纷纷开展了针对各自不同育种目标的人工定向杂交育种工作，比如日本、美国的帚型类桃花品种'照手姬'、'Corinthian Pink'等，日本的菊花型品种'京舞子'和'幸白'，以及乌克兰的抗病远缘杂交品种'Nikitskij Rubin'，中国早花的'粉花山碧'桃、'粉红山碧'桃和'紫奇'，为丰富桃花的枝型、花型、抗性做出了贡献。

针对观赏桃的用途，增强观赏性，延长花期，提高抗性，培育无果桃花新品种是今后桃花育种的主要方向。在种质资源保护的基础上，芽变及实生选种，种内及种间常规杂交育种，结合变突、倍性诱变及现代分子育种等手段，都为桃花新品种的培育提供了途径。

一、观赏桃主要性状遗传规律

整个桃属（*Prunus*）植物是栽培植物中拥有最小基因组的类群之一，而桃的基因组估计只有290 Mbp，每细胞核DNA含量仅有0.61pg（Baird *et al.* 1994），仅相当于模式植物拟南芥基因组的2倍。桃具有杂交方便、童期短（2～3年）、基因组小等优势，是木本植物形态建成、遗传与生理研究的模式植物。

目前已经揭示的有关桃形态特征中符合孟德尔遗传规律的质量性状共有42个（Dirlewanger and Arus 2008）。这些研究成果为今后观赏桃花育种工作奠定了坚实的理论基础，其中在观赏桃花中主要可以应用树型、花、叶、抗性等方面的一些规律。

数量性状在选择中也相当重要，通常为多基因控制，受环境影响，在群体里表现为连续性，比如花瓣数量、节间长度等性状。

(一) 生长型方面

桃花的生长型实际上主要取决于枝条自身的性质，以及枝上的节间密度两个方面。前者决定了树体的外观形态，后者则决定了树体的高度，二者共同构成了观赏桃花丰富的枝型多样性。

枝条自身的性质，即枝性，首先包括枝条本身的形态。与大部分桃花类型的枝条多相对平直不同，枝条自然弯曲的龙游类型虽然在桃花中并不多见，但也是桃花众多枝型中最为独特的一类，具有重要的园艺观赏价值。

其次，枝性包括枝条生长的方向性，向下即为垂枝型，向上斜出或直上则因分枝角度不同而呈现出不同的类型。

分枝角度小，枝条内抱，树冠窄高，树体呈柱状的一类是为帚型桃，分枝角度相对较大的桃花类型则最为普遍。

枝条上的节间密度，或者说是枝上芽与芽之间的距离，是决定树高的关键。节间密度小，即节间距大，树高表现正常，即为最常见的普通型；节间紧密，则节间距小，树体矮小，通常被称作矮化的寿星桃。

1. 垂枝型

分枝角度包括基角和延伸角两个概念。基角指枝条基部与上一级次枝之间的角度；而延伸角则是完全生长后的枝条从顶端到枝条起点的角度（Bassi 2003）。枝条延伸角是定义枝条类型的一个重要参数，特别是对枝条下垂性状的描述。

垂枝型是观赏桃中独特的类型，延伸角可以达到90°以上（Monet *et al.* 1988；沈向等 2008），表现为枝条下垂，树冠伞形，具有良好的观赏特性。

垂枝性状（*pl*）是受单基因控制的隐性性状（Dirlewanger and Bodo 1994），与普通枝型品种杂交，垂枝型表现为完全隐性（Lesley 1957）。

2. 帚型

帚型桃是日本给予世界的一大贡献。早在其江户时代就有帚型桃的记载。这一类型的桃花用于观赏也始于日本（Yamaguchi *et al.* 1987a）。帚型桃分枝角度小，通常不超过35°～40°（Layne and Bassi 2008），使植株看起来呈柱形。

帚型性状（*br*）为不完全显性，当*br*是杂合体时，表型为直立型（*Br/br*）（Scorza *et al.* 1989; Scorza *et al.* 2002）。山崎和雄等（Yamazaki *et al.* 1987a）通过帚型桃与垂枝桃的正反交实验，证实帚型对垂枝型表现为完全显性。Werner等用帚型桃‘Pillar’（*brbr*）和垂枝油桃‘White Glory’（*plpl*）杂交，后代表型分离比例说明帚型性状（*brbr*）对垂枝性状（*plpl*）具有隐性上位效应（Werner and Chaparro 2005）。

3. 曲枝型

小枝自然弯曲的曲枝型桃花品种目前只有‘云龙’桃（Yoshida *et al.* 2000）。Okie曾对这一枝型有所描述，认为曲枝性状是单基因隐性性状（Layne and Bassi 2008）。

4. 矮型

关于矮生型桃最早的研究见于1786年（Monet & Salesses 1975）。Strong和Hooper在1867年描述了矮生型桃的基因型。Connors（1922）从法国矮生型桃品种‘Elberta’的后代中观察到半矮生型植株，认为是隐性特征。目前共有3种矮生桃基因型。Lammerts 1945年从原产中国的寿星桃与直枝型桃的杂交后代F_2中得到了矮生型桃，认定矮生型基因（*dwdw*）为单基因控制的隐性性状。这种矮生型桃被普遍称为"寿星型"（brachytic），以节间短（通常小于10mm）和大叶为主要特征。Lammerts 1945年还报道了另外一种具有矮生特征的灌丛型半矮生桃类型（Bushy），为双隐性基因 bu_1bu_1/bu_2bu_2。Monet和Salesses 1975年报道了一种半矮生桃变异'A72'，基因型和遗传性状都不同于广为熟知的"寿星型"（brachytic）。'A72'为不完全显性的杂合体，表现为半矮生型。Gradziel和Beres 1993年报道了从粘核型果桃品系自然授粉后代中分离出的变异类型'SD22-59'，表现为和'A72'相同的变异特征。Hu和Scorza（2009）通过对'A72'树体和枝条的详细测量及分析，发

图7.1　'A72'的叉形分枝
Fig.7.1　Forked-form of 'A72'

现'A72'(Nn)基因型具有一种独特的叉形分枝（FBR）特征（图7.1），以此特征来划分实生后代，获得的分离比例为$1NN$：$2Nn$：$1nn$。NN与普通型桃没有明显区别，Nn具有明显叉形分枝特征，而nn则为极矮化型。

5. 普通直立型

枝条斜出，分枝角度在40°～70°的范围内的类型，在果桃中的研究甚为详细，由于分枝角度大小的差别，呈现出丰富的变异类型：普通型（Standard：40°～60°）；直立型（Upright：50°）是普通型和帚型杂交产生的杂合体（Scorza *et al.* 1989）；拱型（Arching）和直立型最大的区别，在于1年生枝有一定程度的拱形下垂（Werner and Chaparro 2005）；开展型和开张型基本接近，开张角度为60°～70°。

作为观赏植物，这些差别并不能影响枝型的整体结构，因此视为同一枝型类型，在观赏桃花分类中不再细分，均归为普通直立型。

6. 其他新型枝型

不同生长型之间的相互作用会带来新的类型。根据直立型对帚型、矮型及垂枝型均为显性，且受单一基因控制（Yamazaki *et al.* 1987a）；以及帚型对垂枝型为完全显性的结论，一些混合枝型应运而生。例如利用帚型桃和矮化桃杂交，已经得到了矮型帚桃（Scorza *et al.* 2002）（图7.2矮型帚桃）；半矮化型（'A72'）与帚型桃的相互作用（Hu & Scorza 2009）再次证实了桃花在树型结构上具有良好的可塑性，完全可以通过育种手段，得到预期的更多变异组合枝型

图7.2 矮型帚桃
Fig.7.2 Dwarf pillar

图7.3 35年生的密枝型植株株高只有不到2m
Fig.7.3 Compact, less than 2 m for 35 years tree

（Scorza *et al.* 2002）。细心的研究，审慎的观测和严谨的测定，将决定何种性状和组合才能够更适合未来的观赏需求。

此外，密枝型（*Ct*）（Mehlenbacher and Scorza 1986）因大量节间较短的二级和三级侧枝，使得整个株型形成相对矮小的密集树冠（图7.3）；圆球型（*GL*）（Scorza *et al.* 1989）则是由密枝型和帚型桃杂交得到的中间类型，既有密枝型众多分枝的特点，分枝角度又相对窄小。目前这两个类型中尚未形成专门用于观赏的桃花品种，但其丰满紧凑的树型和密布的花芽所形成的良好观赏效果，已经显示出很好的应用前景，在未来桃花育种中具有很大潜力。

这些不同枝型的桃花所体现出的丰富的观赏特性，为桃花这一传统观赏植物在园林、道路、家庭中得以更为广泛地应用奠定了基础。

（二）花的方面

花是观赏桃最重要的部分。

关于花的大小，铃型花对普通花型为显性（*Sh/sh*）（Connors 1920；Bailey and French 1942；Lammerts 1945）。

关于花色，粉色对红色，粉色对白色，红色对白色，浓粉色对淡粉色分别均为显性（Lammerts 1945），这4对性状的遗传均受单基因控制（Yamaguchi *et al.* 1987a）。

关于花瓣数，即重瓣性（D_1），单瓣花对重瓣花完全显性（Lammerts 1945）。

关于花药的育性，通过对不同桃品种的研究，先后发现了两类花粉不育基因型*ps/ps*（Connors 1926；Blake and Connors 1936；Scott and Weinberger 1994）和ps_2/ps_2（Blake 1932；Werner and Creller 1997）。花粉不育的类型花期偏晚，Connors（1920）认为可能是因为花粉败育降低了花药的膨压，从而造成花期延后而且花期较长。而观赏桃中的'玫粉'、'玫紫'等品种均为花粉败育，众多复瓣品种的雄蕊瓣化程度极高，正是这些特性使得重瓣花类型比单瓣花类型花期晚，而且花期相对较长，从而提高了桃花的观赏价值。

（三）叶的方面

在叶色方面，Black（1937）得出桃的叶色是由一对基因（Gr/gr）控制，红色叶对绿色叶为完全显性的结论。吉田雅夫则通过赤叶寿星桃的培育，进一步得出紫红叶遗传子 R 对绿叶遗传子 r 为显性，并受单一基因控制的结论（Yoshida *et al.* 2000）。

山崎和雄等人（Yamaguchi *et al.* 1987a）用紫叶桃与绿叶杂色花品种杂交，在 F_2 代中得到了紫斑绿叶的品种，而且所有杂种后代中的花朵均为杂色花。用红花或粉花的紫叶桃与绿叶白花的品种杂交，F_2 代中并没有出现白花绿叶的品种，因此认为控制紫色叶的基因与决定花色的基因为连锁遗传。

Connors（1921）曾经研究过叶腺的形状与对白粉病的抗性有一定关联。通常桃的叶腺分为无叶腺（e/e）、球形叶腺（E/e）和肾形叶腺（E/E）3种类型，其中没有叶腺的品种最易感染白粉病，肾形叶腺的品种感染白粉病的几率相对最低。经过观察，桃花中的大部分品种均为肾形叶腺，白粉病的发病率在观赏桃中并不像果桃中那么严重。

寿星桃类通常会有波状叶缘，其他类型则很少见。研究表明，桃叶缘有无波状形成，是由单基因控制的（Wa）（Scott and Cullinan 1942）。

窄叶型（或称柳叶型）桃（Chaparrc *et al.* 1994；Okie and Scorza 2002）是由于叶宽的遗传改变而形成的一种新类型。桃的成熟叶片为 4cm×16cm，而窄叶桃的叶子通常只有一般叶子宽度的一半（Glenn *et al.* 2000），这使得树冠可以获得更多的阳光，减少病虫害的发生。研究表明，窄叶型桃树可以更有效地利用水分，在水分胁迫条件下表现明显比直枝类型桃好。目前还没有窄叶型的桃花品种系列。

（四）抗病性方面

目前在果桃研究中，关于抗病性、抗虫性以及抗线虫的遗传规律比较多。Lownsberry 和 Thomson（1959）发现了抗褐腐病的基因型（Mj），为不完全显性。此外，抗根瘤线虫的（Mi）（Weinberger *et al.* 1943）和抗桃绿蚜的（Rm）也有所报道（Monet and Massonie 1994）。观赏桃中尚未出现类似研究报道。

二、桃花主要育种途径

(一) 杂交育种是桃花育种中的主导方法

作为常规育种手段，杂交育种是培育新品种的主要途径。杂交引起基因重组，杂交后代可以获得来自父母本双方的优良性状，因此具有不同基因型的桃花品种之间、桃花与其近缘种之间有目的的杂交，均有望产生预期的新品种和类型。一般包括品种间杂交和在种间完成的远缘杂交。

1. 杂交技术

去雄：桃花是两性花，自花授粉容易，因此在杂交时需要在授粉前去雄。一般去雄应在大蕾期完成。利用镊子去除花瓣、雄蕊，只留下雌蕊和很少部分的花被。要注意一定要去雄彻底，且不要把花粉碰破。在进行不同杂交组合去雄工作之间，一定要用酒精消毒，以避免花粉污染。

套袋：去雄的花朵不再具有吸引昆虫授粉的能力，所以通常全部进行去雄的植株不再需要额外对之加以保护。然而，如果是育种研究，则必须保证没有由于风或昆虫造成的意外授粉，需要额外用细纱或尼龙纱布、草图纸等做成隔离袋，以防来自旁边正处在花期的其他植株的花粉以及昆虫的介入。

花粉采集及贮藏：花粉采自父本，可以采集露天自然条件下的花粉，也可以采集尚未萌动的枝条，放在温室中根据需要时间进行定期花粉采集。同样在大蕾期采集花蕾，即在花粉尚未打开之前，在室内将花瓣去除，将花粉收集在培养皿内，保存在内底放置无水氯化钙的干燥箱里，使花粉自然散出。干燥的花粉可以在 -30℃冰箱内保存2年，在 -80℃冰箱内则可保存更久。

花粉的贮藏和运输可以打破杂交育种中亲本在时间和空间上的隔离，扩大了杂交育种的机遇。

通常贮藏后的花粉需要进行生活力的测定。用培养基检测桃花花粉的人工萌发，是测定生活力的方法之一，以萌发率的高低鉴定花粉的生活力。用琼脂、糖和水制成一种与柱头相类似的基质。用0.4g琼脂，5g蔗糖，再加蒸馏水至50ml混合，在琼脂完全融化且煮沸后，将烧杯放在一个盛有开水的大烧杯中备用，以免培养基凝固。准备载玻片、玻璃环若干，在玻璃环的一端涂上凡士林并放在载玻片上，再滴一滴蒸馏水在玻璃环中，以保持一定的湿度。将配制好的培养基滴到载玻片上，厚度为2mm，当培养基凝固后，用刀片切成0.5cm×0.5cm的小方块，将做好的发芽床床面向下放在玻璃环上备用，以免干燥。将发芽床取下，蘸取少许花粉条播在发芽床上，每床播3～4行，花粉在100～300粒左右为宜，将播好花粉的发芽床放置25℃的恒温箱中进行培养，2个小时后用显微镜观察花粉发芽情况，计算发芽率。

结果表明，观赏桃的花粉萌发力大部分处于20%～40%之间，远远低于食用桃花粉平均为70%的萌发力（刘会宁，冯义龙 2004；朱更瑞等 1998）。'云龙'桃、'单白垂枝'、'照手桃'、'北京紫'、'粉花山碧'桃等品种的花粉萌发力较高，达到40%～50%左右，这些主要是单瓣型和复瓣型品种，而'京舞子'、'五宝'桃、'簪粉'等，其花粉萌发力相对较低，均在20%以内。

图7.4 授粉
Fig.7.4 Pollination

授粉：柱头开始分泌黏液发亮时即可授粉，通常发生在去雄后的24～48小时内完成。父本的花粉可以用毛笔、蜜蜂棒或手指直接涂抹在已经去雄的柱头上（图7.4）。为了节省花粉，也可用玻璃小棍儿蘸少许花粉抹到柱头上。连续2～3次授粉可以确保授粉成功。

获得实生苗：种子需要在0℃～4℃的冷室进行3个月以上的沙藏以打破休眠，待到生出幼根，即可将实生苗移至温室；而当种子发育不够完全，即种子采自开花后少于110～120天的果实时（Layne and Bassi 2008），种子几乎没有任何营养与能量储备，即使在正常沙藏条件下，也无法生根，则需要人工条件下进行胚培养（详见第八章繁殖）。

图7.5 用降落伞布罩住整株树进行自交
Fig.7.5 Self-pollinated plant covered by parachute

杂交后代的选择：杂种后代性状并未稳定，还需要经过多代选择才能选出优良而稳定的新品种。杂种第一代会发生分离，根据目标进行单株选择。对于那些具有特异性、优良性状明显的植株，可以直接进行嫁接以获得新品种。

2. 自交

即自花授粉，就是用自身的花粉进行授粉，授粉过程本身一般无需人为介入。但需要在开花前，将整株植株或需要进行自交的枝条用防水纸袋或透水帆布完全罩住，以防昆虫进入传粉。在花开之后即可去除所有保护（图7.5）。

自交不仅可以使杂合体的后代群体迅速趋于纯化，还能够使杂合体中的隐性基因显现出来，是人工定向杂交育种工作中的一个重要环节。通过自交与人工定向杂交后代多代的反复更替和多次选择，可以得到新的性状的重组，从而获得新品种。培育一个果桃品种，要经过杂交、初选（选择优良单株进行自交）、复选（逐代系间淘汰、选择）和决选，到最后进行区域试验，确定推广地区，要完成整个育种进程，获得目标性状，通常至少要经过5代、20年以上才能实现（Layne and Bassi 2008）。相比之下，作为以观赏为主的桃花，由于没有结实和果实品质的要求，一旦发现新品种，即可用无性繁殖方法获得大量具有同样稳定性状的植株，自交则主要用于育种研究和鉴定品种的基因型。

3. 远缘杂交

桃的野生近缘种中具有各具特色的优良性状。山桃花期早，耐寒性较好；甘肃桃开花早，具有一定的抗病性；生长在青藏高原的光核桃寿命长，耐旱、耐寒、耐瘠薄。利用这些近缘种培育桃花新品种，可以提高抗逆性。

自然界里存在着自然发生或人工获得的种间杂交种，观赏桃花中应用最广的当属桃与山桃的天然杂交品种'白花山碧'桃，具有山桃早花的特点，使桃花的花期明显提前；Dirlewanger（1996）等的研究表明，桃与山桃的杂交后代有明显的抗白粉病的能力。

'Lyubava'是桃与光核桃（*P. mira*）的杂交后代；'Malenkij Princ'是桃与甘肃桃（*P. kansuensis*）的杂交品种；'Fleur Pompon'是桃与扁桃（*P. amygdalus*）的杂交品种（Komar-Tyomnaya 2007）。远缘杂交后代明显结实性降低，'Lyubava'和'Fleur Pompon'均为不育品种；'白花山碧'桃更是连雌蕊形态都未能形成。无疑，桃与其近缘种进行杂交为培育不结实的观赏桃花新品种提供了很好的方向。

此外，桃能够与山桃、甘肃桃、光核桃、新疆桃进行远缘杂交，并未表现出任何明显的不亲和性，并且还在一定程度上体现出其在抗病性上具有明显超亲性，这也从另一个方面证实了这几个种的亲缘关系比较接近。

4. 杂交培育出的主要品种

北京植物园利用桃花与山桃的天然杂交种'白花山碧'桃培育出'粉花山碧'桃和'粉红山碧'桃2个早花的山碧桃品种（胡东燕，张秀英 2001），不仅保留了'白花山碧'桃的早花习性，同时具有鲜艳的重瓣花，在4月上旬即可开花，比正常桃花的花期提前了将近1周的时间，正好可以和山桃连接，弥补了山桃与碧桃花期之间的断档。

郑州果树所培育的早花品种'探春'（方伟超，朱更瑞，王力荣 2008）和观赏食用二者兼顾的'满天红'（朱更瑞，王力荣，方伟超 2008），前者是以'迎春'为母本，以'白花山碧'桃为父本，在郑州3月初即可开花，需冷量仅为400h；后者则是花有一定观赏性，果实又有一定品质的花果两用新品种。刘佳琴等以寿星桃和'白凤'为亲本，经过5代育种，获得了花果兼顾的'北9-9'（刘佳棻等 2000）。

日本杂交培育出的紫叶寿星桃品种'赤叶寿星'选自单瓣粉花、紫叶的'赤芽'和另一寿星桃的F_2代（Yoshida 1974）。'赤枝垂'（Yoshida et al. 2000）、'白枝垂'、'赤叶寿星'等，还作为亲本不断参与到新的杂交组合之中，帚桃品种'照手红'（Yamazaki et al. 1987b）（'赤枝垂'和'帚桃'的F_2代）和'照手桃'（Yamazaki et al. 1987b）（'帚桃'和'赤枝垂'的F_2代）就是利用'赤枝垂'做亲本，通过与帚桃的正反交得到的新品种；白色花半重瓣的'照手白'（Yamazaki et al. 1987b）选自'帚桃'和'残雪垂枝'的F_2代。最新培育出的白花菊花桃新品种'幸白'（前田克夫等 2005）也是用寿星桃和'菊花桃'杂交而得来的。

美国培育出的'Corinthian Pink'等4个Corinthian 系列帚桃品种（Werner et al. 2000a；Werner et al. 2000b；Werner et al.，2000c；Werner et al. 2001）均选自油桃'NC174RL'和来自日本的'Pillar'（帚型桃）的F_2代实生苗。

（二）选种是获得新品种的简单而有效的手段

由于桃本身属于高度异质体，自然条件下出现芽变的几率相当高。因此，从野生或现有栽培群体中发现符合育种目标的自然芽变单株，进而选育出新品种是得到桃花新品种最为简洁的方法。'凝霞紫叶'、'斑叶'桃（Hu et al. 2005）等就是从现有品种植株中选择出来的芽变变异；'京舞子'是'菊桃'的芽变；'云龙'桃则选自油桃的芽变（Yoshida et al. 2000）。

选种的另一个重要途径是通过人工实生选种，即从天然开放授粉或人工授粉杂交种子的播种实生苗中进行人工筛选。实生变异分离几率大，实生选育是获得新品种的重要途径。'白枝垂'选自于单瓣、红花的垂枝类型'赤枝垂'的天然授粉后代；红花重瓣的'赤花蟠桃'选自粉花重瓣的'八重蟠桃'的天然授粉后代（Yoshida et al. 2000）。淡粉色花、半重瓣花的'照手姬'则是'帚桃'的天然授粉后代（Horikoshi et al. 1992）。美国育成的紫叶、粉花重瓣寿星桃'Bonfire'（Moore et al. 1993）则是用来自日本农林水产省的'筑波2号'的种子而得到的自然实生种。

（三）诱变育种是桃花育种行之有效的手段之一

通过人工物理或化学手段，对桃花的枝条、种子进行辐射或化学激素诱变，方法简便，在变异后代中一些重要性状，如节间长度、花色、叶形、育性等出现频率较高，因此诱变育种与常规育种技术结合是培育桃花新品种的有效手段。

笔者曾于1996年进行过桃花休眠芽^{60}Co辐射育种的尝试，经高接后连续观测数年，并未发现变异现象，主要可能是由于供试材料数量有限，未能达到辐射几率出现的频率。陈青华等（2005）2002～2004年也曾经利用^{60}Co-γ射线25～40Gy对白碧桃、菊花桃、红碧桃、红花垂枝桃及粉花垂枝桃等品种休眠芽进行处理，并得出半致死剂量在34～39Gy之间。

山东农业大学2007年通过对'紫叶'桃开放授粉的种子进行γ射线辐射照射，培育出紫叶、单瓣、淡粉花的早花新品种'紫奇'（沈向等2007），与母本相比，花期提前8～12天。

（四）倍性育种是培育新型无果桃花新品种的重要育种手段

桃是二倍体植物，染色体倍数为$2n=16$。具有单倍体的桃在20世纪70年代以后开始有所报道。单倍体具有株型矮小、小枝细弱、节间短、叶小、花小的特点（图7.6）。单倍体的意义并不在其本身的观赏价值，而在于它的育种价值。Hesse（1971）曾经观察到单倍体桃的花粉萌发率只有7%。Toyama（1974）用秋水仙素加倍单倍体桃，获得了单倍体纯系。Pooler 和 Scorza（1995）观测了单倍体桃的花粉粒，其中不乏能够萌发的二倍体花粉，从而使获得三倍体桃的构想成为可能。Monet等（Layne& Bassi 2008）的观测完全证

图7.6 单倍体的花（右）与小花铃型（中）和单瓣花（左）的比较

Fig.7.6 Comparison of haploid flower (R) with bell-shaped (M) and single form (L) flower

实了这一可能，他们观测到了单倍体桃植株上的果实，并且获得了三倍体、二倍体和非整倍体桃，三倍体植株生长旺盛但结实性很差。这些结果无一不为观赏桃培育不结实的新品种提供了广阔的前景。

(五) 现代生物技术是桃花辅助育种的新途径

现代生物技术为桃花育种提供了新的途径。利用已知品种的特异基因进行分子育种，也是今后桃花育种的方向所在。

Dirlewanger等（1996）用RAPD分子标记，发现桃与山桃的杂交后代中与抗白粉病连锁的基因；用分子标记辅助选择抗PPV病毒的基因也已经成功分离（Pascal *et al.* 2005）。

Dirlewanger和Bodo (1994)发现了垂枝性状(pl)的RAPD标记，遗传距离为11.4cM和17.2cM，但遗传距离稍大，不足以说明是一个精确的标记。李亚蒙通过SSR分子标记技术，发现位于G1染色体52.6cM和65.1cM处的微卫星标记BPPCT020、CPPCT029与垂枝基因的连锁关系密切，其中标记CPPCT029与pl基因共分离，为以后进一步精确定位奠定了基础（李亚蒙 2006）。

三、观赏桃育种趋势

针对观赏桃的用途，增强观赏性，延长观赏期，提高抗性，培育无果桃花新品种 (Hu and Zhang 2008) 是今后桃花育种的主要方向。

1. 丰富枝型、花型、花色，组合不同性状，提高桃花的观赏性

枝型变化丰富是观赏桃的重要观赏优势，充分利用现有的桃花品种资源，选育出新颖别致的树型是桃花育种的目标之一。原有枝型进行组合，形成复合型枝型，是观赏桃花育种的一个趋势。目前已经有这种育种材料存在，比如矮化帚桃，超矮化寿星型，以及密枝型等中间类型。曲枝型目前仅有1个单瓣品种，不同花色的复瓣品种都有可能出现在不同枝型类型之中。

花永远是观赏桃最为重要的观赏特征，在花型、花色等方面的育种，是今后桃花育种的重要目标。根据花色、花型在各类枝型品种中的平行关系（周建涛等 1998），现在已经出现在直枝类型的各类花型及花色都可以在其他各类枝型中出现，因而在此方面的育种应

图7.7　黄色桃花
Fig.7.7　Creamy flower of the peach

该是值得期待的。

花色育种也是观赏桃的一个方向。黄色一直是桃花中几乎不可能的神话。但从我们掌握的资料中可以断言，出现黄色花的观赏桃品种只是时间上的问题（图7.7）。

铃型小花类型以前并未在观赏桃花中受到很好的重视，随着育种工作的推进，在不同枝型类型当中培育复瓣到重瓣的铃型小花类型，也是今后丰富观赏桃品种的一个主要方向。

2. 培育早花、晚花以及色叶品种，延长桃花的观赏期

花期育种是延长桃花整体观赏期最有效的手段。'紫奇'除花色的独特性之外，花期早是其重要的特异性；'粉花山碧'桃、'粉红山碧'桃也具有明显早花的特征。利用现有资源选育早花或者晚花的桃花新品种，对延长桃花品种的绝对观赏周期具有重要意义。

随着近年来紫叶品种的不断涌现，使桃花在花期之外仍有很好的观赏性，特别是紫叶寿星类和紫叶帚桃类品种，无疑增添了桃花作为观赏植物的价值。培育不同枝型与叶色的品种，可以使得桃花的观赏期相应延长到夏秋，甚至冬季。春季观花、夏秋观叶、冬季观枝，延长桃花周年观赏期是育种工作的又一目标。

3. 培育抗病虫害、抗寒、抗旱品种，提高桃花的抗性

桃花的近缘种，如甘肃桃（*P. kansuensis*）、新疆桃（*P. ferganensis*）、光核桃（*P. mira*）以及同属的早花且有香味的梅花（*prunus mume*），早花、抗寒、抗旱的榆叶梅（*prunus triloba*）和毛樱桃（*prunus tomentosa*），抗寒、抗旱、适应性强的东北杏（*prunus mandshurica*）、'辽梅'山杏（*Prunus sibirica* 'Liaomei') 和 '陕梅' 杏 (*Prunus armeniaca* 'Shanmei') 等均可广泛用作远缘杂交的亲本, 通过大规模授粉参与到观赏桃育种中, 增强观赏桃的基因丰富性。培育出具有更强抗性的新品种类型。尽管目前在远缘杂交方面已经有了一些尝试, 但在抗性育种方面仍有很大潜力。

4. 培育不结实的桃花品种是观赏桃育种的主流方向

目前, 在观赏桃的育种目标中存在着两个方向, 一是从枝型、叶、花的观赏性着眼, 注重桃花品种观赏性的选育; 二是花果两用桃的选育。笔者认为, 桃花区别于果桃的主要研究重点就在其观赏性, 这就决定了观赏桃的主流方向, 始终应当是以观赏为主要目的, 果实并非观赏桃的主攻目标。

常年以来, 人们总希望能培育出既能春天看花、又能秋天吃桃的两用桃新品种, 尽管不少机构已经在此方面进行了多年的工作, 也取得了一定的成果, 但实际效果证明, 出于观赏和出于食用的种植目的本身各不相同, 相应的栽培条件和管理手段更有很大差别, 要想二者兼顾, 就很难保障充分满足任何一方的条件, 往往导致各自特征的削弱。花果兼顾始终不是观赏桃的主要方向, 观赏桃的重点还是应该集中在它的观赏特征上, 而果实则是尽可能越少越好, 既可以减少花后结实对树体营养的大量消耗, 也可以有效地避免果实脱落后对园林环境的影响。当然也不完全排除为家庭庭园等特殊市场需求培育的花果两用桃, 虽然可以避免公共绿地人为干预的问题, 但果实要想达到一定的品质, 还需要相当水平的技术要求和格外精心的管理, 这也是一般家庭往往难以保证的。

目前除 '白花山碧' 桃天然不结实之外, 尚没有任何一个品种可以达到无果。'粉花山碧' 桃和 '粉红山碧' 桃已经体现出结实性差的特征; 桃与光核桃的杂交后代品种也有一定的不育性 (与乌克兰桃花专家Komar-Tyomnaya的个人交谈)。因此利用这些桃的近缘种与桃进行杂交, 再利用种间杂交品种进一步进行回交或自交, 都有望获得不育的后代。此外, 单倍体桃的发现也为培育不结实的观赏桃新品种提供了可能。

特别需要强调指出的是, 所有育种工作都是建立在丰富的种质资源的基础之上, 因而, 种质资源的保护是育种工作中的重中之重。野生种及桃花种质资源的收集, 桃花基因资源的保护, 是进一步育种研究的根基所在。

斑叶桃

8 繁　殖

PROPAGATION

北魏（386-534）贾思勰的《齐民要术》是中国果树嫁接记载最早的文献，书中记载"桃、柰桃欲种法，熟时合肉全埋粪地中，至春，既生，移栽实地。"可见，早在1500年前，我国在桃的实生繁殖方面就已经认识到实生会发生变异；在不了解种子层积概念的情况下却采取了相应措施，满足了在北方层积与后熟所需的温、湿度条件。

我国从12世纪就开始了多种桃树嫁接组合的探索，北宋已经对桃、李、杏、梅的属间远缘嫁接、正反嫁接等进行了较为详细的观察和记载。

元代嫁接桃树与实生桃树并行发展。

根据不同的目的要求，繁殖桃花需要采用不同的繁殖方法。有性繁殖通常用于砧木的培育和杂交选种；但桃花品种的主要繁殖方法还是以无性繁殖的嫁接、扦插等方法为主。

一、种子繁殖

通常为了砧木的繁殖，或者是特殊目的的杂交育种后代的实生选育，会采用种子繁殖的方法。

种子的采收通常在8月进行。待果实软化后，去果肉，水洗干净，置于通风干燥处阴干，得到纯净种子，装入纸袋，干藏。如贮藏1年以上，要求冷藏于5℃条件下。

桃的种子果壳致密，具有休眠特性。播种前采用低温层积处理方法打破种子休眠。种子用0.5%的高锰酸钾溶液浸种消毒2小时，后用80℃水浸种，自然冷却后浸种24小时后捞出，置于筛子上沥干水分。混沙（沙子与种子比例：3∶1），其中沙子要求用0.5%的高锰酸钾溶液喷洒后密封消毒2小时后使用。沙子湿度在60%～70%之间，以手握沙子成团，松开即刻能够散开为宜。将种子混合均匀后，置于0℃～5℃处。20天检查1次，约100天后，陆续有种子萌发。这时，改为2天检查1次。等到有1/3种子裂口后，即可播种。也可将混沙直接放入冰箱，100天后有同样效果。温度过高或冷量不足虽然可以正常萌发，但会出现莲座状上胚轴畸变。而过久的沙藏还有可能诱发种子的再度休眠（Guerriero and Scalabrelli 1984）。利用赤霉素（GA_3）等生长素也可以达到打破休眠的目的（Koornneef *et al.* 2002）。

北京地区播种约在3月底至4月上旬。桃种子较大，适宜采用条播。种子播种前3天，浇透水1次，播种前1天再浇水1次，保持土壤有一定的湿度。种子条播前，在畦面上先开一道约为2cm宽、3cm深的小沟，条沟之间宽度为10cm左右。然后进行种子点播，种子间距离为6cm左右。待每条种子播完后，覆土于种子上，覆土厚度2～3cm。

桃花品种的种壳硬度因品种不同而有所差异，出苗情况也有不同。2003年采集的'Corinthian Rose'种子，次年播种1株未出，待到2005年在原播种地长出帚型小苗；而大部分品种播种后都能正常出苗。

二、嫁接繁殖

为了保证桃花品种的特性，嫁接是桃花品种繁殖中应用最为广泛、最简单便捷，成本又低的方法。

（一）砧木的选择

具有不同抗性的砧木能够有效改进植株对积水、干旱、寒冷、高pH值、石灰质土壤等不同栽植条件，以及外寄生性线虫、根腐病等不同病虫害的适应性。

用于果桃生产的砧木品种屡见不鲜，但鉴于观赏桃的研究远没有做到如此精准，但这并不妨碍在观赏桃领域引用果桃所用的砧木品种，改进观赏桃品种对不同栽植条件和病虫害的适应性。'Lovell'，Halfold'，Nemoguard'，'Nemared'，'Bailey'和'Guardian'等6个砧木品种已经成为95%的美国桃树的砧木（Layne & Bassi 2008），这些砧木不仅具有高出苗率和整齐统一的出苗品质，而且在抗寒、抗病、抗线虫等方面各具特色。目前以观赏为目的专门培育、应用的砧木，仅有加拿大培育出的'Harrow Frostipink'、'Harrow Candifloss'和'Harrow Rubirose'3个耐寒晚花的系列品种（Layne 1981）。

在中国，桃花品种的繁殖还主要依赖于山桃或毛桃为最主要的砧木来源。山桃主根较深，抗旱，耐寒性强，不耐湿，适合于微碱性土壤栽培，在北方地区应用最为普遍；毛桃主根较浅，须根发达，在同样耐旱的情况下，具有耐瘠薄、耐潮湿的特性，树龄也比以山桃为砧木的寿命长，广为南方各地采用。山桃和毛桃的种子一般于8月采收。

砧木的播种分为秋播和春播两种。秋播一般在北方地区上冻之前，让种子在田间自然通过休眠；春播的种子则需要层积处理打破休眠，通常用湿沙在2℃～7℃放置100～120天，3～4月播种。

（二）嫁接方法

桃花品种繁殖主要采用的嫁接方法为芽接和枝接。

芽接采用当年生接穗，多用于苗木的大量繁殖。根据进行的时间不同主要分为3种，即夏秋芽接、春季芽接和6月芽接。夏秋芽接，一般是在7月中下旬到9月中，利月休眠芽进行；而春季芽接则是利用叶芽进行，这两种方法嫁接的苗木多数在第二年秋季出圃；而一般在6月进行的芽接，即在皮层尚未剥离，不易切取芽片时，进行带木质部芽接，嫁接苗可以于当年出圃，是为半成苗，可以节省接穗量，用于快速大面积增殖。

枝接（切接或劈接）多在开春前进行，此时地下开始解冻，但地上部树液尚未开始流动。多用于接穗数量很少的引种，以保证品种存活为准。

根据桃花品种枝型本身的特点，嫁接时也应根据其不同特点选择适宜的砧木。在进行垂枝桃品种嫁接时，要采用高接的方法进行繁殖。一般需要2年生或者2年以上的砧木，以达到主干高大、小枝下垂的效果，使垂枝桃类型的品种更富观赏性。近年来也有人将寿星桃高接在2年生以上的砧木上，人为抬高了寿星桃的高度，在庭院栽植中也不失为一种特色（图8.1）。

图8.1　奥林匹克公园的高接寿星桃
Fig.8.1　Higher grafting of dwarf peach in Beijing Olympic Park

三、扦插繁殖

尽管扦插繁殖普遍具有苗型整齐健壮、苗龄寿命长的优点，以前也有过不少关于桃树扦插繁殖的研究报道，但在观赏桃的实际生产中很少采用这种繁殖方法。第一是因为嫁接繁殖简单便捷，成活率高；第二扦插繁殖不仅需要一定的设施建设投入，而且扦插苗生根尚存在问题，而且成苗慢，目前还没有明确的研究指明快速繁殖的基本指数和条件，难以形成批量生产。

关于观赏桃的扦插繁殖，曾经试验采用半木质化扦插（6月）和硬枝扦插（8月及2月）两种方法进行比较试验。结果表明半木质化扦插明显成活率要高。

(一) 硬枝扦插

硬枝扦插的插条一般在秋末、冬初采集（10月和1月），选择25～30cm长的顶部或中部枝条。硬枝插条通常持有诱发不定根所需的营养、能量和激素物质。采穗母株的特性、插条采集的时间，以及生根处理，都是影响硬枝扦插结果的主要因素。

IBA是应用最多的生长素，1000～3000 mg/L浓度的溶液采用速蘸法，10～20 mg/L的稀释溶液中则需要浸泡12～24小时。直接蘸取粉末虽更易操作，但会造成一定药剂损失。底部加热对促进生根有一定作用。为了诱发根的形成，使插条根部温度保持在18℃～20℃几天，同时要保证插条部分仍然是在相对较凉的温度下继续保持休眠状态。

扦插基质一般选用珍珠岩：泥炭（3∶1），以保证良好的透气性和适宜的湿度。

（二）半木质化扦插

一般在7月中到8月中进行，选用尚未木质化的枝条。多在温室内采用全光喷雾方法，可以通过喷洒细密的雾珠提高相对湿度，减少插条的水分散失。冷却系统和适当遮荫可以降低温室内的温度和光强度。与硬枝扦插相同，IBA也是应用最广的生长素，同时NAA，或者两种生长素混合使用，都对促进半木质化扦插生根有很好的作用。激素溶液浓度相对要低，一般采用可溶于水的盐类生长素。过高浓度的生长素会导致组织脱水，从而瓦解插条基部的皮层组织，更易感染真菌和细菌。珍珠岩等渗透性好的基质可以避免对插条的潜在伤害，并提供良好的透气性，促进排水。

将扦插苗从温室移到室外的过程中要特别注意。要逐渐降低扦插苗所处环境的湿度，或放置在遮荫条件下。

用于半木质化扦插的成本明显要高于硬枝扦插。

2005～2006年用‘菊花’桃、‘照手姬’做材料，曾经尝试用KIBA做生根剂，选择600mg/L，1000mg/L，2000mg/L，3000mg/L 4个浓度梯度，采用速蘸法（20秒），全光喷雾采用10s/20m。结果表明以600～1000mg/L 生根率较高。扦插基质采用透水性强的纯珍珠岩，以保证扦插苗不致积水（图8.2）。

图8.2　1000mg/L KIBA处理后的‘菊花’桃和‘照手姬’扦插生根情况
Fig.8.2　Rooted cuttings of 'Juhua Tao' and 'Terutehime' treated by 1000 mg/L KIBA

四、微繁

微繁是一种基于细胞全能性的无性繁殖方法，通过组织培养的手段在实验室完成，所以成本也明显高于其他繁殖方法。目前用于桃花的微繁工作尚没有开展。果桃中有关于微繁的较为详尽的记载和报道，在今后开展桃花微繁工作中可以加以借鉴。

4h光照/2h黑暗的周期，比16h/8h更有效地刺激试管苗根的形成和生长（Morini *et al*. 1990）。其他因素，诸如光周期、光质（Loreti *et al*. 1991）、琼脂类型和浓度（Debergh 1983），以及试管相对湿度等因素都会影响繁殖体的生长（Sciutti and Morini 1993）。增值率在第一次培养中可能会相对较低，但会在随后的次培养中明显增加（Loreti *et al*. 1985）。

五、胚培养

胚培养是将未完全成熟的胚从种子中取出，进行无菌培养（Kester and Heses 1955; Hartmann *et al*. 2002）。这一方法对于改进早花桃花品种育种工作非常有用。采集时胚的大小，决定用何种培养基进行培养，10mm左右的胚可以用SBH培养基（Smith *et al*. 1969），而相对较小的胚（5~10mm）则适合采用MS培养基（Ramming 1990）。

9 园林养护及管理

GARDEN CARE AND CULTURE

直枝桃开心圆冠型修剪

温度、光照、水分是植物生长最重要的因素。桃花喜光，耐旱，有一定的耐寒力。喜肥沃湿润、排水良好的土壤，不耐水涝。和植桃花时应选择阳光充足、排水良好之处，忌高大乔木之下，忌低洼积水之地。

桃花的病害主要以叶部病害为主，其次是枝干病害；虫害从较早以鳞翅目为主的食叶害虫，如"大虫"类的桃天蛾、天幕毛虫等，到近年来逐渐转变为以刺吸、潜叶为主的"小口"，如蚜虫、叶蝉、潜叶蛾、螨类等。

不同于果桃为了获得更多更大果实的修剪目的，以观赏为主的桃花的修剪，则是以满足美观的观赏需求，以完善树型，最大限度地展现其观赏特性，维持最佳观赏效果为主要目的。

一、桃花栽培的基本条件要求

(一) 温度

桃是广泛分布在温带地区的重要树种，也是用于商业生产的李属植物中最耐寒的植物之一（Quamme *et al.* 1982）。通常生长在北纬30°～40°之间，相当于年平均温度为8℃～17℃左右。桃花的适生温度为18℃～23℃。

完全适应环境的桃花芽可以忍受-30℃的低温，而叶芽则能耐-35℃低温（Layne 1984）。花芽在休眠期间，绝对低温低于-18℃，花芽有可能会受到冻害（曲泽洲 1990），低于-27℃时，会冻死大部分花芽。

近年吉林果树所在延边地区珲春县（北纬43°）发现珲春桃，命名为f. *hui-chun-tao* Gu，能抗冬季地温-30℃以下。是我国桃树栽培最北线的一种桃，可能起源于桃与山桃的自然杂交种，品质尚好，是培育抗寒优良品种的重要种质资源（俞德浚 1984）。

Byrne（1986）认为，在低温或晚霜等逆境条件下，品种的耐寒性通过每个节上花芽的增加而得到了增强。Werner等在1988年曾经有过花芽的密度可以有效地通过选择而控制（Werner *et al.* 1988a）的假说。在其试验中，来自美国东北部及加拿大的寒冷地区的品种，其平均每个节上芽的平均数较高（个别可达到1.59），与南方品种有着显著的差异。

花芽在结束自然休眠后，如忽遇短暂高温，耐寒力会显著下降，而当温度再度降低时（即使高于冻害的临界低温温度），也很容易受到冻害，这使得桃在萌芽期和初花时往往很容易遭受晚霜的危害。

桃花能耐夏季高温，在广东、福建等地也可栽培生长。在高温多雨地区，枝条生长时间长，养分消耗过多，积累少，影响花芽分化，花量及开花质量均会受到一定影响。晚花基因型虽然可以避免春寒的危害，但可能不具有足够的耐性抵御极低的冬季低温。小气候，或者是砧木、树龄、以及健康状况都可能影响其耐寒性。晚花品种可以有效地避免春寒带来的晚霜危害。研究表明（Werner *et al.* 1988b），不同的晚花品种在一定的需冷量紧接着的热量积累所达到的基本温度有所不同，而不像一般的品种仅仅只是需冷量的问题。

很多观察支持不同基因型在休眠后开花的需热量不同的假说。目前尚不清楚需热量积累的基本温度与开花早晚的相关性。

桃花在落叶后进入深度休眠，这种休眠状态必须经过一个低温阶段才能打破。需冷量是休眠期能保证花芽正常生长所需要达到的最小低温积温。Weinberger（1950）用低于7.2℃低温的累积来表示不同品种通过这个低温阶段所需的时间。

关于需冷量在生物化学方面的机理并不清楚，但可以肯定的是内源生长调节剂包括赤霉素、生长抑制剂等起了一定作用(Bowen 1971)。

大多数用于商业生产的品种桃的需冷量在650~950h之间，在美国南部加州、佛罗里达和得克萨斯等地，有需冷量仅为150~500h的近40个品种。

王力荣等（2003）按照0℃~7.2℃需冷量模式，以秋季日平均温度稳定通过7.2℃的日期为有效低温累积的起点，以打破休眠所需0℃~7.2℃的累积低温值为品种的需冷量计算，我国重瓣桃花品种的需冷量分布范围从400h到1250h，分布范围广泛，并以900h以上的品种为主（朱更瑞等 2004）。观赏桃花品种的需冷量和始花期的相关系数为0.58，达到极显著水平。早花品种'白花山碧'桃和'迎春'均为450h；而晚花的'菊花'桃（1250h）、'洒红'桃（1214h）、'鸳鸯垂枝'（1188h）、'紫叶'桃（1112h）等均高达1000h以上。'红雨垂枝'（926h）和'寿红'（930h）的需冷量也均接近1000h。Pawasut（2004）等人的研究结果同样说明了'菊花'桃是目前已知拥有最高需冷量（1433h）的桃花品种。

关于桃花品种的耐寒性的研究并不多。李书文（2000）用电导法测定出'白花山碧'桃、'寒红'桃、'菊花'桃、'洒红'桃、'二色'桃5个品种的半致死温度（LT50）均低于－30℃，均较抗寒。其耐寒顺序呈降序排列如下：'白花山碧'桃>'寒红'桃>'菊花'桃>'洒红'桃>'二色'桃。

(二) 光照

桃是喜光植物。充足的光照是桃花健壮生长、保持良好观赏效果的重要条件。

对于不以结果为主要栽培目的的观赏桃品种来讲，充足的光照对保持不同类型品种的形态特征，充分展示其观赏特点就显得更为关键。光照不足会造成花芽分化不良，花芽质量下降，郁闭的光照会造成内膛枝枯死，外围枝条为获取更多的阳光出现徒长，使得花朵仅仅集中在树冠外围，影响整株植株的观赏效果。

(三) 水分

桃原产在高原地带，较耐干旱，但不耐水涝。桃是对水涝和土壤透气状况最为敏感的果树之一（Anderson *et al*. 1984; Alvino *et al*. 1986; Schaffer *et al*. 1992）。种植时应选择阳光充足之处，雨量过多不仅会造成枝叶徒长，严重时，连续3天的降雨可以导致整片桃花无一幸免。积水和排水不良的低洼之地，会影响桃树的正常生长，严重者可造成死亡。

在旱情严重的地区，早春，即萌芽期的灌水极为关键，对于桃花花芽的正常分化、开花的质量和数量、以及病虫害的预防都有重要作用。此时灌水要灌足，一般需要达到80cm左右深度，最好一次灌水充足，不要多次反复灌水，以免造成地温降低。

开花前如果有条件，可以进行补充灌水，能够起到延长花期、保持花色艳丽、花朵大而饱满的作用。

11月下旬的冻水对于桃花是必不可少的。此次灌水也一定要灌透、灌足，以保证越冬所需水分。

果桃一般会由于结实而造成更严重的水分缺失（Berman and Dejong 1996; Naor *et al*. 1999），这一点在观赏桃中无疑不会产生额外的水分负担。

关于桃花品种耐旱性的研究，仅限于'二色'桃1个品种。李书文（2000），通过对其盆栽2年生嫁接苗进行的水分胁迫试验表明，'二色'桃具有很强的耐旱性，土壤含水量达到5.284%，就可基本满足其正常生长。

(四) 土壤

桃为浅根性，根的水平分布明显大于垂直分布。根系扩展程度相当于树冠的0.5～1倍，而深度则只相当于树高的1/4左右。桃的吸收根分布深度通常在土层下50～60cm之内（Layne and Bassi 2008），10～30cm根系吸收水平最高。

桃树对土壤的适应性广，pH值为5～8之间的沙质或黏质土壤上均可栽培，以4.6~6.0为好（汪祖华，庄恩及 2001）。一般在土壤较深的沙壤土上生长旺盛，树体寿命长，土壤过于黏重、排水不良，pH值高于7.5时，会造成树势减弱，出现缺铁等症状，树体易早衰。不同地区可以通过不同的砧木提高植株的抗性，例如北方地区会选用耐寒、稍耐碱的山桃作为砧木，而南方地区则采用有一定耐水湿能力的毛桃做砧木。国外对于桃的砧木的研究很多，也培育出不少适合不同环境条件的系列砧木。例如耐寒晚花的系列砧木品种：'Harrow Frostipink'、'Harrow Candifloss'和 'Harrow Rubirose'（Layne 1981），不仅本身具有良好的观赏性，还可以作为很好的耐寒性砧木在Zone 6a地区使用。

土壤通气状况与根系生长关系密切。排水不良，通气性不好的土壤还会造成根系生长减弱，出现黑褐色腐烂，直至死亡。

（五）施肥

桃树喜肥，但用于观赏的桃花并不像果桃一样对施肥那么要求严格。整体上讲，如果施入过量肥料，反倒会造成枝条徒长，花芽分化不良，影响开花、树型等重要观赏特征的表现。所以适时适量施肥是保持桃花树体强健、减少病虫害侵害、提高观赏效果的关键。

和果桃一样，秋肥对于观赏桃花至关重要。花芽形成在秋季，因此此时重视施肥，会有效地增加花芽的数量和提高质量，以达到更好的越冬储备，在第二年有良好的观花效果。秋施基肥以迟效性的有机肥为主，在当年落叶前施入，要保证氮、磷、钾全面配合。条件允许的话，可以在秋季以环状沟施、放射状沟施、井字状沟施与撒施施入等方式施基肥，但在定植时不适宜施肥，以免苗木生长过旺而抽条。叶面喷肥在大规模的桃花观赏区内一般不予采用，个别精细管理中可以适当喷施尿素、硼砂、磷酸二氢钾进行叶面追肥。

（六）营养元素

共有16种营养元素对桃花的生长至关重要。叶片特征通常用来诊断桃花营养状况。也有一种看法是用休眠的枝条或根进行桃花营养状况早期判断（Johnson *et al.* 2006）。

氮肥不足的临界值已经由原来的200 kg/hm^2（Daane *et al.* 1995;

Tagliavini *et al.* 1996）降低到100 kg/hm²，或者更低（Tagliavini *et al.* 2000）。徒长枝的增多和众多病虫害的产生，很可能是氮水平太高造成的（Lobit *et al.* 2001；Daane *et al.* 1995）。

磷肥能够使组织充实，树势强健，促进花芽形成。健康植株的磷肥临界值通常是在0.14%～0.25% (Robinson *et al.* 1997)之间。高磷还往往会加剧缺锌（Ballinger *et al.* 1966）。缺磷叶片会变成深紫绿色，并出现叶卷和革质。

出现在仲夏的苍白叶色和卷缩叶面是缺钾的表现，随后出现叶缘灼伤状，严重时造成枝条生长受阻，形成"小老树"。健康植株含钾通常在2.0%～3.0%。过多的钾肥会造成其他元素，比如镁元素的不足（Weir and Cresswell 1993）。

适宜钙含量在2.0%或者以上。叶缘变色、落叶，枝条顶尖坏死都是缺钙的表现。

镁含量不足的临界值是0.1%～0.3%（Sanchez 1999）。叶片缺镁的表现是从老叶开始的，斑状脱色和落叶都是从基部叶片开始，严重时会造成生长停滞、花芽形成受阻。

硫含量的理想值是0.17%～0.40%（Leece *et al.* 1971; Robinson *et al.* 1997）。低于0.09%则被认为是缺硫（Finch *et al.* 1997）。缺硫的症状与缺氮很相似，区别在于缺硫是从新叶开始出现，枝条顶端开始的，会出现生长量减少，小叶、黄叶等症状，严重时会沿着叶缘坏死。

叶脉间隙出现变黄，使得缺锌的症状很难与缺锰区别开来。严重时会节间缩短，形成叶缘波状、黄化的莲座状叶，最终导致老叶脱落。最新研究表明，秋季叶面喷施锌肥可以移动到根部（Sanchez *et al.* 2006）。

硼的移动性使其因含量过高而引起的硼中毒变得更为敏感（Gupta *et al.* 1985）。尽管过多的硼很容易从土壤中过滤掉，但仍有大量的硼残存在桃树体内（Dye *et al.* 1984）。小枝和枝条会在春季死亡，或叶芽在春季无法萌发。硼中毒与硼缺乏很难区别。春季或初夏小枝坏死，小叶，皱缩，以及腐烂病和流胶病等造成树皮的粗糙和剥落。叶面喷施以及高浓度的土壤施用补充硼。

在高pH值的重碳酸盐水平的石灰质土中，铁会变得无法移动（Koseoglu 1995; Morales *et al.* 1998），从而造成铁缺乏。缺铁的症状从年轻的叶片开始，最终会波及整株植物。叶脉仍为绿色的网状黄化是缺铁的明显症状，严重时整个叶片会完全失去叶绿素。

采集春天的叶片，或者是对花的分析都能体现与缺铁的良好相关性（Sanz *et al.* 1997a; Abadia *et al.* 2000）。叶面喷施，树干注射，以及土壤施用都可以补充铁的不足。

缺锰的植株老叶会出现叶脉仍为绿色的"人字形"黄化色斑。

在实际栽植应用中，缺素现象对于桃花来说并不构成非常严重的问题，缺硼、缺钾、缺钙、缺镁、缺硫、缺锰、缺铜都不会造成对桃树太大的危害。一旦发现叶片、枝条出现异常，还是应当首先从生长条件（光照）和栽培条件（水分、土壤）加以考虑，然后是是否有病虫害等方面的原因，最后可以考虑是否缺素。

二、病虫害防治

桃栽培历史悠久，分布范围广泛，与人们的生活密切相关。早在公元前我国就有对虫灾的记录。清代《花镜》中"如生蚜虫，以多年竹灯挂悬树梢间，则虫自落。"则记载了应用诱集方法来防治害虫的方法。

针对主要用于观赏的桃花的病虫害研究并不多见（高玲，周凤英 2004；胡东燕，俞思佳 2002年），但大部分桃树本身的病虫害也同样是观赏桃花中普遍存在的问题。针对果桃的病虫害研究，对于观赏桃的病虫害防治有很好的借鉴作用。但由于主要用于观赏桃栽植的植株在其观赏区域内（以北京植物园和杭州西湖景区为代表）基本上会采取相应的控果措施，即花后及时摘除残花的工作，使得观赏桃中的病虫害发生几率明显少于用于果桃生产的栽植区域，即那些直接发生于桃果上的病虫害（Layne and Bassi 2008）在观赏桃花中并未构成严重危害。

桃花病害主要以叶部病害为主，其次是枝干病害。

桃花虫害从较早以鳞翅目为主的食叶害虫，如"大虫"类的桃天蛾、天幕毛虫等，到近年来逐渐转变为以刺吸、潜叶为主的"小虫"，如蚜虫、叶蝉、潜叶蛾、螨类等。

(一) 桃树的主要病虫害

1. 主要病害

（1）桃细菌性穿孔病（Bacterial spot）[*Xanthomonas campestris* pv. *pruni* (Smith) Dye]

（2）桃缩叶病（Peach leaf curl）[*Taphrina deformans* (Berk.) Tul.]

（3）流胶病（Fungal gummosis）——[*Botryosphaeria dothidea* (Moug. ex Fr.) Ces. & de Not.]

（4）腐烂病（Canker）[*Fusicoccum amygdali* Del]

细菌性腐烂病（Bacterial canker）[*Pseudomonas syringae* pv. *syringae* van Hall *P. mors-prunorum* Wormw.]

桃树腐烂病（Perennial canker）[*Leucostoma persoonii* (Nitts.) Hoehn.] or （Cytospora canker）[*Valsa leucostoma* (Pers.) Fr.]

（5）褐腐病（Brown rot）

美澳型核果链核盘菌（American brown rot）[*Monilinia fruticola* (Wint.) Honey]

核果链核盘菌（European brown rot）[*M. Laxa* (Aderh. & Ruhl.) Honey]

果生链核盘菌[*M. fructigenta* (Aderh. & Ruhl.)Honey]

（6）白粉病(Peach powdery mildew) [Sphaerotheca pannosa (Wallr. : Fr.) Lev.]

（7）根癌病（Crown gall）[*Agrobacterium tumefaciens* (Smith and Townsend) Conn.]

2. 主要害虫

（1）蚜虫（Aphids）[*Myzus* ssp.]

桃蚜（Green peach aphid）[*M. persicae* Sulzer]

桃瘤蚜（*M. momonis* Mats）

桃粉蚜（*Hyalopterus amygdali* Blanchzrd）

（2）介壳虫（Scale）

朝鲜球坚蚧（*Didesmococcus koreanus* Borchsenius）

桑白蚧（*Pseudaulacaspis pentagona* Targioni-Tozzetti）

（3）红蜘蛛（Mite）

山楂叶螨（*Tetranychus viennonsis* Zacher）

（4）叶蝉（Leafhopper）

小绿叶蝉 [*Empoasca flavescens* (Fabricius)]

（5）蛀干害虫（Borer）

桃红颈天牛 （*Aromia bungii* Faldermann）

（6）桃潜叶蛾 （Peach leaf-miner） [*Lyonetia clerkella* Linn.]

（7）蛾类 （Moth）

苹果小卷叶蛾 （Codling moth） [*Cydia pomonella* (L.)]

梨小食心虫 （Oriental fruit moth） [*Grapholitha molesta* Busck]

3. 主要线虫

根结线虫 (Root knot nematode (RKN) [*Meloidogyne incognita* (Kofoid & White.) Chitwood]

（二）主要病虫害产生的原因和防治

1. 病害

桃树的病害一般分为真菌类、细菌类、病毒等不同类型。

危害桃树的真菌类病害主要包括以下危害不同部位的类型：褐腐病、缩叶病、白粉病主要危害花、叶和果实；桃树腐烂病和流胶病主要危害树干和枝干；桃树短命综合症 （Peach trees short life） 则主要是针对根部和树冠而发生的病害。几乎所有因真菌感染的病害均具有非常相似的病症，一般也采用相近的防治措施。

危害桃树的细菌类病害主要包括：桃细菌性穿孔病、细菌性腐烂病和根癌病。

危害桃树的病毒在果树中颇为常见，例如近年来果桃种植者及育种者无不谈之色变的李树痘病 [Sharka or Plum Pox Virus (PPV)]（Garcia and Cambra 2007）、以及西红柿环斑病毒 (Tomato Ringspot Virus)、草莓潜隐环斑病毒 (Strawberry Latent Ringspot Virus)、桃树潜在花叶病 (Peach Latent Mosaic) 和桃花叶病 (Peach Mosaic) 等，但目前在观赏桃花中并未发现能成为主要危害的病毒。

下面介绍一些常见的桃花病害。

（1）**桃细菌性穿孔病**　主要危害叶片，对新梢和果实也有一定危害 （图9.1）。恶劣的环境条件，如降水多，伴随高温强风，会导致细菌性褐斑病的感染 (Larsh and Anderson 1948)。细菌性穿孔病受害叶片上，叶斑最初成半透明油浸状褐色小点，然后逐渐扩展为圆形或不规则形四周有浅黄色晕环的褐色病斑，上常有菌脓；当病斑干枯时，便会形成穿孔或脱落，病症从顶芽向下扩展，会导致叶芽和花芽无法正常展开。一旦细菌性穿孔病在桃树

图9.1　桃叶细菌性穿孔病危害状（李菁博 摄）
Fig.9.1　Bacterial spot (Photograph courtesy Jingbo Li)

图9.2　桃树流胶病
Fig.9.2　Fungal gummosis

群中扩散开来，将很难控制（Ritchie 1999）。

　　加强栽培管理，增施基肥，改善土壤条件，增强树势，提高树体自身的抗病能力是防止各类病害的关键。针对穿孔病，应及时剪除病残体，集中烧毁，减少再侵染病原。采用药剂叶面喷施时，在早春发芽前喷波美3～5度石硫合剂，可兼治害虫；生长期可喷70%甲基托布津可湿性粉剂800～1000倍。对细菌性穿孔病可选用链霉素、土霉素、井冈霉素杀细菌性药剂。从芽开裂开始，通常施用3～5次，但要注意适量（Ritche 1999; Brannen et al. 2007）。每次喷施中间至少要间隔7～14天。

　　(2) **桃缩叶病**　主要危害早春新叶，易发生于早春低温多雨季节。春季初发嫩叶会因感病而皱缩，变红，上有一层白色粉状物，随后病叶最终变成褐色，脱落。严重时也会危害新枝和果。

　　没有腺体的类型有很强的白粉病易染性，在果桃育种中已经基本不再使用，尽管这一类型表现出一定程度的抗缩叶病能力（Wickson 1889；Monet 1983）。具有圆形腺体的类型比具有肾形腺体的类型更易染白粉病（Saunier 1973）。观赏桃品种基本都具有肾形叶腺，这也与其白粉病发生几率并不大有直接关系。

　　防治时采用喷施波美3～5度石硫合剂，1∶1∶100波尔多液，以消灭越冬病菌；在花蕾露红时可喷施70%代森锰锌500倍液，70%甲基托布津1000倍液；一般只需1次喷施药剂处理，但一定要把握好最佳时间：可以在晚秋时节90%的叶片脱落以后，或在春季芽萌动之前。

　　(3) **流胶病**　（图9.2）主要危害枝干，树龄大的较为严重。真菌、细菌、枝干害虫、霜寒、冻害等均能引起透明褐色树胶的不断分泌，以后逐渐变硬成为胶块，会减弱树势，从而为其他病虫害的发生创造条件，极大影响桃花的观赏效果。在中国、日本、美国东南部（Pusey 1989）以及澳大利亚（Scorza & Okie 1991）均有报道。

增强树势，提高树体自身抗性，是防止流胶病和各种病害的根本。注意及时、有效地控制枝干害虫，减少不必要的机械损伤，早春喷波美3～5度石硫合剂，也可根据具体情况选择适当的杀虫剂、杀菌剂。建议冬季进行修剪，可以有效地避免灭菌药残存的时期。

(4) 桃树腐烂病 是桃在北方栽植区域内最为严重的病害，在美国大西洋中部地区、欧洲国家（法国、意大利、葡萄牙、保加利亚等）、北非（突尼斯）、南美（阿根廷、巴西等）及日本均有报道（Layne and Bassi 2008）。主要有细菌性腐烂病和桃树腐烂病两种。后者更容易发生于寒冷地区。在美国东南部，桃树腐烂病也是造成桃树短命综合症的原因之一。

细菌性腐烂病主要危害主干枝，从而减弱树势。症状始于小枝和叶子。染病1年生小枝表现为红棕色或淡棕色，椭圆形损害多发生在芽和节间。腐烂病会减少成枝数，对叶芽、小枝、枝干均可造成危害，2～6年生的树比树龄较大的树更易感染细菌性腐烂病。受细菌性腐烂病侵染的植株根部并未完全死亡，会有萌蘖随后萌生，这也是其与感染Armillaria root rot而导致的根部全部坏死的明显区别。

桃树腐烂病在高氮水平下会更容易发生，因此控制氮在中等或偏低的水平，会降低腐烂病的易染性（Layne and Bassi 2008）。早期人工修剪、高线虫居群数量和冬季冻害会增加腐烂病的发生几率。Chang *et al.* (1989) 发现大多数耐低温基因型很少感染腐烂病，品种的耐寒性和发生腐烂病的几率有一定的相关性。Layne（1971）认为这是由于低温伤害是感染腐烂病的前提条件。中国的桃种质资源被认为是抵御腐烂病的重要资源（Biggs and Lilers 1985）。

及时去除染病枝条是减少感染的主要策略；在芽萌动之前和秋季进行杀菌剂的喷施，对腐烂病的控制更为有效，通常不建议在夏末或初秋进行药剂喷施。

(5) 褐腐病 是一种世界范围内桃树的严重病害（Batra 1991）。主要危害花、叶、枝干和果。以胶质形式残存在小枝上。树枝会因为染病形成溃疡，病部会出现褐色的水渍状腐烂，湿度大的条件下，易长出灰色霉斑，在干燥条件下萎垂、经久不落。低温多雨会诱发花及幼果发病。

美澳型核果链核盘菌在北美发生普遍，是引发北美褐腐病的主要病原体，在欧洲尚未发现该种(Batra 1991)。生长季雨量充沛、湿度高的地方发生率高，在冬季寒冷、夏季干燥的区域则不会造成大的危害。

核果链核盘菌是引发欧洲、澳大利亚、南非、智利、伊拉克、中国、日本等世界范围内褐腐病的主要病原体（李世访，陈策 2009），但在美洲褐腐病存在的地区，如美洲地区的分布并不普遍，仅在美国加州有发现，远没有前者危害严重。

果生链核盘菌主要发生在欧洲、巴西、智利、埃及、中国。北美已经基本没有。

加强栽培管理，尽可能保证通风透光，排水良好；结合修剪，清除病枯枝、僵果、落果，集中烧毁，消灭越冬病原；休眠期防治非常关键，发芽前喷施苯并咪唑类杀菌剂，可以抑制病组织菌丝生长，有在开花期保护花朵免受侵染的作用；或在花蕾现红时，喷波美5度石硫合剂加0.3%～0.5%五氯酚钠消毒；花后喷0.3度石硫合剂。发病严重时，每隔10～15天喷1次药，可用速克灵、福美锌、甲基托布津、代森锌等。

(6) 白粉病　危害叶片、嫩枝和果实，很少有花会感染此病。多发生在湿凉的夜间和温暖的白天。最初的症状表现为小枝上出现小块、圆形的白色侵染，最终孢子会连接成串。叶子会出现卷缩，长期侵染通常会导致叶脱色和坏死，严重者会导致落叶。春季新生叶会通过已经染病的芽形成病害。

降低湿度，减少灌溉次数，采用低角度喷头以保持叶面干燥。杀菌剂应在萼片脱落后、果核开始硬化时使用。

2. 虫害

根据果桃上的分类，桃树害虫主要分为直接害虫和间接害虫。

直接产卵或取食于果实的梨小食心虫、苹果小卷叶蛾等为直接害虫。由于观赏桃花在栽植过程中，特别是结实早期，尽可能地会去除果实，所以这类直接害虫在桃花中并不是主要害虫；与之相反，那些以叶、树干和根为主要取食对象的间接害虫，则是危害观赏桃花的重要害虫。

蚜虫和红蜘蛛（螨类）是各地桃花最为常见的虫害。近年来，小绿叶蝉、桃红颈天牛、桃潜蛾、朝鲜球坚蚧等虫害时有发生，危害程度每年有所不同，气候、栽培条件及管理措施是影响其发生的主要因素。

(1) **蚜虫** 危害桃树的蚜虫主要有桃蚜（图9.3）、桃瘤蚜（图9.4）和桃粉蚜（图9.5）。危害部位主要以群集方式吸食嫩枝、嫩叶或花蕾

图9.3 桃蚜及桃蚜危害状（李菁博 摄）
Fig.9.3 Green peach aphids (Photograph courtesy Jingbo Li)

图9.4 桃瘤蚜及桃瘤蚜危害状（李菁博 摄）
Fig.9.4 *Myzus momonis* (Photograph courtesy Jingbo Li)

图9.5 桃粉蚜密集状（周达康 摄）
Fig. 9.5 *Hyalopterus amygdali* (Photograph courtesy Dakang Zhou)

图9.6 小绿叶蝉
Fig.9.6 *Empoasca flavescens*

等的汁液，影响枝叶的正常生长发育，粉蚜危害时叶背布满白粉；桃蚜危害嫩叶，使其皱缩卷曲；瘤蚜危害后叶缘向背面纵卷，变成紫红色。蚜虫的排泄物往往会引发煤污病。蚜虫1年发生十几代，以5月繁殖盛期最为严重。草蛉、瓢虫、食蚜蝇等是蚜虫的天敌。

对蚜虫的防治可以在冬季结合冬剪及时去除有虫卵的枝条；萌芽前喷波美3～5度石硫合剂可以杀死越冬虫卵；抓住萌芽期越冬卵孵化的高峰期、花后桃叶未卷缩前、以及5～6月危害高峰期等这些关键时期进行喷药防治，通常使用吡虫啉、清源保、印楝素、烟参碱等。同时注意保护和利用天敌。

研究证明，抗蚜虫的基因是受单基因显性控制（Massonie 1979）。山桃以及山桃和桃的杂交种对蚜虫都有一定抗性（Massonie 1979; Massonie et al. 1982; Pisani and Roselli 1983）。

对蚜虫的防治通常采用化学喷施药剂的方式，但由于观赏桃花不同于果园栽植的相对封闭性，大多栽植于人群相对密集的生活区或观赏区内，人工喷施化学药剂不仅每年消耗大量工力、财力，而且对环境也会产生不良影响，既对瓢虫、草蛉、食蚜蝇等有益的天敌昆虫造成杀伤，又会直接影响人们的生活和游览，化学物质在土壤及水体中的残留也会对生态系统带来潜在的危害。因此，北京植物园尝试开发了一种人工摘叶控制蚜虫的方法。根据蚜虫以卵在叶芽基部越冬的特点，在秋季桃花自然落叶之前，人工摘除叶片以阻止叶片上的蚜虫在叶芽基部产卵越冬，以减少次年蚜虫基数。试验证明，这一方法不仅显著地减少了桃花上的蚜虫发生量，而且对桃花的花期、花径都并无明显影响，有效地减少了农药的使用，保护了天敌，将农药对生态系统的不良影响降到最小（李菁博等，北京植物园2008年度交流论文）。

(2) **小绿叶蝉** 以成虫、若虫在叶背刺吸新梢和叶片的汁液造成危害，春季开花时即开始出现。危害初期叶出现失绿白点，严重时全叶苍白早落，影响树势和花芽分化。早春先在发叶早的寄主上取食，然后扩展到其他寄主上危害。平时隐蔽于新梢及叶背。每年发生5代以上，以成虫潜伏于枯草、落叶、树皮裂缝等处越冬（图9.6）。

对河南新乡市人民公园的观赏桃园的昆虫群落的调查显示（吕文彦等 2008），小绿叶蝉是主要优势种，危害主要发生在7月中下旬至9月底。其次是在6月中旬至7月中旬危害严重的山楂叶螨和桃粉蚜。

为了提高防治效果，喷药应抓住各代若虫孵化盛期，成虫向桃树迁飞时进行。化学防治通常使用吡虫啉、滑源保、印楝素、烟参碱等。秋冬及时清除落叶，铲除杂草，并集中烧毁，消灭越冬成虫。

(3) 山楂叶螨　以成虫和若虫群集叶背，吸食叶片汁液（图9.7）。叶片受害后大多先从叶背近叶柄的主脉两侧开始出现许多黄白色至灰白色小斑点，其上有丝网，严重时会扩大成大片枯斑，终至全叶灰褐，迅速焦枯脱落，影响树势和花芽分化。在北方年发生5～13代，均以受精雌虫在枝干各种缝隙内、落叶，或干基附近土壤缝隙中群集越冬。以7～8月高温干旱季节为全年发生高峰期。

可以在9月底成虫开始越冬时，在树木枝干上绑草绳诱杀过冬螨，或在休眠期进行防治，即在早春萌芽前彻底刮除主干主枝上的粗皮翘皮，集中烧毁，能够有效地消灭大量的越冬雌螨。喷药防治则应该抓住其抗药性差的第一若螨期，防治效果最好，通常选用1～3度石硫合剂，发生期可喷阿维菌素。近年来，引用了一种胡瓜钝绥螨，在合适的释放时间，选择适当的高度施放适量螨态的螨虫，可以有效控制山楂叶螨对桃花的危害。

(4) 桃红颈天牛　以幼虫蛀食树干和主枝的皮层、韧皮部和木质部（图9.8）。幼虫在横向弯曲伸展的虫道内蛀食，虫粪堵满虫道，有的从排粪孔内排出大量粪便堆积于树干基部，有的从皮缝内挤出，常有流胶发生。虫口密度大时，造成树干中空，皮层脱落，树势减

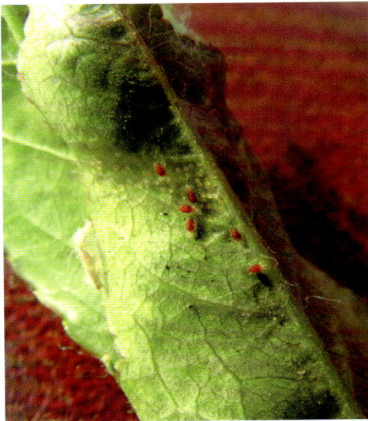

图9.7　山楂叶螨（李菁博 摄）
Fig.9.7　*Tetranychus viennonsis*
(Photograph courtesy Jingbo Li)

图9.8　桃红颈天牛及危害状（李菁博 摄）
Fig.9.8　*Aromia bungii* (Photograph courtesy Jingbo Li)

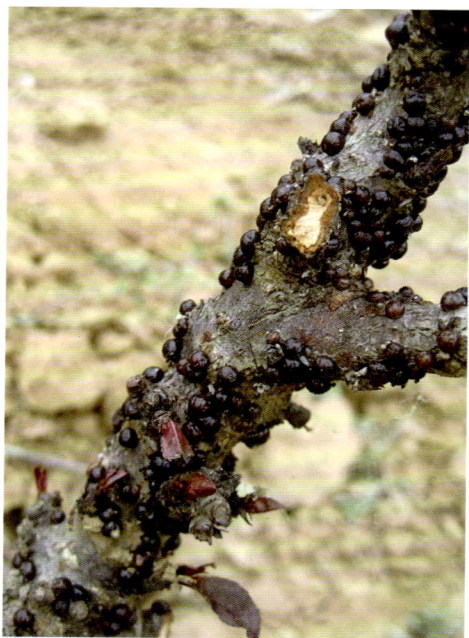

图9.9 朝鲜球坚蚧（付俊秋 摄）
Fig.9.9 *Didesmococcus koreanus* (Photograph courtesy Junqiu Fu)

图9.10 桃潜叶蛾危害状（周达康 摄）
Fig.9.10 Peach leaf-miner (Photograph courtesy Dakang Zhou)

弱，寿命缩短，严重时可引起死亡。2～3年完成1个世代，以不同龄期幼虫在树干隧道内越冬，树液流动后越冬幼虫开始活动危害。

对幼虫的防治，通常采用铅丝从排粪孔顺隧道钩杀幼虫，或在虫孔注药后用泥封蛀孔，毒杀幼虫的方法。对成虫则可以在其羽化前通过树干涂白（生石灰10份、硫磺1份、食盐0.2份、油0.2份、水40份）防止成虫产卵，或在6～7月其成虫期，辅助糖醋液在枝干上进行捕捉。

蛀干虫更容易被那些因为腐烂病、冻害或溃疡感染等而引起的胶质堆积物所吸引，产生侵染。流胶物的化学成分分析显示，怀孕的雌性成虫更容易被这些胶质物所吸引（Gentry and Wells 1982），而胶质沉淀中所含的挥发性物质能够刺激蛀干害虫排卵。

(5) **朝鲜球坚蚧** 以成虫和若虫吸食树干枝条的树液，严重时可使枝条或全株死亡（图9.9）。1年只发生1代，以2龄若虫在枝条上越冬，4月上旬开始活动，重新寻找永久固定场所吸食汁液。防治时针对不同时期采取相应措施。雌虫产卵前，人工剪除虫枝或刮除虫体，集中烧毁；卵孵化盛期喷药防治，可选用吡虫啉，连续喷2～3次效果会更好。另外注意保护其优势天敌日本方头甲、黑缘红瓢虫。

（6）**桃潜蛾**　以幼虫潜入叶肉组织串食，危害部分表皮变白，不破裂，严重时整个叶片都被潜食，枯黄而引起早落（图9.10）。每年发生约7代，以成虫在杂草、树皮缝等处越冬。

防治时注意及时清除并集中烧毁落叶，杀死蛹或成虫；在成虫盛发期和幼虫初孵化期，选用灭幼脲、甲基阿维菌素等喷洒防治。

近十年来桃花越来越受到各地的重视，随着应用范围的不断扩大、应用品种的不断丰富，在养护管理中，尤其是病虫害防治上，也开始出现了可持续发展理念及节约型园林等概念的引入，因此在观赏桃病虫害防治方面进行的有益尝试，必将成为今后观赏桃植保工作的必然趋势。

三、修剪篇

果桃的修剪在其多年栽培的基础上早已广为熟知，但作为以观赏为主要目的进行栽植的桃花的修剪，则是以满足美观的观赏需求，以完善树型，最大限度地展现其观赏特性，维持最佳观赏效果为主要目的。

（一）桃花修剪的根本原则

观赏桃的修剪和果桃有着明显的区别，这首先取决于修剪的目的。果桃是为了获得更多更大的果实，而用于观赏目的的桃花修剪，则应主要关注树体的观赏特性，简单地说就是要使花枝搭配合理，分布疏密有致，花朵大而饱满，而相对不必过多在意传统果桃三大主枝等概念化的模式，根据不同类型、不同品种、不同树龄、不同树势、不同环境、不同栽植地点等具体情况，适树适剪才是桃花修剪应当遵循的根本原则。

桃树是一种喜光植物，改善树体光照条件，调节树体营养，创造良好的通风透光的树体结构，保持树势，是持续保证桃花具有良好观赏效果的重要保证。及时去除枯枝、残病枝、徒长枝、过密枝、内膛枝及萌蘖，减少枝叶的水分蒸腾，改善透光条件，充分满足桃花生长的基本需求；根据观赏特性，不断更新树冠的枝组群，合理调节符合各自枝型的枝条分布，使之排列有序，错落有致，层次分明，结构合理，尽可能地使修剪后的树体均衡协调，富有姿态的美感，是为观赏桃修剪的最高境界。

要修剪桃花，首先要了解桃花的生长习性。桃花的生长方式为合轴分枝方式，即在生长末期会由于顶端分生组织生长缓慢，致使顶芽瘦小或者不充实，造成冬季干枯死亡；或者由于有的顶端形成花芽，不能继续向上生长，而由顶端下部的侧枝取而代之，继续向上生长，每年如此循环向上，均由侧芽抽枝逐段合成主轴，故称合轴分枝。如果放任生长，在顶梢上会有几个势力相近的侧枝同时生长，形成多分权树干。

其次要了解桃花的生物学特性，即桃花芽的类型，干性的强弱，以及桃花枝条的特点。桃花的芽包括花芽、叶芽，单芽、复芽，每一类枝条上都很容易形成花芽；桃树本身的干性弱，不易形成明显的中央主枝，生长迅速，新梢一年能抽生2～3次副梢，树姿较开张，易整型，寿命一般20年。领导枝、干枝、侧枝、花枝、花枝组、背上枝、膛内枝、护干枝等均代表不同的含义，应采用不同的修剪手段。

最后，也是桃花修剪中最为重要的一点，就是要了解不同类型桃花的枝型特征，根据枝性强弱保持原有枝姿，体现出桃花丰富的观赏特性。

(二) 修剪的手法

灵活掌握桃花修剪中短截、疏剪、回缩等最为常用的修剪手法。

短截通常是指剪去1年生枝条上的一部分枝。短截可以提高成枝力，对枝条局部有促进作用，通常剪口下第一侧芽的萌生力最强，常形成长枝，利于扩冠。截不能过狠，必须在有萌芽点处才有可能发出新枝。对于枝条基部不易萌发的休眠芽，可以通过重截刺激，使其萌芽抽枝更新树体。所有针对桃树的修剪，剪口芽必须有叶芽。

短截包括轻短截（1/5～1/6），剪去枝顶很少的部分。截后枝条萌发多，但生长势较弱。中下部容易形成花芽；

中度短截（1/3～1/2）：从枝条的饱满芽处下剪。枝条萌发较少，但生长势较强。中下部形成少量短枝，剪口芽萌发强枝，是利于生长的剪法；

重剪（2/3～3/4）：较重程度的短截。一般从枝条基部半饱满芽处下剪。斜生枝条剪口下萌生枝条较弱，易形成2～3个短花枝，是削弱的修剪；直立枝剪口下可能形成强枝，但可以通过回缩培养花枝组；

极重剪：仅留基部3～5厘米，一般是瘪芽，易抽生几个中短枝，去强留弱可培养结果，是削弱的剪法。

疏剪即疏枝。把1年生枝条或当年生枝条从基部剪除，尽量不留残桩。主要剪除干枯枝、徒长枝、病虫枝、交叉密生枝及不能利用的枝条，以改善树冠内部光照条件，提高光合效能、增加营养积累为目的，从而利于开花。

回缩是更新复壮的手段，目的在于恢复树势，刺激枝条健壮生长和潜伏芽的萌发抽生。适当的压，才能刺激出新的花枝，使得树体趋于圆满，达到平衡树势的目的。主要是针对2年生以上的枝条。

(三) 修剪的时期

桃花的修剪一般分为冬季修剪和花后修剪。根据实际情况和需要也可二者结合进行，相互配合和补充。

冬季修剪。通常在休眠期进行（12月1日～翌年2月中旬），是树冠定型的关键。以整型为主，根据树形要求、品种特性进行剪留，骨干枝要注意长度适宜和留芽方向，要控制好主枝角度；选留侧枝要考虑位置、方向、角度，适当地多保留花枝和侧枝，不宜枝枝过剪，以便维持树体的自然姿态。多采用短截、回缩手法进行修剪。实践证明，在冬剪或花后剪的时候，短截所留部分如遇到背上芽，应该及时顺手掰除，既可以有效减少其成枝后所带来的养分消耗及来年的大强度修剪，又可以省去了运输修剪下来的枯枝所产生的相关经费和工力。

花后修剪。冬季修剪时短截所留的过量花枝，可结合花后疏果同时完成，以减少树木养分的消耗。花后修剪通常是对冬季修剪不能一步到位的植株进行补充修剪和完善。澳大利亚曾经有利用花后修剪的办法，减少桃缩叶病发生的记载，即在花落之后立即进行修剪，剪去枝条的一半，可以使枝条健壮、干净，免受缩叶病的危害。但目前还并不清楚原因，但实际效果很明显（Harrison 1963）。

冬季修剪结合花后修剪是桃花修剪的最佳方案，但由于对工力要求较高，在实际工作中经常难以保障！

果树上常用的夏季修剪——即以疏剪为主，通过处理内膛枝，剪去过密的枝条及病枯枝——在桃花日常管理工作中，在工力充足的情况下，特别是针对观赏盛期的植株，建议使用，能够达到通风透光的目的，为桃花生长创造良好条件。

修剪是持续的反应，很少有人可以仅仅通过一年两年的修剪就能达到最终目的，一般至少要有连续3年的跟踪观察和调整，根据树体的生长情况，逐步检验修剪意图是否达成，是否需要调整。需要调整的部分也往往需要一年一年慢慢地进行，不能急于一次成型。

修剪没有绝对的对与错，最为关键的在于"合理"，只有合乎目的的修剪才能真正充分体现修剪这一手段的效果和重要性。老话说的"因树修剪、随枝作形"（卢戈编 1990年）非常适用于桃花修剪，根据原则"适树适剪"的灵活处理才能达到预期目的。

(四) 各类枝型桃花修剪的基本原则

了解不同类型桃花品种的枝型特性，对于桃花的修剪至关重要。不同枝型、不同品种的桃花萌芽力、发枝力、分枝角度等都各有差别。根据顶端优势的原则，直立枝生长旺盛，斜生枝次之，中等偏强，水平枝缓和，下垂枝较弱。

生长素浓度与枝条生长活力和枝型有关。帚型桃枝条从生长素到细胞分裂素浓度均高于直枝桃（Tworkoski *et al.* 2005）。修剪可以影响当年生枝条的荷尔蒙比例，增强生长，提高生长素含量，但不能影响细胞分裂素的浓度。修剪可以刺激生长和分枝。

1. 直枝型桃花的修剪

直枝桃类桃花是目前园林中应用数量最多的一个类型。枝条直立，冠性开张。直枝桃类桃花的修剪原则是要注意留外芽。

自然杯状形，自然开心形是北京地区应用在直枝类型桃花品种中最为广泛的两种修剪形式。这两种修剪方式基本是沿袭了果桃的指导思想，又丰富了桃花培养花枝组、丰富花量等特殊观赏需求。三大主枝的概念在桃花修剪中并不像果桃修剪中那么严格，特别是对枝干基角等没有特殊要求。但通常苗圃的桃花苗基本上也会在第一年保留三大主枝，第二年在一级主枝先端萌发的枝条中选取5～6个二级主枝，第三年再从二级主枝先端萌发出的枝条中选留7～12个三级主枝，使整个树体枝条三级主枝向四周均匀分布延伸（图9.11）。

对主枝的修剪：

主枝是从树的主干分出来的，向外生长的主枝。一般情况下，每株树保留3～5个主枝即可。修剪时根据生长空间条件，对几大主枝进行定位（即留头）。一般根据芽子情况保留25cm左右，选择离枝条底芽或侧芽1.5～2cm位置剪去，剪口要有一定的斜度，约在35°～45°左右。

对侧枝的修剪：

侧枝也被称作辅枝。侧枝留头长度一般不得超过主枝位置。在主枝两侧适当保留侧枝，增多枝条生长量，以便形成更多的花枝。

对花枝组的处理：

花枝组是由主枝或侧枝上的花枝，经多年修剪培育而形成的一组花枝，既可以保持满树是花的效果，又可以尽量使花分布在树冠内膛，使人更容易接近。花枝组有大有小，在修剪时，去强留弱，不宜枝枝短截，留有一定的生长点即可。

"半边树"的现象在栽植环境不当的情况下时有发生。这主要是由于光照条件、通风情况而造成的枝条强弱各异，大枝营养充足，过于强壮，生长过快；而弱枝则越来越弱，产生偏冠，不仅容易倒伏，也很不美观。在修剪中要注意保持平衡，即改变生长旺盛的枝条的生长方向，或者加大过旺枝的开张角度，使其平缓生长，以减弱生长，"去强扶弱"，缓和树势，平衡生长。

修剪时如果忽略了花芽在枝条上的位置，随意短截则会带来疯长叶、不开花的结果。在树龄尚小的时候强度短截，刺激过强，第二年抽生过壮过长、没有花芽的新枝，形成疯长。应适当保留长枝，扩大树冠，尽量保留中等长度的枝条。

图9.11 自然式修剪的桃花
Fig.9.11 Natural pruning

　　修剪中还应根据不同品种的特点进行修剪，不能一味拘泥于共性修剪。

　　'菊花'桃、'京舞子'等菊花型品种具有树势弱、枝条细的特点，修剪时不能像对待一般直枝类桃花一样，枝枝过剪，应当仅作适当枯死枝、病残枝、过密枝的去除，基本保持品种原有的自然状态。

　　针对发枝多却成枝力低的品种，比如'白碧'桃、'晚白'桃等要截在饱满芽处，以促发长枝，并注意及时疏枝，保持通风透光，减少病虫侵害的机会。而对那些抽枝能力强的品种，比如'红碧'桃、'碧桃'等，则应在相对较弱的芽处轻截，促进抽生形成花芽，以缓和树势。

2. 矮型（寿星类）桃花的修剪

　　此类桃花枝条节间短，着生紧密，生长较慢。基本不需要进行特别的修剪，一般只做去除内膛枝、枯死枝、病残枝、并生枝、交叉枝及过密枝等常规修剪，保持原有树冠圆球形的饱满和匀称。

3. 帚型桃花的修剪

　　此类桃花具有明显的枝条直立、分枝角度小的特点。观测表明，帚型桃的树冠仅相当于直枝型桃花的40%，修剪量也仅相当于直枝型桃花的50%（Bassi *et al.* 1994）。帚型桃一般可以不做整型修剪，尽可能保持原有自然株型。修剪时首先考虑去除并生枝、交叉枝、病死枝外，修剪时留枝的外侧芽，同时同一级次的枝条的高度要保证协调统一，使枝条聚拢不开张，特别注意不要去除过多的护干枝，否则会影响当年开花；而保留一定数量的护干枝，可以达到树体匀称丰满的效果。同时注意切忌去除中央领导枝，否则会刺激萌生过多侧枝，难以恢复原有树型。

4. 垂枝型桃花的修剪

　　此类桃花节间长，枝条细长下垂，营养传输要经历先上传再弯到下垂枝顶的"长途跋涉"，相对比直枝型更为困难，生长较慢。北京地区冬季风大，垂枝桃类稍条现象比较严重，修剪时首先要去掉枯死枝、过密枝、病虫枝、衰老枝，然后采用短截和疏枝的方法进行更新，特别注意剪口一定要留上位芽进行修剪，诱发新枝向外出挑延伸，保持伞形树冠不断扩大，形成丰满树形。

5. 曲枝型桃花的修剪

　　'云龙'桃是曲枝类型桃花的唯一1个品种，具有枝条扭曲、生长势强健的特点。一般仅作适当去除枯死枝及过密枝的工作，不必进行整型修剪，小枝会根据光照情况弯向适当的空间，形成良好的枝型分布。

6. 山碧桃类桃花的修剪

此类桃花因同时具有桃和山桃的血统，树体高大，小枝纤细，树皮光滑，是重瓣桃花中开花最早的类型。为保留其原有姿态和特点，山碧桃类品种通常不宜做大规模修剪，尽可能保留自然的树体结构，只对残枯枝做及时修剪即可。

(五) 不同树龄的桃花有不同的修剪原则

幼年期的桃树是来自苗圃的幼龄树，直到定植后的6～7年均被视为观赏初期。这一时期的桃花枝梢生长旺盛，营养生长大于生殖生长，很容易发枝，分枝角度尚小。修剪时以扩大树冠，增加主干、主枝分枝级次，增加叶片数量等营养体生长为主。在此期间，直枝类品种主干、3大主枝等骨架枝条基本形成并定型，生长过程中容易出现徒长现象，如有可能需要进行夏季修剪，控制徒长枝条。注意随树造型，以保持良好树型为主要目标。修剪时应采取轻剪，不宜过重。

桃树的生命力很强，尤其是幼年和青年期，是为最佳观赏期。此时植株对修剪的反应更为敏感。成年树的生长较好，开花繁茂，修剪时以平衡树势为主，即壮枝轻剪以缓和树势，弱枝重剪以增强树势。这一时期的桃花一般以休眠期修剪为主，花后进行修复修剪。由于观赏桃以观花为主要目的，休眠期短截时，一般实际留枝长度比需要留的长度稍长，在花后再补充修剪到位，以保证观赏效果。

树龄在20年以上的桃花基本进入衰老期，树势逐渐减弱，营养生长小于生殖生长，成枝力减弱，新梢生长量减少，营养缺乏，树势回落。为了延缓衰老，避免过早的树势衰弱，需要采用重剪，刺激隐芽萌发，以达到复壮更新，提高单位保留芽的营养，促进萌发壮枝。

"小老树"现象在桃花中很为常见，这主要是由于修剪过度，伤口过多；或者先天营养匮乏，生长缓慢，修剪不当，如此反复几年就会造成树势极度衰弱，形成"小老树"。

法国曾经有因为采取了成功的修剪措施，而使桃树活到72年仍在继续健康生长的例子（Dowing 1869），可见合理的修剪能够有效地延长寿命，增强观赏性。

　　修剪与综合管理措施有着密不可分的关系。修剪只是一种辅助手段，树体自身的条件、栽植条件及生长环境都在很大程度上影响着树木的生长，修剪不可能解决所有问题！修剪只有充分依靠施肥、灌溉和病虫害防治等综合管理措施，才能显示出其最大的效果，发挥事半功倍的作用。

Peach blossom by Van Gogh （1853-1890）
这是荷兰著名画家凡高绘制的法国阿尔地区桃花盛开的景象，桃花所表现出的春天的繁盛反映出了作者心目中对自己未来生活的乐观憧憬。

10 桃花的应用

APPLICATION OF ORNAMENTAL PEACHES

鲜艳灿烂、繁茂艳丽的桃花是繁荣、幸福、美满、和谐的象征，

自古就是深受我国人民喜爱的传统名花。

桃花无论是作为庭院栽植、行道树、专类园、盆栽或切枝等观赏应用，

还是其药用和美容功能，

在中国都是应用最为广泛的国家。

　　鲜艳灿烂、繁茂艳丽的桃花是繁荣、幸福、美满、和谐的象征，自古就是深受我国人民喜爱的传统名花。我国华北地区四季明显，在经过了严冬的肃杀和萧条之后，艳丽开放的桃花带给人们春天般的温暖与希望，是春回大地的使者；而南方的广东、香港一带，由于粤语中"红桃"与"宏图"谐音，人们形成了一种在春节前要买一枝桃花回家的特殊习俗，寓意"大展宏图"，行"桃花运"的好兆头，桃花在广东及周边地区又被誉为"中国的圣诞树"；我国少数民族地区有庆祝自己特殊节日的习俗，白族"秧歌会"的礼仪植物就是盛开的桃花花枝（陈重明 2004）。桃花在我国大江南北深受喜爱，并广为应用。

　　桃花原产中国，在世界范围内也有所应用。中国、日本及美国是目前培育及应用桃花品种较多的3个主要国家。

　　日本是应用桃花相对比较多的国家，但与其传统名花樱花相比，规模甚小。东京附近的古河综合公园栽有2000株桃花，以每年春天的春日桃林祭而闻名。东京都神代国立植物园也有一个桃花专类园。日本至今

图10.1　古河桃林祭女孩穿和服的照片（吉田雅夫 摄）
Fig.10.1　Girls with kimono in Koga Peach Festival in Japan (Photograph courtesy Masao Yoshida)

还有过女孩节（也叫桃花节）的传统，据说是从中国传过去的。因为过去的女孩节是旧历三月三日，正好桃花盛开，桃花节就因此而得名。在日本，桃花节是女孩子最高兴的日子，她们要穿漂亮的和服，并且邀来自己最亲密的伙伴，尽情地在桃花林下欢度节日（图10.1）。

1998年对日本桃花资源的调查显示，桃花除在茨城县、神奈川县有相对较多应用之外（所用品种也不过20个），在东京、京都、大阪、横滨等地虽尚可见到，却当属极为罕见。相比之下，日本应用帚型桃的比例要远比其他国家大，在街道、公园、苗圃和花卉市场中，出现频率最高的就是各色的帚型桃。

影响桃花在日本的受喜爱程度，一方面是历史原因，日本有众多关于樱花的历史传说，赏樱花的习俗更为悠久；另一方面，日本民族普遍更喜欢颜色稍微柔和的淡粉色，相对而言，艳丽的桃花远不如樱花被广为接受和喜爱。

桃花在美国并不普遍，尽管在苗圃中也可以见到一些品种有售，但市场需求却相当有限。主要原因是寿命短，制约着桃花的广泛应用，此外果实过多、病虫害严重也是限制其在私人庭院、公园绿地中栽植的主要因素。

尽管日本某些地区也有赏桃的传统，美国也有将漂亮女孩称作"桃儿"的习俗（Georgia peach），但无论是重视和喜爱程度，还是桃花应用的形式，中国还是世界上桃花应用最为广泛的国家。

桃花最早应用于园林始于汉代。桃花源、桃花坞、桃花潭、桃花径、桃花洞与黄山的桃花峰、桃花溪，华盖山的桃花圃均为古代的桃花胜地。据《花史》记载："古田县黄檗山多桃树，下有桃坞、桃湖、桃洲，春日景色不减武陵，下有桃溪，春风微和，夭桃夹岸，亦胜地也。"俨然一处"桃花世界"。《西湖志》也有记载：包家山多桃花，花开之时云蒸霞蔚，宋时有匾名曰"蒸霞"，西湖栖霞岭也因岭上桃花烂漫，色如凝霞而得名。此外，江郎山（今浙江省江山市）、武夷山、锡山、简州（今四川简阳一芾）、河阳（今云南澄江县）等地自古也是桃花胜地。我国南北各地以桃花为名的景点不胜枚举，仅"桃花源"一名就分别有湖南桃源县、安徽黟县、浙江宁波以及江西、四川等地。这也从另一个角度体现了桃花在中国应用之广泛。

桃花的应用主要从其观赏和药用两个方面体现。

一、观赏

由于桃的分布范围较广，在大部分地区主要还是作为室外观赏植物栽植，伴以少量的室内盆栽、盆景，以及切花、切枝应用。

(一) 庭院栽植

早在晋代，陶渊明在《归田园居》中就曾描绘过"榆柳荫后檐，桃李罗堂前。"的栽植方式，可见将桃花用于庭园栽植已经有千余年的历史。

孤植桃花可以体现个体美，以繁茂取胜 (图10.2)。

群植、片植或成林种植于园林、道观等公共场所的桃花，能够更好地渲染空间，增强花色对人的视觉冲击力，烘托出烂漫春色。

不同花色的桃花相互搭配更能够相互映衬，体现出春色盎然的意境 (图10.3)。利用不同枝型的桃花与环境相协调，可以创造出完美的景观效果。例如垂枝桃种植在水边更能体现出枝影婆娑的魅力 (图10.4)；而与周围色叶树、常绿树搭配种植在一起，则共同构成一幅色彩丰富的、生动的春景图 (图10.5)。庭院内高接的寿星桃，则将平日只有低头才能发现的枝叶繁茂、树型圆整丰满的寿星桃，抬高到人的视线之内，体现出一道郁郁葱葱的夏日风景。

桃花与不同花木组合栽植，可以营造出格调不同、色彩鲜亮的绚丽春色景观。最为著名的传统栽植方式就是桃柳组合。柳树枝条纤细，树影婆娑；桃花烂漫芳菲，花团锦簇。桃红柳绿，相互对比、烘托，和谐中更显出桃花的娇艳和明媚，突出春季生机勃勃的特殊美感。这一搭配利用桃花花期与柳树展叶大致同时的特点，产生了物候现象重叠之美(图10.6)，是中国园林中重要的春季景观，并已形成古老的造景模式。西湖的苏堤、白堤堪称这一手法运用的代表作；在南京玄武湖、扬州瘦西湖、北京颐和园西堤等中国著名的园林中，更是得到了广泛的应用，并且深受人们的喜爱。

(二) 行道树

枝型窄高的帚型桃特别适合做行道树、隔离带 (图10.7)，可以春天看花，夏秋作为高篱，起到隔离、遮挡的效果。

帚型桃作为行道树和隔离带种植，在国内外均有很好的范例。日本代代木体育场外和杨树间隔种植'照手桃'(红、白、粉各色均有)，形成良

图10.2　北京植物园盆景园外湖边最大一株桃花，开花时节可以同时容纳7～8组人同时照相，而不至于互相干扰

Fig.10.2　A huge specimen of ornamental peach planted in Beijing Botanical Garden

图10.3　不同花色的桃花配植在一起体现春意盎然的意境

Fig.10.3　Differently colored ornamental peach clustered together is a striking way to display the prosperity of spring

图10.4　北京植物园绚秋苑水边垂枝桃

Fig.10.4　Weeping peaches planted next to the lake in Beijing Botanical Garden

图10.5 北京植物园树木园垂枝桃与常绿树和其他早春花木共同组成一幅生动的春景图
Fig.10.5 The weeping peaches with conifers planted in arboretum of Beijing Botanical Garden

图10.6 碧桃园桃红柳绿
Fig.10.6 Peaches with weeping willows --the most famous traditional planting style of ornamental peaches

图10.7 北京植物园南门外'照手红'做成的隔离花篱
Fig.10.7 Flowering fence of 'Terutebeni' in the south gate of Beijing Botanical Garden

图10.8 日本代代木体育场外的各色帚桃
Fig.10.8 'Terutebeni', 'Teruteshiro', and 'Terutemomo' planted around Yoyoki stadium in Tokyo

图10.9 重庆用'照手红'作绿化分车带
Fig.10.9 'Terutebeni' is beginning to be planted along highways, driveways as street trees in Chongqing

图10.10 亚特兰大植物园温室外的'Corinthian Pink'
Fig.10.10 Purple-leaved 'Corinthian Pink' planted in front of the greenhouse in Atlanta Botanical Garden

好的行道树格局（图10.8）。在我国重庆市也开始大规模使用'照手红'作为城市道路分车带及街道绿化美化使用，收到了很好的效果（图10.9）。

紫叶品种的帚型桃丰富了桃花的观赏期，树型优美、叶色独特的'Corinthian Pink'在庭院中往往可以独树一帜，成为盛夏时节的一道亮丽风景（图10.10）。

树型圆满、高大的帚型桃具有良好的遮挡作用，春季艳丽的花篱、夏秋繁茂的叶篱、甚至冬季丰满的枝干也很具观赏性。

(三) 专类园

中国应用桃花建设专类园的历史由来已久，唐代贞观年代就已经有桃花用于专类园使用的记载。《太平御览》卷九六七引《唐书》说："唐国贞观十一年，献金桃、银桃，诏令植之于苑囿。"唐代长安大明宫，就栽种了大量的桃树。皇帝为游赏之便，还专门辟有桃花园。

桃花专类园凭借丰富的桃花品种，集中栽植展示桃花品种的多样性，是为桃花应用的最好形式。

桃花原产中国，有着丰富的种质资源和深厚的文化内涵，近年来随着社会经济和人民生活水平的不断提高，桃花在我国逐渐受到广泛的重视，并得到迅速的发展和多方面的应用。湖南桃花源和桃江、上海龙华、上海南汇、杭州西湖、广州石马、成都龙泉、江西庐山、北京西山、兰州安宁等地以种植大量桃花或举办一年一度的桃花节，而成为中国十大桃花观赏胜地。以后安徽芜湖、广西柳州、四川遂宁、广东清远等地纷纷举办各种形式的桃花节、桃花会，踏青赏花已经成为人们借助赏花观景推进经济发展，创造良好的社会效益和经济效益的一种手段。其中不乏大批以果桃栽植、农家游园活动等为主的旅游经济行为，但桃花作为一种烘托气氛的背景植物，每年早春盛开的时候，赋予了这些地方极高的观赏性。

近20年来，随着人们对桃花文化内涵理解的不断深入，以及桃花栽培品种的不断丰富，北京、杭州、湖南桃花源、昆明郊野公园等地相继形成了以栽植观赏桃为主题的真正意义上的桃花专类园。把桃花与文化交流，以及植物科学知识普及结合起来，为桃花应用注入了新的科技内容。

北京植物园将展示桃花品种与弘扬中国传统花文化相结合，在桃花的栽植手法上创新，利用吟咏桃花的石刻诗文和桃花文化相结合的科普形式，体现桃花的文化内涵，为人们认识、了解桃花资源提供了生动的园地。同时与其他春天的花配植，营造出繁花似锦的景象（图10.11），形成了自己独特的魅力。在经过20余年的桃花引种、选育、应用等方面的持续努力，"北京桃花节"现在已经成为北京植物园的品牌，也是中国最为著名的桃花收集展示中心，每年春

图10.11 桃花和郁金香等早春花卉配植
Fig.10.11 Ornamental peaches with early spring bulbs in Beijing Botanical Garden

天吸引着众多市民和周边地区的人们到植物园观赏桃花，成为北京市民春季一项长盛不衰的传统活动。

(四) 盆景、盆栽

由于桃花本身寿命的限制，作为桃花盆景的栽培并不多见。1998年4月在日本埼玉县大宫市（Omiya-shi Saitama-ken）的金蔓青园见到的百年桃花盆景（图10.12），是由日本著名的盆景大师加藤秀男（Hideo Kato）制作的。

更多的桃花盆栽形式还是以北方地区春节盆栽催花为主（图10.13）。催花应选用自然花期较早的品种，如'绛桃'、'合欢二色'桃等。选择健壮植株，在落叶后将盆桃放在7℃左右的低温环境，使其完成充分的低温休眠，春节前40～45天将其移入温室，从5℃～10℃到20℃～25℃逐渐提升温度，要特别注意不能迅速增温，否则会导致叶芽萌发，只长叶不开花。

(五) 切花切枝

桃花本身不仅备受人们的喜爱，整株桃花插放在容器里，更是中国新年祈求健康长寿和吉祥幸运必不可少的装饰。在广州，人们在新年的时候，甚至将桃花从近根部整个斩断，直接将开满桃花的树枝放在家里作为"中国的圣诞树"。2008年曾创造出一株桃花卖价8200元的天价。

图10.12 1998年4月在日本埼玉县大宫市 (Omiya-shi Saitama-ken)的百年桃花盆景
Fig.10.12 Bonsai of ornamental peach in Omiya-shi Saitama-ken, Japan, made by Hideo Kato

图10.13 盆栽催花'合欢二色'桃
Fig.10.13 Potted ornamental peach for the Spring Festival

　　桃花姿态多样，枝型、枝色优美，花色艳丽，花型各异，作为折枝瓶插的历史由来已久。宋代陈骙《南宋馆阁录》中《碧壶桃花图》一篇就表明瓶插桃花已是当时馆阁重要的生活图景。清代陈淏子的《花镜》中也详细叙述了切花水养的步骤和方法。桃花切枝瓶插催花主要与桃花品种和着花密度有一定关系。例如早花品种'白花山碧'桃、'合欢二色'桃和'绛桃'，需冷量较低，着花密度分别为0.8/cm、1.2/cm、0.75/cm，均属于中等，催花相对容易，又能达到较好的观赏效果。在20℃～25℃室温条件下，'白花山碧'桃大约需要18天，后2个品种则需要约25天左右即可开花。

二、药用及美容

　　桃树浑身是宝。除了深受喜爱的的桃子作为水果食用之外，桃仁、桃胶、桃根、桃叶、桃花，甚至桃树的干皮也都具有一定的药用价值。

　　桃仁很早就有药用记载。成书于公元2世纪左右的《神农本草经》记载其性味苦，是活血祛瘀的良药。现代中药研究表明，桃仁可以治疗高血压及慢性阑尾炎等症。药理研究还证明，桃仁在一定程度上可以改善肝脏表面的血液微循环，可抗凝血和抗血栓形成，还有一定的消炎作用。临床还可用于跌打损伤，瘀血肿痛，血燥便秘和妇女经血不调。将桃仁和决明子一起用水煎服，有缓解头痛的功效；将混入桃核汁液的牛奶涂抹在额头上，也可以起到类似的缓和放松的作用。桃仁与醋一起煮沸，对斑秃部位头发的再生还会有意想不到的效用。

桃树所分泌的树胶叫桃胶，味甘苦，能生津止渴，益气活血，可以用于糖尿病、痢疾、石淋、血淋等症的治疗。《本草纲目》（1578）中有"治恶痔作痛，桃根水煎浸洗之"的记载。

民间还有将桃根加入少许红糖，捣烂外敷治疗骨髓炎之法。另有报道说，用桃树枝干受到物理创伤后流出的树液与款冬、红酒及藏红花粉混合，对咳嗽、喉咙嘶哑或失声有一定功效。

桃叶味苦辛，具有祛风、除湿、清热解毒、杀虫、消炎作用，常用于治疗头风、湿疹、疟疾等症。取鲜嫩桃叶，捣烂敷患处，可以治疗脚癣、手癣。桃叶煎冲剂浓缩成膏，外敷疮疖可以很快收敛。意大利至今还有一种很流行的说法，即用鲜桃叶涂抹瘊子，然后埋在树下，瘊子便会随着埋下的叶子腐烂而掉落。

桃树的干皮和叶一样，不仅有缓和镇痛的效力，还有利尿、化痰的作用。水中加树皮和干叶，可以有效清理胃表面积累的附着物。对于慢性支气管炎患者，一茶匙的用量即能有止咳平喘的功效。

桃花盛开在阳春三月，桃花的娇美常会令人联想到生命的丰润。古人曾用"人面桃花相映红"来赞美少女娇艳的姿容，其实桃花确实有美颜作用。对大多数女性来说，桃花不失为经济方便的美容佳品。

《本草纲目》记载：桃花味苦，性平。有活血悦肤、利尿通便、消肿化瘀止痛的作用。桃花应用于美容的历史源远流长。桃花最早被作为女性的美容用品至少要追溯到南朝时期（420～589）。明·董斯张《广博物志》卷四十三："陶真君曰服三树桃花尽，则面色如桃花。""陶真君"即南朝梁名医陶弘景。唐代药物学家和养生学家孙思邈在《千金方》中记载："桃花渍酒服之，好颜色，治百病。"中国现存最早的药学专著《神农本草经》里也谈到桃花具有"令人好颜色"之功效。《国经本草》记载：采新鲜桃花，浸酒，每日服用可使容颜红润，艳美如桃花。

利用桃花美容的简单方法，就是将新鲜的桃花捣烂、取汁涂于面部，轻轻按摩片刻；也可以将桃花阴干后研成细末，用蜂蜜调匀涂敷脸部，每晚敷面，次晨洗去，连续使用也能使皮肤艳丽光彩。

桃花中富含植物蛋白和呈游离状态的氨基酸，容易被皮肤吸收，对皮肤干燥、粗糙及皱纹有一定的缓解作用；桃花含有山柰酚、香豆精、胡萝卜素及维生素等成分，其中山柰酚有较好的美容护肤作用，能够扩张血管，疏通脉络，润泽肌肤，改善血液循环，防止黑色素在皮肤内慢性沉积，有效地预防黄褐斑、黑斑和雀斑，从而达到面色红润、皮肤润泽光洁、富有弹性的美容效果。

民间还流传着多种桃花护肤、美容、瘦身的偏方及食谱。

《普济方》（15世纪）中记载的"桃花白芷酒"，即在三月三采花苞初放或开放不久的新鲜桃花，与白芷共同放于瓶中，以上等白酒浸泡，密封1个月后即可服用。开瓶每日服用或将酒倒少许在手掌中，两掌搓至手心发热，来回揉擦面部，数月后可使皮肤白皙红润，有除病益颜之效。

号称"隋宫增白方"的偏方是用桃花、冬瓜仁和橘皮共研成极细的药末，饭后用温糯米酒送服，有活血化瘀、去斑增白、润肤悦色之功效。

《千金方》（652）中也曾记载："桃花三株，空腹饮用，细腰身。"

以桃花烹制美食也不失为滋润皮肤、补益身体的良方。以桃花、猪蹄、粳米等为主料的桃花猪蹄粥，有活血润肤、益气通乳、丰肌美容、化瘀生新之功效，特别适用于脸有色斑的哺乳女子。产后服用此粥，既可通乳、去体中瘀血，又可去脸部色斑及滋补身体。

此外，桃花用于药用的偏方在民间也有很多。桃花阴干，捣成末，温清酒送服可以治脚气；同样将阴干的桃花捣为细末，用蜂蜜调制成膏，则能缓解皮肤瘙痒；收集未开的桃花，阴干后与桑葚等份研成末，用猪油调和，涂敷在先用灰汁洗去疮痂的秃疮患处，数次即可有效；采集早上带露桃花，与醋同捣绞汁涂抹 [《圣济总录》（12世纪）] 对背疮痛有一定效果；采集鲜桃叶捣烂敷在患处，可以用来治湿疹。

桃花的应用广泛而源远流长，从庭院栽植到盆栽入室，从赏花观枝到利用桃花美容护肤，桃花无疑已经深入到了人们的生活中……

《芥子园图说》（1662年第一次出版）
"施朱施粉色皆好，倾国倾城艳不同，疑是蕊宫双姊妹，一时俱肯嫁东风。"——宋代邵雍赞美桃花的诗篇《二色桃》。

11 幸福与吉祥的象征
——中国文化及艺术中的桃花

THE HAPPINESS AND LUCK FLOWER
——ORNAMENTAL PEACHES IN CHINESE
CULTURE AND ART

桃原产中国，分布广泛，
其花色艳丽，果实鲜美的独特植物属性，
很早就与古人的生活有着密切联系，
并通过历代文献记载和文学作品得到了充分的表现。
伴随着我国特有的桃文化的产生、发展，
人们对桃花的认识和利用也在逐步演化和升华。
这一代表着春天，
代表着美丽女性，
代表着理想境界的桃花，
在中国文化中具有独特的地位，
千年来深得我国人民的喜爱。

一、桃文化溯源

桃文化的产生起源于远古时代的桃图腾（王焰安 2003）。桃的药用功能所赋予桃的神力以及多籽所赋予桃的旺盛繁衍力，都是其成为图腾的重要原因。

《典术》中记有"桃者五木之精，故能压伏邪气"。《本草纲目》中则清楚地记载有：桃的叶、茎、根、皮、胶、仁、花都有治疗多种疾病的功能。而在原始初民的眼中，疾病和由此引起的死亡往往具有一种不可战胜的神力，而桃以其祛除各种疾病的天然药用功能，成为可以战胜疾病和死亡之神的"神上之神"，从而赢得了古人的信任、敬仰，甚至是神秘崇拜，使桃本身更具神化意义。汉代王充在其《论衡·订鬼篇》中引古本《山海经》称："沧海之中，有度朔之山，上有大桃木，其屈蟠三千里。其枝间东北曰鬼门，万鬼所出入也。上有二神人，一曰神荼，一曰郁垒，主阅领万鬼。恶害之鬼，执以苇索而以食虎。于是黄帝乃作礼，以时驱之，立大桃人，门户画神荼、郁垒与虎，悬苇索，以御凶魅。"的记载，这里所记将神荼、郁垒二神的画像与虎符共刻于桃木之上悬于门户的做法，就是后来桃符、桃板的最早来源。南朝梁代宗懔在《荆楚岁时记》中记有，正月初一，"造桃板著户，谓之仙木"，并贴画神荼、郁垒，"插桃符其旁，百鬼畏之"。隋代《玉烛宝典》记曰："元日造桃板著户，谓之仙木……即为今桃符也。其上或书神荼、郁垒之字。"由此已经反映出桃木可以驱鬼的民间意识的形成。五代后蜀，人们开始于春节时在桃符板上书写联语，以后渐渐改为写在红纸上，贴于门上，后来便发展为现在大家熟知的春联，从而成为中国传统习俗中一项重要的文化内容。

日本的桃文化与中国一脉相承。日本现有的最古的史书《古事记》（严绍璗 1993）中即有"桃太郎的传说"。在日本，几乎无人不知那个从桃实中跳出，力大无比，征服了三个恶鬼的桃太郎，他不仅是生命的象征，更给人们带来了祥和、幸福。《日本书纪》（星川清亲 1981）在叙述同一神话故事之后注曰"此用桃避鬼之由也。"从中透露出日本民族对桃的崇拜和信仰。至今，日本列岛从北到南，还有在新年里用三四片桃叶，围以稻草，中间置一蜜橘作为饰物，置于车头的习俗，表现出日本民族对驱邪避害、祝福的追求（严绍璗1993）。

桃的神力和桃果自身的鲜美，使得桃一向被古人视为世界上最鲜美的佳果，被奉为仙人的享有物。《汉武故事》有记载为证："西王母种桃，三千岁一为子。……武帝欲种之，王母笑曰：此桃三千年一著子，非下士所植也。"《西游记》则更进一步将这一观点推向了高潮：孙悟空因未被王母邀至蟠桃会而大闹天宫，同时他也因偷食了王母的仙桃而得以长生不老。据考，桃是我国祖先最早栽培的珍果，在春秋战国时代珍贵的"五果"（唐代王冰注）（枣、李、杏、栗、桃）中，惟有桃以其艳美的果形、果色可与长寿老人的鹤发童颜相媲美。现在我国民间还始终保持着选用"寿桃"为长辈祝寿的习俗，借"仙桃"的神圣、珍贵，祈盼人们健康长寿。

《本草纲目》中曰："桃，性早花，易植而子繁，故字从木，十亿曰兆，言其多也。"《山海经·海外北经》："夸父与日逐走，入日。渴欲得饮，饮于河渭。河渭不足，北饮大泽。未至，道渴而死。弃其杖，化为邓林。"其中大片的"桃林"（即邓林所指）即是希冀自己的氏族人丁兴旺，子孙能在桃林环境中繁衍如桃之"易植而子繁"。正是旺盛的生命力和繁衍力，才使得桃这一古老的树种长盛不衰，始终如一地得到人们的喜爱。

二、桃花文化的发展

人们在关注桃果的同时，也产生了对桃花的关注。鲜艳烂漫的桃花带给人们无限美好的遐想，并逐渐形成了一种蕴涵着深厚的社会文化和艺术内涵的桃花文化，成为中国传统花文化的重要组成部分。

(一) 象征春天的名花

古人认识到桃花之美，可以溯源至三千多年前，《诗经》中《周南·桃夭》："桃之夭夭，灼灼其华。之子于归，宜其室家。桃之夭夭，有蕡其实。之子于归，宜其家室。桃之夭夭，其叶蓁蓁。之子于归，宜其家人。"这一篇章为我们展现出了一幅桃花盛开时热烈和壮观的繁盛景象。正如朱熹的评注："'夭夭'少好之貌，'灼灼'华之盛也"（宋·朱熹《诗集传》），古人对桃花的细致观察，不仅突出了桃花色彩的艳丽，更显现出桃花姿态的美艳。

桃花早春开放，独特的时空优势和生物学特性，使得桃花成为象征春天的名花。

相比牡丹、菊花、荷花等久负盛名的传统花卉，桃花不仅分布广泛，而且花期较早，在清明前后即可开花，时值万物复苏的季节，在经过寒冬的肃杀之后，"争开不待叶，密缀无枝条"（苏轼）的桃花在相对单调的环境映衬下，鲜艳的花色更为夺目，旺盛的生机最能代表生命和活力。

相比同样在春天开放的蔷薇科春日花卉梅花和杏花，桃分布更为广泛，开发和利用也较早。在《诗经》六篇涉及"桃"的作品中，《周南·桃夭》和《召南·何彼秾矣》写的即为桃花；而涉及"梅"的四篇作品或是描写"梅子"或是描写"梅树"，无一关涉梅花；《诗经》中没有描写"杏"的篇章。就视觉效果而言，盛开的桃花具有鲜明的色彩，相比暗香的梅花和色淡的杏花，鲜艳抢眼的桃花更是"占断春光是此花"（唐·白敏中《桃花》），很早就得到了上古时代先民的关注和喜爱，在春秋时代就已经走入了文人的视野和文学的表现领域，当之无愧地成为春季的月令之花。

（二）女性美的象征

在中国古代"天人合一"的传统文化背景下，桃花是春天和健康、青春、美丽女子的象征，从而奠定了中国文学传统中桃花与女性关系的基础。同样是一首《诗经》里的《周南·桃夭》，借以桃花"灼灼"花色的艳丽，比喻初出嫁的姑娘。这种鲜艳的色泽和怒放的花朵所呈现出的娇嫩，正与青春、美丽、健康的待嫁女子极为契合。旺盛的生命力和蓬勃的生机与活力，是桃花和青春女子的共同内涵，形成了桃花文化的雏形。这一象征意蕴在不同的时代条件下，又衍生出不同的文学和文化内涵，"桃花流水"、"人面桃花"、"桃花与女性"等成为这一内涵的经典显现，逐步形成了独具特色的桃花文化。

继《诗经·桃夭》篇中美丽的新娘形象之后，春秋时代的息妫较早被形容为"桃花夫人"。以后，便有以"桃"称颂美貌的女子的习俗。晋代大书法家王献之的美妾及其妹妹分别名为"桃根"、"桃叶"；宋代名相寇准的爱妾亦名"茜桃"。现在民间还有将面貌姣好的女孩称作"桃"的习俗。

女子的青春期称为"桃李年",女孩子使用的胭脂被称作"桃花粉",女子两颊白中透粉的淡彩被称为"桃花妆",《妆台记》中"隋文帝宫中梳九真髻红妆"谓之"桃花面","桃羞杏让"也常常用来比喻女子装扮的艳丽。

日本每年三月三是"女孩节",因为正是桃花盛开的时节,所以俗称"桃花节"(高国藩 1993)。到了这一天,家家都要在花瓶中插上桃花,女孩子们身穿美丽的和服,与家人一起观赏桃花,欢度节日。三月三的"桃花节"始于1000年前的平安时代(749—1192),若再追其起源,则可回溯到中国周朝"三月上巳祓于水滨"的古老习俗(史丽华 1993)。

(三) 理想生活的象征

桃花鲜丽而烂漫,明艳的色彩营造出一种热烈鲜活的氛围,能够让人感受到春天明丽阳光的照射和暖融融的切肤之感,因此常常能给人以温暖、悠远的想象和安慰。浪漫文人历来就用"桃花"装扮着幻想中的理想世界,无论是幻想中的祥和天地,还是憧憬着的怡然仙境,莫不是桃花遍布、流水纵横的世界。

晋代陶渊明的《桃花源记》所描写的"忽逢桃花林,夹岸数百步,中无杂树,芳草鲜美,落英缤纷"的理想世界,以灿烂至极的夹岸桃花林为背景,点缀着他心中怡然自得的天地,不仅为后人留下了一则尽人皆知的桃花典故,更为人们描绘了一幅和平、安宁、幸福、美好,没有世俗烦恼的理想世界的图画。后世王维、王安石的《桃源行》、韩愈的《桃源图》等,众多文人雅士,多次借用桃花源的概念塑造了自己心目中的理想天堂。

现在"桃花源"已经成为远离尘嚣、宁静和平、幸福怡然的理想世界——世外桃源的代名词,唤起人们对理想境界的向往和追求。

晋人干宝在《搜神记》中记载了汉帝刘晨与阮肇"入天台山,远不得返,行十三日,甚饥",远望山上有桃树。"子实熟",于是赶紧跑去采食充饥。接着刘、阮在一条小溪边上遇见了两位绝色佳人——两位仙女,邀至她们家中用饭,膳罢又有一群少女持桃上供,然后酒酣作乐,庆贺这一仙凡艳遇。后人则在这一记载的基础上加以想象,刘、阮天台山遇仙而发生的人仙之间美好爱情的故事，便成为了人们想象中的爱情桃花源的象征。

源自《诗经·周南·桃夭》对桃花原始单纯的生命的欣喜，唐代崔护《题都城南庄》中"去年今日此门中，人面桃花相映红。人面不知何处去，桃花依旧笑春风。"所描绘的故事，同样以美艳的桃花衬出了少女光彩照人的面影，又设置了春华艳遇的美好背景，在赞美美好境界和美好事物的同时，又增加了一种对昔日美好情感的追忆、欣赏、珍惜和留恋。这与《桃花源记》及天台遇仙的故事中偶然与不经意之间邂逅美好事物，而一旦再去刻意追求，却又往往再不复得的人生体验，有着异曲同工的意味。而其中"人面桃花"无一不是美好事物和可遇不可求的幸福和好运的象征。

三、古典文学作品中的桃花之美

从观赏园艺学的角度，一种植物的观赏价值通常以其"色"、"香"、"姿"、"韵"来体现。桃花的美在其色，在其姿，以及由此表现出的韵味。

桃花树态优美，花朵浓密，色彩妩媚。阳春三月，嫣然成林的桃花，云蒸霞蔚、如绮似锦，更别具一番动人之美。

(一) 桃花自身的花色、姿态之美

宋·陆佃《埤雅》卷十三"释木"中写道："俗云'梅华优于香，桃华优于色'。"桃花以其独具的花色之美，成为我国重要的园林观赏树种。

"桃之夭夭，灼灼其华"形容的就是桃花的花色之美。

唐代诗人吴融在其题为《桃花》的诗句中"满树和娇烂漫红，万枝丹彩灼春融"，就是抓住了桃花的红艳，描绘出满树的桃花宛若天工所开，烂漫得融入人间春光的景象。古典诗词中对常见桃花的红色描写，往往采用借喻的方式，用"红雨"["桃花乱落如红雨"（唐·李贺）]、"乱红"["满院桃花，尽是刘郎未见。……浓睡起，惊飞乱红千片。"(苏轼)]、"飞红"["燕子风高，小桃枝上花无数。乱溪深处，满地飞红雨"（南宋·周紫芝）] 等词语描写桃花花色之红。

"初桃丽新彩，照地吐其芳"（梁简文帝），写出了初发桃花的粉嫩鲜丽。明代文徵明在一首《粉红桃花》中即言："温情腻质可怜生，泡泡轻韶入粉匀。新暖透肌红沁玉，晚风吹酒淡生春。窥墙有态如含笑，对面无言故恼人。莫作寻常轻薄看，杨家姊妹是前身。"将一枝细腻柔嫩，没有大红大紫强烈刺激色彩的出墙粉桃，活脱脱地呈现在众人面前。粉色桃花的形象由此臻于完整，它的美才得到了真正的认可。

色彩素雅的白色桃花，在宋代开始备受青睐。宋·王十朋的《千叶白桃》"洗尽夭夭色，冷然众卉中。却将千叶雪，全胜几枝红。"渲染出白色桃花的冰清玉洁。

桃花花色变化丰富，"桃花一簇开无主，可爱深红爱浅红"（杜甫）所描绘的二色桃屡见不鲜。唐·元稹《桃花》以"桃花浅深处，似匀深浅妆"写出了桃花或深或浅的红色如美人之淡浓相宜的妆容；宋·邵雍《二色桃》中"施朱施粉色俱好，倾国倾城艳不同"，不仅极言二色桃花出类拔萃的美，也写出了桃花轻盈的姿态和妩媚、鲜妍的动人之处；宋·周必大《以红碧二色桃花送务观》"碧云欲合带红霞，知是秦人洞里花"中"碧云"、"红霞"渲染出这二色桃花之不同凡俗的美；元·方回《二色桃花》"阮郎溪上醉腮融，蓦忽深红又浅红"的诗句写出了二色桃花深红、浅红的色彩错落之美。而绯碧两色桃花在宋代范成大"碧城香雾赤城霞，染出刘郎未见花"诗句中的描写，更是得神仙居所碧城的金碧交辉与赤城红霞之神韵，才幻化出如此脱俗超凡的世外桃花。

"千朵秾芳倚树斜"（唐·白敏中《桃花》）、"烂漫东风态绝佳"描绘出了桃花的姿态之美。

"敷水小桥东，娟娟照露丛。"唐代诗人温庭筠笔下的桃花带露含羞，摇曳无依，在洒满露水的草丛中独自开放，它的美好娇艳的姿态楚楚动人。

"凭伊几点清明雨，催出新装试小红。"清代诗人沈荣俊将初试春装的桃树，比作天真活泼的姑娘，穿上红色的春衣，美丽可人，逼真地刻画出春桃娇美的姿态。

明·王衡在《东门观桃花记》中写道："桃花醉面垂垂，傍水洗妆，不见头额……"突出刻画了桃花娇羞可人的姿态。

唐代诗人皮日休在《桃花赋》中更是尽谴历代美女比之桃花，"或歌或舞，或幽柔或娉姿，俯者若想，倚者如瘦"，形态各异，婀娜多姿。

"桃李虽丰开，菁萼满芳枝"（王维），"宠光蕙叶与多碧，点注桃花舒小红"（杜甫）及"始见洛阳春，桃枝缀红掺"（韩愈）等诗句中所描绘的含苞桃花微小，颜色淡红，与盛开桃花之烂漫妩媚和落花之满地红芳相比，极易被忽略，而以"菁萼"、"小红"、"掺"等字眼极为形象地表现了桃花似开未开或初开的特征，初开桃花的粉嫩莹润呼之欲出。

不同品种各异的花瓣形态，也在古代诗词中得到了竞相吟咏。"一花五出尚可饮，何况重重叠叠开。"（宋·陶弼）和"异姿夺众妍，姝萼同一状。重绯杂褐袭，送彩迷下上"（元·刘诜）的诗句写出了千叶桃花的重葩叠萼、美丽绝伦的姿态；明·杨基的《千叶桃花》以"春色千重与万重"、"剪裁宁不费春工"来描写其优美的姿态，每一个花瓣都似乎是一层春色，而这些花瓣又似乎是春工匠心独具剪裁出来的。

（二）桃花的传统栽植方式

桃花最早应用于园林应始于汉代，《西京杂记》（晋·葛洪）中记载道："汉武帝修上林苑，诏群臣献奇花异果……"共有7个桃的品种；《太平御览》967卷引晋代王子年《拾遗记》则明确说："汉明帝时，常尝献巨核桃，此桃霜下结花，降暑方熟，常使植于霜林园。"

尽管魏晋南北朝时期桃已多应用于园林，但对桃花本身的物态之美观赏则极少涉及。直至唐代私家园林的兴起，桃花作为观赏植物的价值才开始通过品种、形态、色彩及与整体景观的搭配逐渐体现出来。

栽植地点的不同，气候条件的不同，以及与不同植物之间的相互配植，都能体现出桃花在不同环境中气质各异的美感。

"君阳山下足春风，满谷仙桃照水红"（唐·陆希声）所描写的山谷中的野生桃花，有着不可遏制的野性生机和旺盛之势；"半红半白无风雨，随分夭容解笑人"（明·王周）中则体现出庭园桃花的另一种柔美多情之态；而杜牧在《题桃花夫人庙》"细腰宫里露桃新，脉脉无言度几春"，精心营造出的氛围则表现了庙宇中的桃花独具韵味的凄美和幽怨；"推窗绿树排檐入，临水红桃对镜开"（元·赵孟頫）和"水底红云迷醉眼，樽前降雪点春衣"（陆游）则将临水而立的桃花在水中的倒影写出了其朦胧色彩的美；"夹岸桃花蘸水开"（宋·徐俯《春游湖》）和"夹岸桃花锦浪生"（李白）的诗句描绘了"夹岸"的桃花的明丽。

桃花性喜阳光，温度越高，开放越快，也更为繁盛。晴日艳阳下的桃花尽情绽放，色彩炫目。李白"桃花开东园，含笑夸白日"的诗句写出了阳光下灿然开放的桃花的骄人情态；王安石的"春风过柳绿如缲，晴日蒸红出小桃"则描绘了丽日淑景下桃花的明媚姿容。桃花花瓣薄而嫩，沐浴雨露的桃花更加润泽、剔透，颇具一种含蓄、温柔之美。"春烟带微雨，漠漠连城邑。桐叶生微阴，桃花更宜湿"（唐·皎然）则表现了与晴空丽日下张扬与热烈的桃花完全不同的温润可人的阴柔之美。

无论是大面积的群植、片植，还是与其他不同花木组合，桃花都能体现出其不同寻常的魅力。

桃花的美感来自于其艳丽的花色和浓密的花朵，成林的桃花更能加强花色的冲击力。韩愈《桃源图》"种桃处处惟开花，川原近远蒸红霞"表现了遍布川原的桃花灿然齐发，云蒸霞蔚般的壮丽气势；明仁宗朱高炽《桃园春晓》"碧桃千万树，鲜妍如锦绚"，则写出了满园桃花共展绚美的壮观。"锦"、"霞"也就成了后来描写桃花的现成比喻。

与不同花木组合，可以凸显桃花独特之美，呈现风格不同的春色。桃红柳绿，竹遮松荫，菜花丛中可称得上是桃花配置应用最多的手法。

宋代许彦周《彦周诗话》云："春时秾丽，无过桃柳。"说出了桃花的花期与柳树展叶大致同时的特点。杜甫"红入桃花嫩，青归柳叶新"；白居易"绿丝萦岸柳，红粉映桃楼"；王维"桃红复含宿雨，绿柳更带朝烟"；刘禹锡"桃红李白皆夸好，须得垂杨相发挥"等诗句都是咏桃柳红翠相映。这一幅幅俗中见新的春景图，和谐中更显出桃花的妩媚娇艳，更突出了春季勃勃生机的特殊美感。这一搭配已成为中国园林中重要的春季景观，并已形成古老的造景模式。西湖的苏堤、白堤堪称这一手法的代表作，在南京玄武湖、扬州瘦西湖、颐和园西堤等中国著名的园林中，更是得到了广泛的应用，并且深受人们的喜爱。

"竹外桃花三两枝"（苏轼）、"窥窗映竹见玲珑"（韩愈），

"种竹交加翠，栽桃烂漫红"（杜甫），"村桃拂红粉，岸柳被青丝"（南北朝·庾信）桃花与竹搭配的诗作以碧绿清新、秀丽淡雅的竹子为背景，点缀几枝桃花，或"竹遮松荫"，与松共植，则更能突出桃花的艳丽、优雅，营造出更高层次的造园意境和文化内涵。

桃花开时，正值田间黄花正旺，红艳的桃花与黄嫩的菜花开于原野，俨然一派村田野趣。峨眉山上报国寺的楹联"一溪红白桃李树，四野青黄菜花田"描绘了这一景色。桃花与菜花的搭配在园林中应用时虽会受到一定的限制，但在自然风景区中也不失为一种新的尝试。

四、千年来广受喜爱的桃花

桃花自身艳丽多魅的观赏特性和幸福、吉祥的美好象征，使得桃花在中国传统文化中占有特殊的地位，无论是皇宫贵族，文人墨客，还是平民百姓，对桃花的喜爱伴随着中国文化的发展长盛不衰。

从数量上看，据不完全统计（渠红岩 2005），在《全唐诗》中常见的观花植物的咏物诗中，咏桃诗数量位居首位（195首），比唐代备受尊崇的牡丹多出58首；在《全宋词》中有关植物意象出现的单句次数统计中（渠红岩 2005），桃位居第四位（1711次），在观赏类花卉中，仅次于被宋人高度评价的梅花与荷花。仅从这两部中国传统文化中最重要的诗词文集中的统计数字，就能清楚地反映出桃花受到的关注和喜爱程度。

以桃花为景名的游览胜地在我国数不胜数。同名异地现象广为所见，由此从另一个侧面也可以看出，桃花在我国栽植之广和人民对它的喜爱程度。桃花源因《桃花源记》而闻名于世。通常认为是在湖南桃源县，其间有桃花观、秦人古洞、遇仙桥等众多古迹。但安徽黟县也自称是五柳先生梦中的理想境地，汇集了众多古桥、古塔、古碑、古牌楼、古祠堂，甚至在其藏有的清代同治年间的陶氏宗谱37代列祖中，东晋诗人陶渊明被列为"五世祖"。陶渊明祖籍江西，故也有称桃花源在江西。古武陵管辖多县，其中包括今重庆的一些地区，所以四川也有可能是当年的桃花源。甚至还有日本人声称真正的桃花源在日本。尽管并无实据，却也说明桃花深得各地人民的喜爱。

　　台湾段木千先生编著的《中外地名大辞典》中带有"桃"字样的地名达92条之多，其中有17处以"桃花"为名；谷向阳编纂的《中国楹联大观》中记载的有关桃花的楹联多达50多处，遍布我国江、浙、川、闽、冀、鲁、徽、桂、黔、湘、赣、津、陕、甘、台等近20个省份，再次体现了桃花在我国栽植之广和受欢迎程度之深，同时也从另一个角度表明了桃花在中国文化中的广泛和重要意义。

　　从桃花栽植的广度上看，桃花自汉代应用于宫廷园林开始，赏玩桃花之风在唐代宫苑中成为一种时尚。大明宫苑更为皇帝游赏之便而专门辟有桃花园（杜甫《奉和贾至舍人早朝大明宫》），"明皇于禁苑中，初有千叶桃盛开，帝与贵妃日逐宴于树下，……御苑新有千叶桃花，帝新折一枝插于妃子宝冠上，曰：'此个花尤能助娇态也。'"（《开元天宝遗事》，王仁裕）。这也是对桃花的审美认识由重果实利用向以观花为主转变的表现。

　　在宫廷赏桃风气的影响下，"移桃"、"种桃"、"赏桃"成为唐宋文人雅士的普遍行为，杜甫（"纷纷桃李枝，列处总能移"）、白居易（《种桃歌》："命酒树下饮，停杯拾余葩"）等著名诗人都有亲自栽桃的经历；南宋周密的《武林旧事》（孟元老等，1957）卷二《赏花》则记载了当时杭州皇城赏花的盛况："起自梅堂赏梅，芳春堂赏杏花，桃源观桃……至于钟美堂赏大花为极盛"，说明桃花的栽培和欣赏渐渐趋于普遍，被越来越多的人所接受。

　　桃花林是道教的代表景观，道士僧人也纷纷大片成林地栽种桃树，花开时节，如霞似锦的寺院桃花成为一道独特的风景。刘禹锡在《元和十一年自郎州召至京，戏赠看花诸君子》中就用"紫陌红尘拂面来，无人不道看花回"描写出了这一盛况。

　　随着花卉栽培技术的提高，种花和赏花之风逐渐向民间普及。宋·吴自牧《梦粱录》卷十九《园囿》曰："嘉会门外有山，名包家山。……山上有关，名'桃花关'，旧扁'蒸霞'。两带皆植桃花，都人春时游者无数，为城南之胜境也。""桃花关"的命名也许更让我们看到了南宋杭州包家山桃花欣赏的盛极一时。

　　桃花越来越成为寻常百姓生活中相当普遍的花卉之一，这可以从人们购买桃花的热情中体现出来。《梦粱录》卷二《暮春》曰："是月春光将暮，百花尽开，如牡丹、芍药、棠棣、木香……千叶桃、绯桃等花，种种奇绝。……买者纷然。"

在中国传统文化的概念里，桃花既是欲望的象征，又是超脱的象征。对于桃花的褒贬之声始终贯穿着桃花文化的发展。

桃极为常见、花期短暂、花色艳丽的特性，因有悖于某些"以少为贵"、"花淡为雅"的审美观点而被赋予了"俗物"与"妖客"《三柳轩杂记》（元·程棨）："花为五十客。桃为妖客"的内涵；《搜神记》中记载了汉帝刘晨与阮肇天台仙凡艳遇的故事，又被演绎为现代版的"桃花运"，使得桃花与美色、艳遇及风流韵事之间形成了特殊的连带关系。另一方面，桃花开花无语而下自成蹊（辛弃疾《一剪梅》）的现实，又被视为"花德"而大加褒扬；长满艳丽桃花的桃花源，成为人们理想生活的世外桃源。人们对桃花的认识始终呈现出一种矛盾状态，但这并没有阻止人们对桃花的喜爱。

唐代诗人皮日休在其《桃花赋》中"花品之中，此花最异。以众为繁，以多见鄙。自是物情，非关春意。若氏族之斥素流，品秩之卑寒士。……其花可以畅君之心目，其实可以充君之口腹。匪乎兹花，他则碌碌，我将修花品，以此花为第一"的措辞尽显其对桃花的喜爱之情，称桃花为"艳外之艳，花中之花，众木不得，融为桃花"，而"台隶众芳"，颇有为桃花正名之意。

桃花盛开在仲春，是古时男婚女嫁的季节，"然则桃之有华（花），正婚姻之时也。"成为色欲的象征。然而远古时期的神话传说，又使桃花具有了仙境的意义，成为无欲的净土。这种相反相成的含义，浓缩为"桃花源"意象，融合了陶渊明幻想的理想社会模式和《幽明录》所描写的美好情爱的精神天地，成为人们心目中幸福吉祥的美好象征。

桃花明丽的色彩和旺盛的生命力，以及由此衍发的、以桃花为承载的桃花文化的形成与发展，使得这一代表着春天，代表着美丽女性，代表着理想境界的桃花，在中国文化中具有独特的地位，深得千年来我国人民的喜爱。

ORIGIN, HISTORY, DISTRIBUTION AND DEVELOPMENT

Origin and History of the Peach

The ornamental peach [*Prunus persica* (L.) Batsch.] is a small deciduous tree of the Rosaceae (Rose Family) that can be extensively grown in temperate and subtropical areas (from USDA Cold Hardiness Zone 5 to Zone 9).

The common peach, a fruit plant, is well known, though the ornamental peach, on which this book focuses; is, as its name suggests; a plant cultivated for its ornamental value. The ornamental peach usually has fully double flowers. In addition to the pink flowers of the common peach, the ornamental peach also has red, white, or variegated colored blooms. Purple-leaved is also one of its ornamental varieties. The ornamental peach has a number of different tree shapes, such as pillar, weeping or dwarf; compared with the normal upright growth habit of the common peach.

The Latin name of the peach (*P. persica*) is misleading as to the origin of the peach. As everyone knows the species epithet "persica" denotes Persia, some of today's European countries and Asian minor; hence people believe that peach originated from Persia. In actual fact the peach originates from China and was commonly spoken of in the literature of China several hundred years before the Christian era.

By far the best way of establishing the origin of a plant is to trace its natural distribution and literature record. Peach stones have been excavated from archaeological sites in China dating back 6000 to 7000 years (Wang & Zhuang, 2001), and ancient wild peaches, which are freestone types with shell-pink flowers, have been found in Shaanxi and Gansu (Hedrick, 1917). The fallacy that the peach originates from Persia, written in nearly all horticultural and botanical works for 2000 years is now disposed to the peach being China's great gift to the world. The specimen of peach collected from Persia was a cultivated peach plant introduced from China. In the *Shi-King*, or *Book of Poetry*, a collection of ancient Chinese poems all of which were written before the Sixth Century B.C.; the brilliance of peach blossoms were described and praised:

Graceful and young the peach-tree stands;

How rich its flowers, all gleaming bright!

This bride to her new home repairs;

Chamber and house she'll order right. ......

Pink peach blossoms on the branches of peach trees were described in this poem making it clear that these flowers were already appreciated for their ornamental value in early times. The cultivation of the peach has been traced back to the time of the Zhou Dynasty (1066-221BC) 3000 years ago.

A statement of the first known dates of peach-culture in various countries is strong proof that its cultivation began in China. De Candolle concluded in his writings that the peach has never been truly wild in Persia (Hedrick, 1917).

The peach did not reach Greece until Alexander's expedition and was first mentioned by Theophrastus in 332 B.C., it did not reach Rome until after the beginning of the Christian era (Hedrick, 1917).

Distribution of Peach

In China

The cultivation of the ornamental peach originated in Shaanxi and Gansu. One group progressed eastward to Hebei, Henan, and Shandong, along the Yellow River, southward to Hangzhou and Shanghai and then further south to Guangzhou, Taiwan. The second group moved southward to Hunan and Sichuan and a third to the west via the Silk Road, to Xinjiang and then to Persia.

In the Wider World

The peach traveled along the Silk Road to the Middle East and then to Europe. It was known to Theophrastus in Greece in the 4th century BC (Haw, 1987) and reached Italy somewhat later, in the first century AD as Pliny mentioned (Li, 1959).

Greeks and particularly Romans spread the peach throughout Europe and England beginning in 300-400 BC. Peaches came to the new world with explorers of the 16th-17th centuries and were brought to Florida by Spaniards or the French in the late 1500s (Jacobson, 1996) (Fig. 1.1, P. 5).

Moving east, the peach spread to Japan around Yayoi era (200BC-300AD) (Hoshikawa, 1981) .

Cultivation History of the Ornamental Peach

In China

China has a long history of ornamental peach cultivation and a written record that dates back 3000 (Tong, 1983).

Peaches began to be planted in the garden as ornamental plants during the Han Dynasty (206 BC -220 AD) as mentioned in Wei Feng (*The Odes of Wei*) (Ruan, 1150). It is clear to see that the cultivation and popularity of ornamental peaches. *Xi Jing Za Ji* (a miscellanea of history and anecdote in Chang'an, now is Xi'an in Shaanxi province) (Ge, 1256) listed 7 peach cultivars during the Han Dynasty.

In the Tang Dynasty (618-907), the ornamental peach has been widely cultivated. In the book of *Quan Fang Bei Zu* (an encyclopedia of botany, which is name as the first botanical encyclopedia in the world) (Chen, 1256), 'Jiang Tao' (deep red, semi-double flowers), 'Fei Tao' (deep red, double flowers), 'Bi Tao' (pink, double flowers), 'Bai Ye Tao' (double flowers), 'Ren Mian Tao' (pink, double flowers) are described. Additionally, *Kai Yuan Tian Bao Yi Shi* (a historiography book) (Wang, 950), stated that ornamental peaches with double flowers were planted in the Emperor's gardens.

Ornamental peach cultivars increased dramatically during the Song Dynasty (960-1279). Thirty cultivars were recorded in the book, *Luo Yang Hua Mu Ji* (Luoyang Flower and tree collection) (Zhou, 1082). New taxa, 'Erse Tao' (two colored flowers on one branch) and 'He Huan Er Se Tao' (bicolor flowers), 'Bai Bi Tao' (semi-double white flowers), and 'Zi Ye Da Tao' (Red leaves, red flowers) were bred during this time.

'Aitao' (dwarf type) was initially described in the book, *Ru Nan Pu Shi* (a book of cultivated plants history in Runan, Henan province) (Zhou, 1620). In an additional volume, *Qun Fang Pu* (a book detailing the history of cultivated plants, morphological characteristics and cultivation) (Wang, 1621), 21 cultivars of peach were noted. Among them, 8 cultivars were for ornamental use only. Three new introductions were 'Ri Yue Tao' (red and white color in one flower), 'Mei Ren Tao' (double, light red flowers) and 'Yuan Yang Tao' (bicolor flowers). In addition the first scientific classification of the peach was documented in a famous Chinese pharmacopoeia, *Ben Cao Gang Mu* (Li, 1596), which targeted the peach for its fruit use. Based on flower color which ranges from red, white, to bicolor, a simple classification system also made its first appearance in this book.

The Qing Dynasty (1644-1911) was the period of great prosperity for peach development. Twenty-four peach cultivar were illustrated in the book *Hua Jing* (a book on horticulture, including flower culture theory and technique) (Chen, 1688) and 8 of them were ornamental peach varieties. 'Mo Tao' (deep purple flower, almost black color) was the most valuable cultivar among them. This cultivar is still widely cultivated in Shanghai, Jinhua and Xinxian County of southeastern China today. Fragrant ornamental peach cultivars were also noted and published in the county annals of Taiping (Chen, 1728). A new ornamental peach cultivar, which had similar striped petals similar to chrysanthemum rays, was recorded in the county annals of Gaoyao, Guangdong Province (Chen, 1728).

With the development of new cultivation and breeding techniques during the last century, more and more ornamental cultivars were produced. In 1949, Huang Y. and Huang D. (1985) described 11 ornamental peach cultivars; using flower color and duration as their cataloguing system. Among them, 'Wu Se Tao' (red flowers with white strips or dots), 'Sa Jin Tao' (double, tricolor of red, light red, and white), and 'Chui Zhi Tao' (weeping type with red, light red, deep red, and white colors), became new in the records.

In Japan

The peach was introduced to Japan during the Yayoi era (200BC-300AD) (Hoshikawa, 1981). From the Nara (629-759) to the Heian (749-1192) era the major purpose of peach in Japan was as an ornamental garden variety as the fruit were of poor quality (Yoshida, 1994).

Many ornamental peach cultivars were developed in Japan, especially during the Edo era (17th to 19th) such as: 'Kanhi' (red double, early flowers), 'Genbei' (variegated pink and white flower), 'Genpei' (semi-flower, pink or red striped in white flowers), 'Kyousarasa' (single, white and pink), 'Yaguchi' (dark rose-pink, double flowers, a good cutting flower), 'Kanpaku' (white double, early flower) (cutting flower), 'Genpeishidare' (single flowers, pink or red stripes on white flowers, weeping), 'Zansetsushidare' (double white flowers, weeping), 'Houkimomo' (semi-double flowers, pink or red stripes on white flowers, upright form, narrow crown), 'Kikumomo' (peach pink, double flowers, narrow petals, looks like chrysanthemum), 'Sagamihidare' (double red flowers, weeping), 'Hagoromoshidare' (semi-double pink flowers, weeping). Single, double red, pink and white dwarf cultivars were all ancient cultivars recorded in Kadan-chikin-sho (Flower bed embroideries) (Ito, 1695). Most of these cultivars are still extensively cultivated today. 'Houkimomo' was probably the first record of a column type ornamental peach in the world (Werner *et al.*, 2001).

In Western Countries

The ornamental peach was introduced to western countries approximately three hundred years ago

(Everett, 1967). England, France and Belgium are probably the most important countries who have cultivated ornamental peach in gardens since its initial introduction.

The first cultivar of ornamental peach recorded in Europe was 'Duplex' (light pink, semi-double flower), which is known to have been cultivated in France since 1636 (Jacobson, 1996). 'Alba' (single white flower) was first named in England in 1829 (Krussmann, 1986). 'Pendula' (weeping peach) was described in Europe in 1839 (Jacobson, 1996). The plant was imported by Siebold from Japan (Krussmann, 1986). 'Camelliiflora' (very large, densely double with ruffled petals, dark red) was catalogued during 1858 in Belgium (Jacobson, 1996). 'Alboplena' (pure white, densely double flower) was introduced to England in 1849. 'Rubro-plena' (semi-double red flower) was introduced to Europe in the 1840s from China and was first described in 1854 (Jacobson, 1996). 'Dianthfolia' (semi-double, narrow and acuminate petals, dark red striped) was first described in 1858 in Belgium (Jacobson, 1996) and is known to have been introduced from Japan by Siebold (Krussmann, 1986). 'Versicolor' (numerous, densely double flower, sometimes totally white or with red or rust colored stripes) was cultivated before 1863 in France (Krussmann, 1986). 'Purpurea' (purple-red leaves, single, pink flower) has been cultivated in Belgium since 1873 (Krussmann, 1986).

A majority of original ornamental peach cultivar present in the United States were introduced from European, Japan and China. 'Klara Mayer' (abundantly double bright pink-red flower, about 4cm wide), was introduced to America around 1890 by Spath nursery of Germany (Jacobson, 1996). 'Magnifica' (semi-double, crimson flower) was introduced by Veitch nursery of England (Jacobson, 1996). 'Peppermint Stick' (pink with red stripes on white flower) was introduced in 1933 by W. B. Clarke nursery of San Jose, CA (Jacobson, 1996); though some ornamental cultivars were later developed in the US. For example, 'Chrysanthemum' (large, light pink, chrysanthemum like flower) was discovered in South Carolina (Jacobson, 1996) and 'Foliis Rubris' (red-leaved, single, pale pink flower) was bred in Mississippi during the 1860s and then exported to Europe, where it was renamed as *P. persica* 'Royal Redleaf' in 1873 (Jacobson, 1996).

Development of the Ornamental Peach

In China

Today, ornamental peach varieties are commonly seen in gardens and public planting and annual peach blossom festivals are held at various locations during the month of peach bloom in China.

There are 14 attractive cities famous for their peach plantation in China (Fig. 2.2 Famous sites for the ornamental peach in China.P.16). Some focus on the fruit peach as with Shanghai Nanhui, Gansu Lanzhou. Guangdong Shima's ornamental peach is cultivated for Chinese New Year as the ornamental peach is so in demand as "the Chinese Christmas tree" in southern China. The rest are famous for fabulous displays of blossom and variety that have become tourist attractions, such as Jiangxi-Lushan, Sichuan-Chengdu, Shanghai-Longhua, Hunan-Taohuayuan and Taojiang (Gu *et al.*, 2007), Hubei Yichang (Gao, 2002) and Wuhan (Chen *et al.*, 2006), Shandong Taian (Zang *et al.*, 1998), Yunnan Kunming, Zhejiang Hangzhou, and Beijing Botanical Garden. Hangzhou is famous for its ornamental peach and pendulous willows planted along the lake edge; the scene is widely regarded as paradise on the earth. Many new cultivars were developed from the above named different climatic zones and habitats. Twenty-one cultivars released from Hangzhou and Shanghai (Hu *et al.*, 1998) was of a different phenotype compared with the cultivars bred and cultivated in Beijing.

Most new ornamental peach cultivars are bred and selected in above places. *P. persica* × *davidiana* 'Fenhua Shanbitao' (double pink flower, early blooming) and *P. persica* × *davidiana* 'Fenhong Shanbitao' (semi-double, pink-red flower, early blooming) are two early blooming cultivars created by the crossing of 'Baihua Shanbitao' with 'Hehuan Erse Tao' (the earliest double pink flower cultivar) and 'Jiang Tao' (the earliest double pink flower cultivar), respectively (Hu and Zhang, 2001). 'Zi Qi' is an early blooming purple leaf cultivar derived from the γ-ray irradiated seeds from open-pollinated 'Zi Ye Tao' (Shen *et al.*, 2007).

In Japan

The selection and breeding of ornamental peach varieties has been very successful in Japan during the last 30 years and has resulted in a great many new cultivars. For instance:

'Kyoumaiko' (red double, narrow petals, chrysanthemum like flower) was a mutation of 'Kikumomo' and 'Unriumomo' (single pink flowers, twist twigs) was a nectarine mutation (Yoshida, 1985).

'Shrioshidare' (single white flowers, weeping) was an open-pollinating seedling selected from 'Akashidare' (single red flowers, weeping) population (Yoshida *et al.*, 2000)

'Akabanabantou' (red double flower, flat fruit) is a selection from an open-pollination seedling population of 'Yaezakibantou' (dark peach pink, semi- double flowers, and flat fruit, edible) (Yoshida *et al.*, 2000).

'Red dwarf' (single pink flowers, red leaves, dwarf form) was selected from F_2 generation of 'Akame'. They are dwarf peaches which were bred by Yoshida and Seike in 1974. One of the parents --'Akame' is an insect and disease resistant cultivar with single pink flowers and red leaves (Yoshida, 1985).

'Saiwai Howaito' (Maeda *et al.*, 2005) is the selected seedling from a cross between 'Kikumomo' and double white dwarf. White chrysanthemum like petals and dwarf tree size are typical characteristics of this cultivar.

'Variegated Leaves' (green leaves with red variegated, single pink flower) is a selection of F_2 generation of 'Akame' and 'Houkimomo' (Yamazaki *et al.*, 1987a).

The column variety of ornamental peaches was bestowed from Japan to the world. The Japanese made good use of it and bred several new cultivars with the same growth habit. 'Terutebeni' (semi-double red flowers) is a selection of F_2 generation of 'Akashidara' and 'Houkimomo' (Yamazaki *et al.*, 1987b); 'Terutemomo' (semi-double peach pink flowers) is a selection of F_2 generation of 'Houkimomo' and 'Akashidara' (Yamazaki *et al.*, 1987b); 'Teruteshiro' (semi-double white flowers) is a selection of F_2 generation of 'Houkimomo' and 'Zansetsushidare' (Yamazaki et al., 1987b). 'Terutehime' (semi-double pale pink dense flowers) is an open pollinated seedling selected from 'Houkimomo' (Horikoshi *et al.*, 1992).

In Western Countries

The Ornamental peach has been present in European gardens for the last three centuries (Bean, 1950). However, cultivar variation has rarely been reported since the beginning of the 20th century.

The United States Department of Agriculture began to collect peaches from China in 1899 (Werner and Okie, 1998) and although the ornamental peach was not the main purpose for their introduction, some cultivars were introduced as stock plants and future breeding lines. For example, 'Pi Tao' (large rose like double flower, very late) and 'Shau Thai Tao' (double) were all introduced from China in 1925. 'Red Weeping' (double red flowers, very late) was received from China in 1931 and was listed in the pedigree of 'Jerseypink' (large rich pink flower with extra petals, late blooming) (Gofferda *et al.*, 1992) and other ornamental peaches released by Layne at Harrow, Ontario Canada (Layne, 1981). 'Harrow Rubirose' (red double flower), 'Harrow Candifloss' (pink double flower), and 'Harrow Frostipink' (light pink) are three very hardy rootstock cultivar released by Layne at Harrow.

Nurserymen also introduced many new ornamental cultivars to the US. W.B. Clarke nursery of San Jose, CA introduced at least 10 cultivars. Although some of them are long extinct commercially ('Double Maroon' and 'Edward H. Rust'), the 'Early Double White' and 'Versicolor Weeping' are still produced commercially today. 'Dwarf Mandarin' was also introduced in 1940 from China by Clarke nursery. The cultivars name, 'Dwarf Mandarin', indicates its origin. To our knowledge, this cultivar was the first dwarf ornamental peach ever recorded in the US. Several flower colors are available (white, pale pink, pink, red, and variegated pink and white) and they are all double (Jacobson, 1996).

'White Glory' (pure white, single to semi-double flowers in early spring, weeping) is an ornamental nectarine released in 1984 (Werner *et al.*, 1985).

Four column type ornamental peach cultivars were released by Werner *et al.* (2000 and 2001). All of these were the selections from F_2 generation of 'NC174RL' nectarine and a selection of Japanese 'Pillar' peach. 'Corinthian Mauve' (double mauve flowers) and 'Corinthian White' (double white flowers) were green leaf cultivar. The other two, 'Corinthian Rose' (double rose flowers; dark purple leaves) and 'Corinthian Pink' (double pink flowers; light purple leaves) (Werner *et al.*, 2001) all had purple leaves.

Other types of ornamental peach such as those with red-leaves; have been successfully bred and released in the US. The dwarf cultivar 'Bonfire' (dwarf, double pink-red flowers with dark red leaf) (Moore *et al.* 1993) was selected from open-pollination seedling population of F_2 of 'Tsukuba No.2' (single flower, red leaf, upright). Two red-leaved weeping cultivars were released by Moore *et al.* (1993). 'Crimson Cascade' (dark red double flowers, dark red leaf) and 'Pink Cascade' (dark pink double blooms, dark red leaf) were the result of crossing a green-leaf, weeping, double-flowered, unnamed New Jersey selection with Rutgers Redleaf - a standard-sized tree with red leaves used as a peach rootstock.

2 TAXONOMY AND PEACH RELATIVE SPECIES

Taxonomy Studies

At the species level, the name of peach was always problematic. Different botanists and taxonomists have different treatments.

The first scientific name of the peach species was *Amygdalus persica* L., published by Linnaeus (1753). In 1768, Miller renamed it as *Persica vulgaris* (Wu and Raven. 2003).

The earliest taxonomic treatments divided *Prunus* (broad-sense) into four genera (Linnaeus 1737), *Amygdalus, Cerasus, Prunus* (narrow-sense), and *Padus*. Later Linnaeus (1754) reduced the number to two, *Amygdalus* and *Prunus* and in 1825 De Candolle recognized five genera: *Amygdalus, Persica, Prunus, Armeniaca*, and *Cerasus* placing both *Padus* and *Laurocerasus* into *Cerasus*.

In 1865 the species was subject to different treatment by Bentham and Hook, who treated *Prunus* as a single genus separated into seven sections: *Amygdalus, Armeniaca, Prunus, Cerasus, Laurocerasus, Ceraseidos* and *Amygdalopsis*. Koehne (1893) treated these sections as subgenera.

Amygdalus, with other five genera *Armeniaca, Cerasus, Laurocerasus, Padus*, and *Prunus* (narrow-sense) are often treated at the subgenus or sectional level within the genus *Prunus* (Wu and Raven, 2003). The three cherry genera (*Cerasus, Laurocerasus*, and *Padus*) are more closely related to each other than they are to the other genera in the Prunoideae. During the last thirty years, most taxonomic works had viewed *Prunus* (broad sense) as a single genus. Recent phylogenetic studies based on molecular data showed that none of these three cherry genera were monophyletic and their separation is probably not justified (Bortiri *et al.*, 2001). The differences between the two subgenera: *Amygdalus* and *Prunus* is as follows: subgenus *Amygdalus* has collateral buds with two flowering buds in the lateral, one leaf bud in the middle and young leaves conduplicate in bud. Whereas subgenus *Prunus* has a single auxiliary bud and young leaves wrapped in bud. There are not many other significant differences between these two subgenera so it seems more reasonable to use the genus *Prunus* and not divide it into the two subgenus *Amygdalus* and *Prunus*.

Prunus persica (L.) Batsch. was published in 1801; this is a legitimate name based on International Code of Botanical Nomenclature (ICBN) (Geruter *et al.*, 2000) that has been widely accepted within the horticultural field in western countries (Bean, 1910; Brickell and Zuk, 1997; Fernald, 1987; Harrison, 1959; Huxley and Griffiths, 1999; Krussmann, 1986; Rehder, 1927). With this in mind, *Prunus persica* has been adopted during the explanation of further study presented this book.

In the broadest category, species; or intraspecific categories a great deal of contradictory information on the ornamental peach has been published. Different countries, geographical locations, taxonomic interpretations and generic name changes associated with *P. persica* have resulted in numerous nomenclature changes. Some varieties and forms of *P. persica* must be changed under various taxonomic treatments.

Key to the Peach and its Relatives

There are four important related species to *P. persica*: *P. davidiana*, *P. kansuensis*, *P. ferganensis*, and *P. mira*. All of them have fleshy mesocarp that does not split when ripe. Many are interest for their disease resistance traits or potential as rootstocks.

Table 2.1   Key to peach and its related species

1. secondary veins of leaf blade continuing to margin; not netted; endocarp surface
    deeply furrowed, sparsely pitted ....*P. ferganensis*
1. secondary veins of leaf blade netted at margin
  2. endocarp sparsely shallowly furrowed ....*P. mira*
  2. endocarp furrowed or pitted
   3. Sepal outside pubescent
    4. winter buds pubescent; endocarp large, furrowed and pitted ....*P. persica*
    4. winter buds glabrous; endocarp furrowed, no pitted...*P. kansuensis*
   3. Sepal outside glabrous; endocarp globose, furrowed and pitted ....*P. davidiana*

P. persica (L.) Batsch.

The term "peach", as used in this chapter; refers to *P. persica*.

P.persica is a diploid species (2n=16), which is a small tree with a spreading canopy usually 2-3.5m in cultivation, with the potential to grow up to 8m. Winter buds pubescent. The leaves are lanceolate, glabrous and serrate, broadest near the middle. Flowers are generally pink with petals that can be large or small, showy or non-showy. Ornamental peaches contain fully double flowers, having many petals that are rose-like, chrysanthemum-like or peony-like in appearance. Colors range from light to dark pink to white, or to red. The fruit is a drupe, pubescent or glabrous; the stone endocarp is large, deeply pitted and furrowed.

P. davidiana (Carr.) Franch.

This species is native to China. As a fully developed specimen it may grow to a tree of up to 10m, with upright slender branchelets. Winter buds glabrous. The leaf is narrow ovate-lanceolate, broadest near base; base cuneate to round. Sepal outside glabrous; petals may be pink, white or red; endocarp surface is deeply pitted.

This species is often used ornamentally because of its distinctive glabrous reddish cupper-colored bark (Fig. 3.5, P. 23). Twisted columnar (Fig. 3.4, P. 23) and weeping forms (Fig. 3.3, P. 23) of *P. davidiana* are also often used in ornamental planting. This species is cold hardy, blooms early and is commonly used as peach rootstock in northern China. Good resistance to leaf curl and powdery mildew is reported in *P. davidiana* varieties (Pisani and Rpselli, 1983). *P. persica* × *P. davidiana* hybrid varieties have also been noted as being resistant to *M. persicae* (Massonie, 1979; Massonie *et al.*, 1982; Pisani and

Roselli, 1983).

P. kansuensis Rehd.

P. kansuensis is native to Gansu, Shaanxi, and Sichuan which are primarily mountainous areas. The tree grows to 7m with slender glabrous branchlets; winter buds are glabrous; the leaf is ovate-lanceolate, broadest from middle; base broad cuneate; sepal outside pubescent. Petals may be white or pale pink. Stamens are shorter than petals; style longer than stamens. Endocarp shallowly furrowed, no pitted.

The hybrid of this species with peach (*P.persica*) is readily made resulting in a vigorous variety, intermediate in characteristics (Meader, 1939; Grasselly, 1974).

P. ferganensis (Kost. & Rjab) Kov. & Kost.

P. ferganensis is cultivated in the Fergana Valley region (south central area of the former USSR) and western China.

P. ferganensis is sometimes classified as a subspecies of *P. persica*, from which it differs in having elongated, parallel leaf veins and continuous parallel grooves on its stone.

P. ferganensis trees may grow to 8m tall, with glabrous, smooth branchlets. Winter buds pubescent; leaves are lanceolate; base broadly cuneate to round; margin crenate. Sepal outside, green with reddish tinge. Petal are pink, stamens shorter than petals, style shorter than stamens; endocarp flatted globrose, surface deeply furrowed and sparsely pitted.

P. mira Koehne

P. mira is a shrub form native to southern Tibet; Nepal, and Northern India (Fig. 3.8, P. 25).

Trees may grow to 10m tall, with green, slender branchlets; winter buds glabrous. Leaves are lanceolate; base broadly cuneate, margin shallowly crenate, entire near apex; sepal outside glabrous; stamens much shorter than petals; endocarp surface smooth or shallow furrows.

Readily hybridized with peach, F_1 plants are sterile (Komar-Tyomnaya, 2008).

BIOLOGICAL AND PHENOLOGICAL CHARACTERISTICS

Most peaches lie between latitudes 30°and 45°North and South (Hesse, 1975). In common commercial fruit peach plantings the average specimen lifetime is limited to 12-15 years at most. This is due to the cultivar either becoming obsolete or it sufferings loss of productivity. Since fruit is not the major purpose of planting in ornamental peach cultivars, they much lower relative nutrient loss than the fruit peach and may live for 20-30 years in "normal" conditions.

Biological characteristics include the basic morphological characteristics, such as growth habit, branch, bud, flower, leaf, fruit, etc. Growing development characteristics and the phenological characteristics of blooming, leaf expansion and fruit setting are also included.

Morphological Characteristics of the Ornamental Peach

The ornamental peach presents great diversity in its morphological characteristics.

Growth habit

A great diversity of growth habit is presented across ornamental peach varieties. In addition to the standard type as with the common peach tree; there are also twist type with twisted twigs and branches; weeping type with pendulous branches; pillar type with narrow crowns, and dwarf type that is short in stature with short internodes. This diversity increases the authentic ornamental value of peach.

Branches

One-year-old shoots of pink or red flowers are usually red-green, turning dark grey-silver when older. 'Fei Tao' is one exception of red flower cultivars, having green branches and white flowering cultivars always have green branches.

Buds

Buds are always present at the base of leaves. There are two kinds of buds present, vegetative buds which are usually small and pointed and flower buds that are larger and plump. In the common peach each node usually displays three buds: two lateral flower buds with one vegetative bud in the middle (Fig. 4.2, P. 29), but in ornamental peach varieties up to three or four flower buds can be found (Fig. 4.4, 4.5, P.29). On occasion only one flower bud is present in addition to vegetative buds (Fig. 4.1, P. 28). There is variability in setting flower buds where genotypes with high bud set pass characteristics to their progeny with high heritability (Weinberger, 1944).

Flowers

Peach has hermaphroditic, perigynous flowers. The reddish-green calyx is gamosepalous and falls after initial swelling of the fruitlet. The color of inner surface of the calyx is related to the color of the fruit flesh: white-green in white-fleshed varieties and yellow to deep orange in yellow-fleshed fruits.

Different from the normal single flowers (less than 10 petals)of common peach, the ornamental peach flowers have semi double flower (10-40 petals) right up to the double flower (up to 40 petals).The numbers of stamens transformed into petals, thus conferring the impression of a double flower. The various flower types and colors give the peach significant ornamental value. Besides normal single type and bell-shaped flowers, mei-flower, rose, chrysanthemum and peony represent the four major flowers types presented in ornamental peach. The number of petals varies from 5 to 80 with the size of flowers varying from 2cm to 8cm.

Flower colors range from monad pink to shades of pink and red or pure white. There are also cultivars with variegated bicolor or tricolor flowers which provide a colorful ornamental display in the springtime.

The length of stamens compared with the length of petals, the length of pistil (s) compared to stamens, the color of anthers and the stamens or calyx petaloids are all different between the cultivars and represent important characteristics when identifying cultivars.

Leaves

Most of ornamental peach cultivars have green leaves, though some first appear brilliant red in the springtime turning bronze red in the autumn. Variegated leaf cultivars are also bred and selected (Hu *et al.*, 2005; Yamazaki, *et al.*, 1987). The leaf blade may be smooth, wavy, crumpled or curled. The leaf base, leaf apex, and leaf margin also present differences.

Fruits

Fruit production begins from the second or third year of life. Fruit is not major ornamental feature of ornamental peach cultivars; on the contrary, the less fruit the better the ornamental effect. Lower fruit production reduces the loss of nutrient during fruit setting, increasing the strength and health of the specimen. The shape of fruit where fruit presents and the pubescent or glabrous nature of the skin are all the characteristics sought in ornamental peach cultivars.

Roots

The root system of peach varieties develops within 50-60cm in depth, depending on soil type. The horizontal distribution of the root is larger than vertical distribution, which makes recovery easy post transplantation.

Growth and Development Characteristics of Ornamental Peaches

The peach originated from the north-western plateau of China. Sites that receive full sun and have good drainage and sandy soil are the best for peach planting. The peach is barren-resistant and cannot endure water logging for long time. It demonstrates cold hardiness and survives well outdoors in northern China during winter.

The ornamental peach is an insect-pollinated and self-fertile species. It is hysteranthous. Initial flowering buds present usually around July to September. There are four stages as follows: physiological differentiation stage, morphodifferentiation stage, dormancy stage, and sexual cell formative stage.

The peach flower bud differentiation is centripetal development (Jiang, 2003). There are five periods in the physiological differentiation stage, these are as follows: flower bud differentiation initiation phase, sepal differentiation phase, petal differentiation phase, stamen differentiation phase and gynoecia differentiation .

The natural leaf falling phase is the middle of November, after which varieties fall into deep dormancy. A period of low temperature is needed to break this dormancy. Chilling requirement refers to the minimum duration of low temperature (under 7. 2°C, usually) accumulatively (Weinberger, 1950). Utilizing the standard cultivar bloom times as an indication of the chilling requirement of unknown genotypes is a simpler and cheaper method of variety identification (Scorza and Sherman, 1996). According to the model of 0°C -7.2°C chilling requirement, Wang *et al*. (2003) measured the chilling requirements of ornamental peach cultivar in China and found requirements ranging from 400h to 1250h with the majority requiring up to 900h (Zhu *et al*., 2004). The early blooming cultivars 'Baihua Shanbitao' and 'Yingchun' (Wang & Zhuang, 2001) all need 450h; while the late blooming cultivar 'Juhutao' (1250h) , 'Sahong'(1214h), 'Yuanyang Chuizhi' (1188h), 'Ziyetao' (1112h) all require up to 1000h. Pawasut (2004) presented similar chilling requirements of 'Juhuatao' (1433h).

Phenological Phases of Ornamental Peaches

The anthesis is the primary period for appreciating a majority of ornamental peach cultivars.

During the flowering bud swelling stage: scales' become loose, 25% of the flowering buds begin to swell.

Calyx stage: (Fig. 4.8, P. 32 'Terutehime'): scales split, 25% of the calyx is shown.

Pink stage: (Fig. 4.9, P. 32 'Fenhua Shanbitao'): calyx spilt and petal shown. When petals are shown in >80% of the flowers buds this is known as the Balloon stage, the best time for appreciation is approaching (Fig. 4.10, P. 32 'Jiangtao');

Early bloom stage (Fig. 4.11, P. 33): 5%-25% of the flowers open;

Full bloom stage (Fig. 4.12, P. 33): Up to 50%- 75% of the flowers open, the best time for ornamental peach;

Late bloom stage (Fig. 4.13, P. 33 'Sahong Tao'): 5% of the flowers dropping, the beginning of late bloom stage, 95% of the flowers dropping at the end of this stage.

The duration of each period differs from climate to climate and cultivar to cultivar.

The blooming phase of ornamental peach usually begins with the pink stage through to the full bloom stage. Flowers may still be present on the tree until the end of late bloom stage. Some cultivar such as 'Unriumomo', have special ornamental value even when no petals, only stamens remain. (Fig. 4.14, P. 34 'Unriumomo'). Under "normal"conditions the full anthesis of a cultivar may last from 10 to12 days depending on temperature, wind, etc.

From the flowering of the first cultivar which is usually begins during early April in Beijing; to the last one, which comes into bloom during the first week of May; the total anthesis of all cultivars of ornamental peach can last almost one month. This ensures that the ornamental peach is the most famous and important ornamental plant of the spring season.

4 SYSTEMATIC STUDIES ON THE ORNAMENTAL PEACH

Morphological Studies

Ornamental peach varieties present a number of shapes and growth habits including standard, columnar, dwarf, weeping and twisted. The flowers, which appear during early spring before the new leaves unfold, are available in single, semi-double, and double forms. Their colors range from pure white, pink, red to bicolor or tricolor. Leaf color ranges from green to red or even purple.

Current problems of describing ornamental peach cultivars

The ornamental peach was cultivated in China for thousands of years before being introduced to western countries approximately three hundred years ago (Everett, 1967). Nowadays, it is widely grown and is well adapted to countries across Asia, Europe, America and Australia where it is planted for its outstanding ornamental value. Many new cultivars have been introduced by the nursery trade; however nomenclature and classification of ornamental peach varieties within literature are contradictory and confusing. A total of 267 taxa of ornamental peach have been documented (Hu, 2004) though the legitimacy of some nomenclature and descriptions are questionable. Some ornamental peach cultivars have the same name (homonyms) in literature, but have very dissimilar morphological descriptions (Krussmann, 1986; Moore et al., 1993). For example, there are two cultivars named 'Crimson Cascade', however, one presents green leaves and the other red. Others have different names but share the similar morphological characteristics (Jacobson, 1996), 'Cardinal', 'Magnifica', and 'Russel's Red' all have large double red flowers. Different countries may also have different names (synonyms) for the same taxa. For instance, cultivars with chrysanthemum like flowers are named of 'Chrysanthemum' in the US, 'Kikumomo' in Japan, and 'Ju Hua Tao' in China. All these three have a particular meaning in English, Japanese, and Chinese which seems to be the same, though these three are presented as three different cultivar names within botanical literature. Clearly the definition of an accurate description of the morphological characteristics of each ornamental peach cultivar the basis of classification. Without a precise morphological description one cannot be absolutely sure of the authenticity and reliability of a cultivar. Too many descriptions include only "white double flower" or "pink single flower, red leaf" etc., which cannot possibly define the distinctness of each cultivar. To standardize morphological descriptionand nomenclature and set up standard guidelines for description, feature by feature and explanation of each grade of ornamental peach is the only meaningful way forward with regard to the exploration and breeding of ornamental peach cultivars in the future.

Criterion that should be followed whilst setting up descriptive categories and grading standards

1. Description of color

Descriptions of color are somewhat subjective and may differ from individual to individual. Even where the same individual is categorizing color, different countries may recognize or represent colors differently. Unification of color description and comparison is necessary for the standardization of color identification. The use of a universal "standard color chart" developed for use in the horticultural field would provide objective and accurate model codes and make the identification of color, and therefore the identification of cultivar a much more standardized procedure.

2. Description of quantitative characteristics influenced by environmental factors

Tree size, flower diameter, and petal size are all quantitative traits highly influenced by cultivation conditions. Without any control cultivar definitions and descriptions, varieties lack evidential support for their individuality. Therefore, several years' constant observation and comparison of the control cultivar in order to define rates and variables of each trait and produce detailed descriptions of morphological characteristics is necessary and important when defining and naming new cultivars of ornamental peach.

3. Illustration of qualitative characteristics

Growth habit is a crucial characteristic to be observed and described in the classification system of ornamental peach cultivars. Other equally important characteristics are flower type, petal shape and bud shape. Clear illustration is absolutely the best way to present this information and is a much more practical approach to adopt when presenting detailed analysis of important variety traits.

Morphological characteristics of ornamental peaches and their description

The most recognized feature of the ornamental peach cultivar is the flower. Flower type, color and shape at every stage from bud through to fading are equally important. The development of pistil, sepal and calyx are particularly useful details to observe and describe as they remain constant under varying conditions. Another equally distinguishing point is the period of flowering, whether early or late.

1. Flower

Shape of flowering bud: There are five forms of flowering bud (Fig 5.1, P. 37): oblong, ovate, oval oblate-round and round. The shape is related to the number of petals and is therefore, to some degree, a secondary characteristic. Single-flowered cultivars usually have ovate-oblong buds, semi-double cultivars ovate to oval buds and double-flowered peaches have mostly oblate to round buds.

Petal number: Single-flowers commonly have 5 petals and always less than 10. Semi-double-flowers have approximately 10-40 petals, double-flowers have up to 40 petals. Semi-double and double flower cultivars have petaloids. Petaloids are stamens that have more or less developed into petals. Usually they are of imperfect shape and filament to anther can be seen halfway up or at the end of the petaloid. Petaloids appear in the heart of the flower because they are derived from stamens.

Petal shape: Petals may be spatulate, lanceolate, elliptic, oval, and nearly round (Fig. 5.3, P.38). Shapes related to flower type, relatively. A spatulate petal is the typical form of a bell-shaped flower and ovate to oval petals occur as the predominant petal forms of single flowers. The petals of chrysanthemum-type flowers usually are lanceolate whereas rose and peony cultivars present inner petals are usually pletaloids from stamen. Ovate-oblong and ovate are typical petal forms of rose and mei-flower type, respectively. The characteristics of one flower may vary for example the outer petals of a peony or rose flower are usually wider and less elongated than the inner ones.

Flower type: The flower type of ornamental peach is a distinguishing characteristic in each cultivar. There are six flower types: single, mei-flower, rose, peony, chrysanthemum, and bell-shaped (Fig.5.2, P. 38). Single type flowers have less than ten petals arranged in a single layer, resembling a flat saucer. Mei-flower and rose type are both semi-double flowers, the former has flat flower with 3-5 layers of regular petals, nearly overlapping; the latter has cup-shaped flowers with outer petals that curl outwards and multiple layers of inner petals that curl inwards; approximately 25-40 petals in total. The Chrysanthemum type is also a semi-double flower with long, narrow, strip-shaped petals, around 30-40 in total. Peony type petals are round or semi-round in shape with commonly more than 40 wrinkled and rumpled petals present in each bloom; some may reach 60-80 petals in total that are crisscrossed in the heart of the flower causing the ovary to be invisible from above. Bell-shaped flowers have spatulate petals that curl inwards; with long stamens that extend beyond the flowering buds.

Flower color: Color is an important aspect in the appearance of ornamental peach specimen. Ornamental peach flowers are darker when in bud, becoming paler or lighter as they expand. White-flowering peaches show a slight pinkish hue when in bud whereas pink cultivars have dark pink or almost red flowering buds. Flowers, when in bud show the fully colored rear side of the outer petals; when the flower opens, the inner petals and inside of the outer petals are shown; which are much less saturated in color. When determining the color of ornamental peach flowers, a clear distinction should be made between the color of flowers in bud and the color of the same expanded flowers. The best time to examine flower and petal color being when flowers are fully expanded. Petals and stamens may become darker during the blooming phase, stamen become darker after pollination and petal color deepens once they begin to whither. To avoid any misunderstanding regarding difference between hues, colors in descriptions should be given as numbers referring to the *Color Chart*, edited by the Royal Horticultural Society. Colors detailed within this volume include number references. Numbers appear in brackets after the color for example, pink (RHS-65D). The diverse colors of ornamental peach (Fig.5.4, P. 39) are grouped as white, pink, red, and variegated with bicolor or tricolor flowers.

Pistil: There are four types of pistil in ornamental peach. (1) The perfect pistil - a fully developed pistil with a mature stigma, style and ovary. Most single-flowered and semi-double-flowered cultivars present perfect pistils. (2) Imperfect pistil - a degenerated or even malformed morphological pistil as presented in 'Baihua Shanbitao' (Fig. 5.9, P. 40). (3) Multiple pistils (polycarpous) - Double flowered varieties have one to five pistils (Fig. 5.11, P. 40). Usually as a result of an abnormally high temperature during initial development of flower buds which leads to increase in gynoecium development. This in turn may eventually give rise to common, double or triple fruit phenomena in ornamental peach. (4) Long pistil - The length of the pistil with stigma is longer than the flowering buds. (Fig. 5.10, P. 40) Observing the characteristics of pistil has proved to be useful in identification of cultivars. The presence or absence of pistil and its length is directly related to the fruit setting making it an important aspect for observation with regard to the ornamental peach classification system.

Stamen: Ornamental peach present three colors of anthers: yellow, orange, and white. White flowers present yellow anthers and most pink or red flowers have orange colored anthers. White anthers usually belong to sterile pollen. Observing filament length is another useful tool in cultivar identification. A majority of cultivars have filaments that are shorter than petals, though bell-shaped flowers usually have long stamens reaching out of their flowering buds and 'Baihua Shanbitao' presents a similar filament length to petals. Filaments color directly is related to flower color; white flowers have white filaments, pink flowers usually have very pale pink filaments and red flowers have light red filaments. Whatever the color, filaments becomes darker after pollination.

Flower size: Flower size helps to identify ornamental peach cultivars. Cultivars producing bell-shaped flowers present the smallest blooms, with an average diameter of approximately 2cm. All other flower varieties have average diameters of more than 3.5 cm with some cultivars producing blooms reaching 5-8 cm in diameter. It is impossible to judge flower size by observing only one or two flowers over a time span of 1-2 years. Several years of constant and systematic observation of typical samples are the only way to produce stable data that defines cultivar.

Calyx: The common calyx form is triangular ovate. Single flowers have 5 calyxes, however, semi-double flowered cultivar, for instance 'Hehuan Erse Tao'; has 5 calyxes (Fig. 5.6, P. 40). Most semi-double and double flowers have two layers, 10 calyxes, with a similar or smaller inner layer of calyx (Fig. 5.7, P. 40). Calyx numbers becomes irregular in petaloid cultivars (Fig. 5.8, P. 40). Calyx color is related to flower color with pink and red flowers usually have red calyx and white flowers producing green calyx. Variegated flowers have calyx color depending on the color variation included in the bloom, white flowers with red or pink stripes or dots have variegated red green calyx (Fig. 5.5, P. 39).

Time of flowering: The period of flowering is the most obvious characteristic by which one is able identify ornamental peach cultivars. There are three stages of flowering. Usually hybrids with *Prunus davidian*a pedigree bloom early, at the end of March or early April. Late blooming cultivars burst into flower during the last ten days of April with some lasting into early May. The flowering period for most ornamental peach cultivars is during the last ten days of April (Times and dates mentioned here refer to the Beijing area, these are variable dependant on location. However, the successive order of cultivar blooming is consistent. Besides those otherwise specified, explanations all refer to the Beijing area). Changes caused by the influence of unusual temperature or humidity and transplantation cannot be considered appropriate when identifying flowering periods in peach cultivars.

2. Fruit and stone

Ornamental peaches, as a rule, are fertile plants. Undeveloped pistil cannot form "normal" fruits, which is particular to the ornamental peach. The primary purpose of planting ornamental peach is not for it's fruit but for it's year round ornamental value. As fruit is not considered an important ornamental feature it should be limited or removed from specimen. For cultivars presenting fruit sets, fruit or stone characteristics such as oblate or round, pubescent or glabrous, could be optional references to be used during detailed cultivar description.

Shape of fruit: oblate, round, obovate, oval and conoid are five shapes presented in ornamental peach fruits (Fig 5.12, P. 42).

Stone Shape: oblate, oval, obovate and elliptic are four typical stone varieties (Fig. 5.13, P. 42).

3. Growth habit

Growth habit: Ornamental peach presents five types of growth habit:

The twisted type - a unique twisting branch pattern (Fig. 5.14, P. 42).

The weeping type - pendulant branches.

Pillar type - upright twigs with narrow crotch angle (Fig. 5.15, P. 43).

Dwarf type - small in stature with short internodes

Standard type (most common) - tilted twigs and branches (Fig. 5.16, P. 43).

Twig color is closely linked to flower color, for instance white flowering cultivars most commonly have green twigs. Pink or red flowering cultivars present red-brown on the upper side facing towards sunlight. Cultivars producing variegated flowers present twigs of a very similar color pattern to their flower color. Red stripes or dots on twigs can be used as a guide to flower color though one known exception must be mentioned; the red flowering cultivar 'Fei Tao' has green twigs.

4. Leaf

Leaf shape: Ornamental peach cultivars exhibit three types of leaf shape: oblong-lanceolate, lanceolate and ovate-lanceolate (Fig. 5.17, P. 44).

Leaf color: The leaves of most ornamental peach cultivars are green though purple and red leaved cultivars also exist. The variegated leaf, which has purple stripes or spots present on a green leaf, can also be found within ornamental cultivars. (Fig. 5.18, P. 44).

Leaf margin: There are three types of leaf margin of ornamental peach varieties: finely serrate, coarsely serrate, and crenulate (Fig. 5.19, P. 44).

Leaf base: Three types of leaf base are found across ornamental peach cultivars: attenuate, cuneate and broad cuneate (Fig. 5.20, P. 44).

Leaf apex: Leaf apex descriptions of ornamental peach fall into two categories: acuminate or acute (Fig. 5.21, P. 44).

Approximately 19 morphological characteristics of ornamental peach are illustrated above. 40 characteristics are needed to define a cultivar and these form the basic data collected when investigating, describing and registering a cultivar (Table 4.1). The characteristics in American patent and Japanese patent are referenced below.

Grouping characteristics are a major criterion for identifying ornamental peach cultivar. Growth habit, flower type, flowering period, leaf color and the absence or presence of fruit set are five characteristics useful for the grouping of all ornamental peach cultivars.

Qualitative characteristics are those that are expressed in discontinuous states (UPOV, 2002). All states are necessary in order to describe the full range of the characteristics and every form of expression can be described by a single state. The absence or presence of pistil and fruit set are pseudo-qualitative characteristics that are, as a rule, not influenced by environment.

Quantitative characteristics are those whose expression covers the full range of variation possibilities from one end of the spectrum to the other (UPOV, 2002). The expression can be recorded on a one-dimensional, continuous, discrete or linear scale. For instance, flower size, tree size and the number of flowers are all quantitative characteristics.

In the case of "Pseudo-qualitative characteristics", the range of expression is somewhat continuous though varies in more than one dimension and cannot be adequately described by simply defining the two ends of a linear range (UPOV, 2002). Growth habit, flower type and flower color are all pseudo-qualitative characteristics.

Table 4.1   Morphological characteristics for ornamental peach cultivar

No.	Type	Characteristics	Rate
1	QL^Z	Bark: texture	glabrous/rough
2	QN^Y	Tree: size	very small/small/medium/large/very large
3	PQ^X	Growth habit*	standard/pillar/dwarf/weeping/ twist/ combination
4	PQ	Flowering shoot: color	green/red/green with red spot or strip
5	QN	Flowering shoot: flowering bud density	sparse/medium/dense
6	PQ	Flowering bud: shape	ovate-oblong/ovate/oval/oblate/ round
7	PQ	Flower type *	single/mei-flower/rose/peony/chrysanthemum/bell-shaped
8	QN	Flower size	very small/small/medium/large/very large
9	PQ	Petal: color	white/lightpink/pink/dark pink/red/red&white/white&pink/ white&pink&red (RHS)
10	PQ	Petal: shape	spatulate/ lanceolate/ovate-oblong/ovate/oval
11	QN	Petal: size	very small/small/medium/large/very large
12	QN	Petal: number	single/semi-double/double
13	PQ	Calyx: color	green/red/green with red spot or strip
14	QL	Calyx: petaloids present or absent	non-petaloids/petaloids
15	QL	Calyx: number	5/more than 5
16	QN	Stamens: position relative to petals	below/same level/above
17	QN	Stamens: number	sparse/medium/dense
18	QN	Stamens: petaloids or non-petaloids	non-petaloids/petaloids
19	PQ	Anther: color	orange/yellow/white
20	QL	Pistil	absent/present
21	QN	Stigma: position relative to anthers	below/same level/above
22	PQ	Pistil: number	0/1-2/more than 2
23	QN	Pedicel: length	short/medium/long
24	QN	Beginning of flowering period *	very early/early/medium/late
25	QN	Duration of flowering period	short/medium/long
26	PQ	Leaf blade: shape	oblong-lanceolate/ elliptic-lanceolate/ ovate-lanceolate
27	PQ	Leaf: color *	green/red/green with red spot or strip
28	QN	Leaf: length	short/medium/long
29	QN	Leaf: width	narrow/medium/broad
30	QN	Leaf blade: length/width ratio	small/medium/large
31	PQ	Leaf blade	smooth/undulate/crumple/curled
32	PQ	Leaf margin	finely serrate /coarsely serrate/crenulate
33	PQ	Leaf: base	attenuate/cuneate/broad cuneate
34	QN	Leaf: apex	acuminate /acute
35	QN	Petiole: length	short/medium/long
36	QL	Petiole: nectarines	absent/present
37	QL	Fruit *	absent/present
38	QL	Fruit: pubescence	absent/present
39	PQ	Fruit: shape	round/oval/conoid/elliptic/oblate
40	PQ	Stone: shape	oblate/oval/ obovate /elliptic

*Grouping characteristics; ^{Z}Qualitative characteristics; ^{Y}Quantitative characteristics; ^{X}Pseudo-qualitative characteristics.

Classification System of Ornamental Peach

Based on morphological characteristics five classification systems have been published (Zhang, 1991, 1993, 1997a; Chen 1993; Hu *et al.*, 2003). In the most recent system, ornamental peach cultivars were classified into 2 Branches, 5 Groups, 11 Forms and 55 cultivars (Hu et al., 2003), based on the exploration, collection and research in China, Japan, and the USA. Morphological characteristics, with consideration of ornamental peach evolution, were reflected in this updated system. Under the premise of pedigree of *P. persica*, growth habit and the type of developed flower were major criteria in this classification system.

Principles of classification

1. Species is the premise

Most cultivars of ornamental peach are descendant from *P. persica*. Along with the hybridization with *P. davidiana*, *P. kansuensis*, *P. mira*, and *P. amygdalus*, in recent years in a number of countries, the pedigree of ornamental peach cultivar has become the first criterion of ornamental peach systematics.

A typical example of hybridization is the natural hybrid of *P. persica* and *P. davidiana*, 'Baihua Shanbitao'. Compared with the pollen parent, it has double white flower, which is of increased ornamental value. Compared with female parent, it is early blooming, about one week to 10d earlier.

According to the updated regulations of ICNPC (Brickell *et al.*, 2004), the David group (the hybrids between *P. persica* and *P. davidiana*) was presented instead of the series or branches presented in older classification systems. Although ornamental hybrids of *P. mira* and *P. kansuensis* were reported (Komar-Tyomnaya L.D., 2007), they will only be included in updated keys once their pedigree has been identified and detailed morphological information collected.

2. Ornamental peach evolutions

Growth habit, flower type, and petals number are all reflexive of evolutionary rules. Simple to complex and primitive to evolutionary are the basic rules for ornamental peach classification.

Growth habit is the most distinguishing characteristics and classification criterion besides the pedigree of ornamental peach. Varying growth habits have emerged during different periods; the standard upright type is the most primitive type and has been the most commonly found form since peach cultivation began. The dwarf type was first recorded in the 14th century (Zhou, 1620) and the pillar and weeping type have records dating back to the 17th (Ito, 1695). The twisted variety was first reported in the 20th (Yoshida *et al.*, 2000).

The wide range of petal number, from simple single petal to semi-double and double; makes flower type the inferior criterion to growth habit for definition of cultivar.

3. Easy to apply

Ornamental peaches are used primarily for their ornamental value in landscape and private gardens. Flower and leaf color are the characteristics most sought by consumers and are the primary influence in the selection and placing of peach species in ornamental planting.

Hierarchy of classification

1. Species (hybrid)

According to current documentation of ornamental peach cultivars, only one hybrid group, the David group (*P. persica* × *P. davidiana*) exists. The cultivars in the David group have characteristics inherited from both *P. persica* and *P. davidiana*, which can be easily noted when compared to the cultivars with pure *P.*

persica pedigree.

Reports on hybrids of *P. persica* × *P. mira*, *P. persica* × *P. kansuensis*, and *P. persica*×*P. amygdalus* (Komar-Tyomnaya, 2007) have proved that it is necessary to put species/pedigree (hybrids) as the first and most important criterion in ornamental peach classification system. After systematic identification the Mira group, Kansu group and Amygdalus group may appear in new classification systems adopted in the near future.

2. Group

As clearly stipulated in Article 3.1 of ICNPC regulations (Brickell *et al*., 2004) "group" is an official grade based on certain similarities of cultivar and plant individuals or plant categories. The standards of composing and maintaining a cultivar group varies dependant on the purpose required.

A cultivar group may be designated if there are two or more cultivars in existence with similar properties in a genus, species, hybrid genus, hybrid species or other nomenclature class. Prior to the introduction of regulations of 'Variety', other names such as "convar", "sort", "type" or "hybrids" were used by scholars, finally the name "cultivar group" was introduced and shall be used from this point forth.

As a result of this change, the "series" (or "branch") terminology used in the old classification system was replaced. Ornamental peach cultivars are now divided into six groups: Standard, Dwarf, Pillar, Weeping, Twisted and David. Each group has its apparent distinction and all cultivars in each group present differenced form the next.

3. Cultivar

Cultivar is the basic classification unit for horticultural plants. According to ICNCP regulations, cultivar must have distinct, stable and consistent characteristics and these distinct characteristics must be able to be propagated by appropriate methods of reproduction (sexual or asexual) and remain consistent. Characteristics can be in morphology, physiology, cell, or biochemistry. Each cultivar may be defined as an independent cultivar once the characteristic(s) can be successfully maintained by propagation.

Classification system

To summarize, updated ICNCP regulations have replace the old orders of "series" and six groups based on growth habit and pedigree were presented. These six groups are: Standard, Dwarf, Pillar, Twisted, Weeping and David. Though flower type is considered an inferior criterion to pedigree and growth habit when identifying cultivar it is still considered as one of the major criterion for observation.

Table 4.2   Key to groups of ornamental peach cultivars

1. Rough bark, medium or late bloom.
   2. Twigs upright or tilted.
      3. Crotch angle around 35°-70°, tilted twigs.
         4. Crotch angle around 40°-70°.
            5. Long internodes length ....I Standard Group
            5. Short internodes length, <10mm ....II Dwarf Group
         4. Crotch angle around 35°-40° ....III Pillar Group
      3. Crotch angle around >70°, pendulant twigs ....IV Weeping Group
   2. Twigs twisted.. V Twisted Group
1. Smooth bark, slender twigs, early flower .... VI David Group

Major characteristics of each group

Standard group: Standard group is the most common type with tilted twigs and long internodes and the widest range of variation. Six types of flower all appear in this type: bell-shaped, single, mei-flower, rose, chrysanthemum and peony.

Dwarf group: Small in stature with short internodes. A dense and compact shape with long narrow-leaves is common characteristic of this group.

Pillar group: Upright twigs with narrow crotch angle are typical characteristics of this group.

Weeping group: The pendulous twigs of this type make its form resemble an umbrella.

Twisted group: A unique growth habit with twisted twigs.

David group: A hybrid of *P. persica* and *P. davidiana*. Its early flowering and smooth bark are related to *P. davidiana*. The absent pistil or presence of underdeveloped pistils are a common characteristic of cultivars in this group.

A parallel relations hip exists between flower type, flower color and the number of petals in each group (Chen, 1997; Zhou and Li, 1998). Cultivar with similar flower types and color appear in each group. Great potential for new variations of known ornamental peach cultivars exists. For example, there is no pillar type cultivar with chrysanthemum flowers and according the parallel relationships noted above it is possible that it may be found or bred in the near future. This relationship will likely provide guidance for future germplasm collection and the breeding of the future.

Systematics Studies on Ornamental Peach

For taxonomic purposes, best results are achieved by combining evidence from as many different organizations as possible including morphological evidence, pollen, isozyme and molecular analysis.

Palynology studies

Pollen characteristics are controlled by genes (Heslop-Harrison, 1968; Wang and Zhou, 1990) and can not be influenced by environmental conditions. Based on NPC classifications when scanning electron microscope (SEM), pollen grains of peach have three (N3) zonally arranged (P4) compound apertures (3 colp-orate, C5) (Heslop-Harrison, 1968). Pollen grains are large in size and prolate ($P/E \leq 2\mu$m) or perrolate ($P/E > 2\mu$m) in shape (Fig. 5. 22, P. 52).

Based on the pollen characteristics and genetic information of 22 ornamental peach cultivars the surface sculptures of ornamental peach pollen can be divided into three groups: stripe-like cavity, stripe-like, and strip-like cavity-like . Upright, dwarf and weeping cultivars belong to each group respectively. These results indicate that cultivars with similar growth habits exhibited similar pollen characteristics and should be placed as a group in the ornamental peach classification system (Zhang *et al*., 1997b).

The pollen grains of 39 cultivars of ornamental peach (*P. persica*), which belong to upright, dwarf, weeping, pillar, twist-twigs, and David groups (hybrid with *P. davidiana*), were collected and examined in Beijing Botanical Garden, China in the 2007 and 2009 growing seasons.

The exine sculpture of ornamental peach pollen grains

There are three pollen types of ornamental peach cultivars. Typical exine patterns and descriptions of each cultivar are shown in Fig. 5.23, P. 52.

Width of ridge varies from 0.27 ('Shouhong') to 0.52 ('Erse Tao') μm. Distance between furrows varies from 0.2 ('Teruteshiro') to 0.57('Erse Tao') μm. Pore diameter varies from 0.2 ('Xiayu Shouxing') to 0.44 ('Jiang Tao') μm. Pore density varies from 0.19 ('Dan Shoufen') to 1.66 ('Corinthian Rose').

Pore type: exine glabrous, multiple, short branching of ridge, irregular, furrows and ridges were all inapparent.

Striate and pore is the most common type of pollen grain found in ornamental peach cultivar. The striate and pores were presented together. Observation of striate indicates that there are two varieties - regular to irregular.

The Striate variety is striate with very few or inapparent pores, commonly presenting wider ridges and narrow distances between striate.

Erdtman (1978) regarded that the evolution of exine sculpture in angiosperms progresses from smooth to pores and then to striate. The exine sculpture of six groups of ornamental peach support this rule. Each group has cultivars at different evolutionary levels.

All three cultivars in the David group have striate and pores type pollen. Both 'Fenhua Shanbitao' and 'Fenhong Shanbitao' share the same pollen parent as 'Baihua Shanbitao'. The similarity of pollen exine sculpture revealed the close relationship between these three cultivars (Erdtman, 1978).

As the only cultivar with twisted twigs, 'Unriumomo' is pores type, and therefore is not at a developed stage of evolution. The single petal of 'Unriumomo' is primitive based on flower petal numbers and cannot assert the typical exine sculpture of this cultivars type (Fig.5.24, P. 52).

Six weeping cultivars examined were all striate and pores type (Fig. 5.25, P. 54), with the exception of 'Daiyu Chuizhi' which was regular striate and pores, all cultivars -, 'Lv E Chuizhi', 'Yuanping Chuizhi', 'Danbai Chuizhi', 'Zhufen Chuizhi', and 'Crimson Cascade' presented irregular striate and pores. Records of weeping peach date back to the 17th -19th centuries (Ito, 1695), indicating that this cultivar is in a well developed stage of evolution.

The five cultivars representing the pillar group can be divided into three types (Fig 5.26, P. 54), 'Teruteshiro', 'Terutehime', and 'Terutebeni' present regular striate and pores type pollen, whereas 'Corinthian Rose' presented pollen with the following characteristics: irregular striate and pores. 'Terutemomo' was observed as striate type. All of these are in higher states of evolution which correlates with post 17th century records of the pillar variety (Ito, 1695).

All six dwarf cultivar present striate and pores type pollen (Fig 5.26, P. 54). 'Dan Shoufen', 'Shoubai', 'Shoufen', and 'Shouhong' are all irregular striate and pores with the exception of the regular striate and pores of 'Dan Shoubai'. The first description of dwarf peach was formed in the Ming Dynasty (1368--1644), which, though later than the upright variety still places it in a relatively high state of evolution.

Nineteen upright cultivars display diverse exine sculpture, which present all four types above (Fig 5.27, P. 55). 'Jiang Tao' and 'Bi Tao' are pores type; 'Juhua Tao' is striate type and all others are striate and pores. The differing degree of evolutionary development presented within upright peach cultivars illustrates their long history and development.

Biochemical studies

Results of isozyme of POD (peroxidase) studies (Hu, 1999) on 35 ornamental peach cultivars supported that growth habit should be considered very highly in the classification criterion of ornamental peach cultivar. All dwarf cultivars shared one unique band, implying that they have closer relationships to each other than other cultivars with differing growth habits. All weeping cultivars generated more bands than all other growth habits, with one shared band within this group of cultivars. All cultivars originating from *P. persica* shared three unique bands, which may be the unique bands for their species.

Molecular studies

In recent years molecular biology has become increasingly popular in ornamental plant study. In ornamental peach, three molecular markers, RAPD, ISSR, and AFLP have been applied for analyzing the genetic relationships of ornamental peach cultivars. A DNA fingerprinting library of 51 ornamental peach taxa has been established using 6 primer combinations of AFLP data. By using SSR methods, the *pl* gene was found that co-segregated with CPPCT 029, providing a basis for further accurate location.

1. Molecular marker

In contrast to morphological markers, molecular markers detect genetic differences at the level of the organism's DNA (Debener, 2002), which is never changed by growth, development processes or environmental conditions. This could offer immediate help with cultivars identification and ultimately lead to an improved classification system. In addition, many molecular techniques may be able to elucidate the hierarchy of relationships among cultivars (Culman and Grant, 1999).

DNA markers are able to distinguish different individuals or cultivars through 'DNA fingerprinting' techniques. These techniques involve the generation of a series of bands (DNA fragments), which are separated by size. Bands are generally not characterized fully but are defined by the presence of a particular target sequence of DNA (Grant and Culham, 1997).

RAPD: 10 primers selected from 36 random primers on 37 ornamental peach taxa were amplified with RAPD for the analysis of the genetic relationships of ornamental peach (Zhang *et al.*, 1999). Results indicated that the dwarf group had much closer relationships than the weeping group and bicolor cultivars. As for the hypothesized hybrid *P. persica* 'Baihua Shanbitao', RAPD results showed a much closer relationship to *P. persica* than *P. davidiana*, supporting the former hypothesis (Zhang *et al.*, 1997b).

22 primers selected from 200 arbitrary primers on 9 red-leaf cultivars, 8 dwarf cultivars, and 10 ornamental peach cultivars were amplified with RAPD (Cheng, 2003a, Cheng *et al.*, 2002, and Cheng 2003b). Molecular checking indexes were presented within the dwarf group and red-leaf group.

ISSR: Ten primers selected from 100 available primers of 16 ornamental peach taxa were applied with ISSR (Hu *et al.*, 2006). A total of 132 useful markers between 300 to 1400 base pairs were generated from 10 ISSR primers (UBC818, UBC825, UBC834, UBC855, UBC817, UBC868, UBC845, UBC899, UBC860, and UBC836). Among them, 62% of bands were polymorphic markers. The average number of markers for each taxon was 80. The agarose gel image showed on Fig. 5. 30, P. 61. Two hybrid cultivars grouped to a cluster with 69% bootstrap support. The results demonstrated that the pedigree of *P. davidiana* has been involved in the breeding and selection of ornamental peach cultivars and these cultivars do have apparent genetic distance from pure *P. persica* cultivars. The ISSR marker system is a useful technique for revealing groups (different pedigree and different growth habit) and genetic relationships of ornamental peach.

Based on 132 useful ISSR markers, pairwise distances ranged from 0.030 to 0.402. The peach is a self-fertile and naturally self-pollinating fruit species with very low genetic variability (Hesse, 1975; Scorza, *et al.*, 1985; Scorza and Okie, 1991). Its genetic base was strikingly narrow (Scorza, *et al.*, 1985). AFLP analysis (Aranzana *et al.*, 2001) also indicated the low level of genetic diversity in fruit peaches. Results from ISSR analysis also confirm the narrow genetic diversity among ornamental peach cultivars.

AFLP: A total of 275 useful markers ranging in size from 75 to 500 base pairs were generated using 6 EcoRI/MseI AFLP primer pairs (Hu *et al.*, 2005a). Among them, 265 markers were polymorphic. The number of markers for each taxon range from 90 to 140 (mean=120) (Fig. 5.28, P. 57). Based on the results of the AFLP data, ornamental peach are primarily derived from *P. persica*. However, species *P. davidiana* was also involved in ornamental peach cultivars breeding and development, demonstrating that the growth habit of ornamental peach should be important in ornamental peach systematics.

Based on 275 useful AFLP markers, genetic distances ranged from 0.044 to 0.404. This is not surprising given that genetic distances among the cultivars derived from different species are higher than those among the cultivars within a species. Two distinguish clades were recognized in the UPGMA tree generated using PAUP (Swofford, 2002), one being accessions of *P. davidiana*, and the other containing cultivars of *P. persica* (Fig. 5.29, P. 58).

P. davidiana was apparently an out-group to *P. persica*. This relationship suggested that *P. davidiana* was genetically distanced from ornamental peach taxa derived from *P. persica*. Clade *Davidiana* had three taxa derived from *P. davidiana* (*P. davidiana* var. *alba*, *P. davidiana* var. *rubra*, and *P. davidiana* 'Fastigate') and supported by 100% bootstrap value. The genetic distance within these three taxa is 0.16, while the average distance of these three taxa to all other taxa originated from *P. persica* is 0.339. Obviously, these three taxa are relatively close to each other when compared with other ornamental peach taxa derived from *P. persica*.

The *P. persica* clade consisted of four subgroups, Clade PR (Red-leaved clade), Clade PT (Twisted clade), Clade PU (Upright clade), and Clade PG [Growth habit clade (included fastigiata, weeping, and dwarf)].

Clade PR included two red-leaved taxa ('Zi Ye Tao' and 'Zuoshuang'), with 100% bootstrap support. *P. persica* 'Zuoshuang' is a mutation from normal red-leaved cultivar 'Zi Ye Tao', and was selected by Beijing Botanical Garden in 2001. Compared with normal red-leaved cultivars, 'Zuoshuang' has unique purple and green bicolor leaves. The smallest genetic distance from this clone to 'Zi Ye Tao' was 0.12. Both morphological characteristics and genetic distance supported that 'Zuoshuang' is a new cultivar (Hu *et al.*, 2005b).

Clade PT consisted of only one cultivar 'Unriumomo' (single, pink flowers). It is the only documented ornamental peach cultivar with twisted twigs. The plant was selected from a mutation of nectarine seedling (Yoshida *et al.*, 2000). It is independent to all other ornamental peach taxa in UPGMA tree. The average genetic distance to other *P. persica* taxa was 0.237. It is possible that this cultivar originated independently and may be an important germplasm source for further ornamental peach breeding. Other growth habits, such as pillar, weeping, dwarf, and standard, were grouped to their relevant clade; again demonstrating that growth habit is a major classification criterion in ornamental peach systematics.

Twenty taxa were clustered into Clade PU, which had the common morphological characters and upright tilted twigs. Among them, 18 out of 20 upright ornamental peach taxa in this study were all in this clade.

Within Clade PG, there were two subgroups, which were Clade PGD (Dwarf clade) and Clade PGM (Mixed clade of pillar, weeping, and dwarf). All seven fastigiata cultivars and 10 weeping cultivars used in this study were clustered into PGM subgroup.

The cultivars with different flower color and form overlapped in each clade. There was not a distinguished cluster of the same flower color or form from this AFLP result. It is possible that flower color and form may not be in a higher hierarchy in ornamental peach systematic.

Based on the results of the AFLP data, the growth habit of ornamental peach should have in important place in ornamental peach systematics and the number of petals should also be taken into consideration. The large number of fragments amplified from ornamental peach was generated by AFLP markers and provided detailed information about genetic relationships between the ornamental peach cultivars.

2. DNA Fingerprint of ornamental peach

DNA markers are able to distinguish different individuals or cultivar through "DNA fingerprinting"techniques. The techniques involve the generation of a series of bands (DNA fragments), which are separated by size. The bands are generally not characterized fully but are defined by the presence of a particular target sequence of DNA (Grant and Culham, 1997).

By using 6 primer combinations of AFLP data a DNA fingerprinting library of 51 ornamental peach taxa has been established. This molecular data may assist in the identification of and support the legitimacy of ornamental peach cultivar. This indicates clearly the potential of molecular markers for identifying cultivars and verifying the origin of ornamental peach breeding (Hu, 2004).

Both 'Fenhua Shanbitao' and 'Fenhong Shanbitao' were derived from 'Baihua Shanbitao' (Hu and Zhang, 2001). Morphologically, all three cultivars have smooth bark, slender and long twigs and all flowers early in the season. There were 61 bands in six primer combinations shared by these three cultivars. 'Baihua Shanbitao', which is hypothesized natural hybrid between *P. davidiana* and *P. persica*, has 7 unique bands to each of the parents (M-CAT/E-ACC-65, M-CTC/E-ACC-83, M-CAT/E-ACT-121/130/162/269, M-CTC/E-ACT-352). Both 'Fenhua Shanbitao' (M-CTC/E-AGG-84) and 'Fenhong Shanbitao' (M-CAT/E-AGG-234) all have their unique band. AFLP data states that these three cultivars belong to a hybrid group.

'Unriumomo' is the only ornamental peach cultivar with twisted twigs. Its unique growth habit can be represented shown though its 16 unique bands generated from 3 primer combinations (M-CAT/E-ACC62, -77, -133, -187, M-CAT/E-ACT68, -75, -96, -120, -190, -219, -330, and -360, M-CAT/E-AGG62, -77, -133, -187). Molecular data fully supports that this is a legitimate cultivar.

Compared to other dwarf cultivars, 'Xiayu Shouxing' (double, bicolor of pink and white) has a unique band pattern at M-CAT/E-ACT270-274-277. This AFLP data indicate that this is also a legitimate cultivar.

'Wubao Chuizhi' is a weeping type ornamental peach with rose pink and pale pink double flowers. Its unique band at M-CTC/E-AGG-368 should be regarded as a molecular evidence to support its cultivar status.

Molecular markers may be used not only to distinguish genotypes but also to provide information about genetic relations between genotypes.

'Red Dwarf' is a red-leaved dwarf cultivar, released by Yoshida and Seike (1974) in Japan. This plant originated from F_2 of a red-leaved, upright growth type and a green leaved, dwarf type peach. 'Red Dwarf' has a unique band pattern M-CTC/E-AGG-336-36-447. 'Bonfire Patio' a cultivar with similar morphological descriptions to 'Red Dwarf' is an open pollinated seedling from red-leaved upright type cultivar and released by Moore *et al.* (1993) in the USA. It has very similar AFLP bands pattern to 'Red Dwarf'. These two cultivars also have very similar band patterns to 'Beijing Zi', whose morphological characteristics are quite similar to the parent of 'Bonfire Patio'. Results from this AFLP data have verified the origin of these two red-leaved cultivars. Obviously, AFLP markers are a powerful tool for detecting the breeding source of ornamental peach cultivars.

Molecular data can be of tremendous assistance when identifying cultivars and verifying the origin of ornamental peach breeding as it provides supporting evidence of distinctness. However, it is clear that molecular data is not being able to completely substitute morphological characteristics. Some key morphological characteristics continue to be essential because it is difficult to establish firm enough links between cultivars simply from the limited molecular data available.

3. Genetic distance of weeping characteristics

Standard and weeping DNA bulks were built by BSA method and 19 SSR markers covered the whole linkage group were found closely linked to the weeping gene. CPPCT 029 is co-segregated with the pl gene (Li, 2006), which is about 242-309bp, with same band form as its parent individual. This is a basis for further accurate location.

CULTIVARS ILLUSTRATION

Standard Group

The standard group is the most common group of ornamental peach, with tilted branches of around a 35°-50° crotch angle. Cultivars in the upright group exhibit great diversity of flower type, flower color and leaf color. Flowers which appear in early spring before the new leaves expand, are available in single, semi-double and double forms. Their colors range from pure white, pink, and red to bicolor and tricolor. Leaf color ranges from green to red (or purple). A key to the standard group based on flower type, color, leaf and resistance to adversity is detailed below.

Table 5.1   Key to ornamental peach cultivars of Standard Group

1. Tilted twigs, flower less than 10 petals, green or red leaves
  2. Small flower size, around 2cm, bell-shaped, green or red leaves
    3. Light pink flowers, green leaves.. 'Rui Guang'
    3. Dark purple flowers, red leaves .... 'Ha Lu Hong'
  2. Flowers size bigger than 3.5cm, flat saucer-like flower, green or red leaves
    4. White flowers, green leaves.. 'Alba'
    4. Pink, red, or variegated (bi-color or tri-color) flowers, green or red leaves
      5. Pink flowers, green or red leaves
        6. Pink flowers, green leaves
          7. Pollen fertile .... 'Dan Fen'
          7. Pollen sterile .... 'Mei Fen'
        6. Pink flowers, red leaves... 'Beijing Zi'
      5. Red or variegated flowers
        8. Red flower .... 'Dan Hong'
        8. Pink and white flowers .... 'Kyosarasa'
1. Tilted twigs, semi-double or double flowers, green, red, or variegated leaves
      9. Semi-double, 15-40 petals
        10. 3-5 layer petals, regularly arranged like overlapping
          11. White flowers, green leaves..'Bai Bitao'
          11. Pink, red, or variegated (bi-color or tri-color), green, red or variegated leaves
            12. Pink flowers, green or red leaves
              13. Pink flowers, red leaves... 'Zi Qi'

13. Pink flowers, green leaves

   14. Tolerant to bacterial spot

      15. Cold-hardy, tolerant to brown rot, powdery mildew

         16. Pink as 55C of RHS color chart, eglandular .... 'Harrow Frostipink'

         16. Pink as 55C of RHS color chart, reniform glands .... 'Harrow Candifloss'

      15. Tolerant to cytospora canker ....'Jerseypink'

   14. No specific disease tolerant, flat or round shape fruit

      17. Round shape fruit, early blooming, 5 calyxes .... 'Hehuan Ersetao'

      17. Flat shape fruit, medium blooming, 10 calyxes .... 'Yaezaki bantou'

12. Red, or variegated (bi-color or tri-color) flowers, green, red or variegated leaves

      18. Red flowers, green, red or variegated leaves

         19. Red flowers, green leaves

            20. Pollen sterile, flat fruit .... 'Akabana bantou'

            20. Pollen fertile

               21. No specific tolerant to disease .... 'Jiang Tao'

               21. Cold-hardy, tolerant to brown rot, powdery mildew ...... 'Harrow Rubirose'

         19. Red flowers, red or variegated leaves

            22. Red leaves.. 'Zi Ye Tao'

            22. Variegated leaves, red spot or stripe on green leaves .... 'Zuoshuang'

      18. Variegated flowers, green or red leaves

         23. White, pink, red, or variegated, green leaves .... 'Peppermint Stick'

         23. Pink with red flowers, red leaves....................................... 'Ningxia Zi Ye'

10. Reflected or narrow petals, 30-40 petals

         24. Cup-shaped flowers, broad ovate petals

            25. Pink or variegated flowers

               26. Light pink flowers .... 'Renmian Tao'

               26. Pink with rose pink flowers .... 'Erse Tao'

            25. Red flowers, early blooming .... 'Hanhong Tao'

         24. Narrow petals, chrysanthemum-like

            27. Pink flowers, around 30 petals ....'Kikoumomo'

            27. Red flowers, around 36 petals ....'Kyoumaiko'

9. Double flowers, more than 40 petals, round shape

         28. White flowers... 'Wanbai Tao'

         28. Pink, red or variegated flowers

            29. Red flowers

               30. Pollen fertile, green or red twigs

                  31. Red flowers, green twigs .... 'Fei Tao'

                  31. Bright red flowers, red twigs .... 'Hong Bitao'

               30. Pollen sterile .... 'Mei Zi'

         29. Pink of variegated

1. 'Rui Guang'

Origin: Unknown.

Characteristics: Branches upright, twigs reddish; flower bud ovate, short pedicle; calyx 5, purple with green hint, pubescent in margin; saucer like bell-shaped flower, petal 5-10, light pink, spatulate, flower diameter 2cm; medium flower density; leaf blade green, ovate-lanceolate; fertile, with edible fruit.

Blooming period: Middle of April (P. 68).

2. 'Ha Lu Hong'

Origin: Unknown.

Characteristics: Upright, twigs reddish; petal 5, ovate-oblong, bell-shaped flower, dark pink (67D); flower diameter 1.5cm; filament pink; pedicle very short; calyx purple; dense flower; leaf blade purple, ovate-lanceolate.

Blooming period: Middle of April (P. 68).

3. 'Alba'

Origin: First named in England in 1829 (Krussmann, 1986).

Characteristics: Upright, green twigs; flower bud oval, petal 5, oval, white (155D), 2.4cm long, flower diameter 5.1cm; stamens 51.3, filament 1.43cm long, white; anthers yellow; one pistil, shorter than filament; medium flower density (flower bud 1.22/cm), pedicle 0.57cm long; calyx 5, green; leaf blade elliptic-lanceolate, green, 11.2cm × 3.05cm , L/W ratio 3.67; leaf margin finely serrate; petiole 0.78cm long, 1-2 reniform gland; drupe green, pubescent, 3.3cm × 3. 2cm, round; elliptic stone, 2.57 cm ×1.7cm, glabrous in the surface.

Blooming period: Middle of April (P. 69).

4. 'Dan Fen'

Origin: Unknown.

Characteristics: Upright, large; bark brown yellow, flower bud narrow ovate, petal 5, 6-9 sparsely, oval, pink (65C), 2.37cm long, flower diameter 5.1cm; stamens 38.7, filaments 1.5cm long, whitish pink; anthers yellow; one pistil, a little bit longer than filament; medium flower density (flower bud 0.92/cm); pedicle 0.57cm long; calyx 5, reddish, pubescent in margin; leaf blade green, elliptic-lanceolate, 12.22cm × 3.74cm, L/W ratio 3.27; leaf margin finely serrate; petiole 0.8cm long, 2 reniform glands; drupe green, pubescent, 4.7cm long × 4.74cm, round; stone 2.9cm × 2.1cm, elliptic, smooth on the surface.

Blooming period: Middle of April (P. 69).

5. 'Mei Fen'

Origin: A pollen sterile mutant found in Beijing Botanical Garden.

Characteristics: Bark grayish brown, twig green; flower bud narrow ovate, petal 5, oval, pink (69A), 2.13cm long, flower diameter 4.57cm;staments 40, filament 1.37cm long, white; anthers white, 1 pistil, almost the same length as filament; medium flower density (flower bud 0.8/cm), calyx 5, reddish, pubescent in margin; pedicle 0.57cm long; leaf blade green, elliptic-lanceolate, 14.02cm × 3.65cm, L/W ratio 3.84, leaf margin finely serrate; petiole 0.73cm long, 2 reniform glands. No fruit through several years' observation.

Blooming period: Middle of April (P. 70).

6. 'Beijing Zi'

Origin: Unknown.

Characteristics: Bark grey-brown, twigs reddish; flower bud narrow ovate, petal 5, ovate, pink (62D), 1.17cm long, flower diameter 3.63cm; stamens 39.3, filament 1.1cm long, whitish pink; anthers yellow; 1 pistil, almost the same length as filament; medium flower density (flower bud 1.2/cm), calyx 5, reddish, pubescent in margin; pedicle 0.57cm long; leaf blade red in new, fainted in summer, elliptic-lanceolate, 13.02cm × 3.14cm, L/W ratio 4.15, leaf margin finely serrate; petiole 1.06cm long, 2-3 reniform glands; drupe red on exposed side, 3.98m × 3.78cm, round, stone 2.23cm × 1.76cm, oval, smooth on the surface.

Blooming period: Middle of April (P. 70).

7. 'Dan Hong'

Origin: Unknown.

Characteristics: Bark grey-brown, twigs reddish; flower bud narrow ovate, petal 5, ovate, red (59D), 1.8cm long, flower diameter 3.8cm; stamens 40, filament 1.77cm long; anthers yellow; 1 pistil, shorter than stamens, medium flower density (flower bud 0.98/cm); calyx 5, reddish; pedicle 0.5cm long; leaf blade green, lanceolate, 11.82cm × 2.14cm, L/W ratio 5.52, leaf margin finely serrate; petiole 0.54cm long, 1-2 glands; drupe green, 4.7cm × 4.4cm, oval; stone 2.3cm × 1.9cm, oval, smooth on surface.

Blooming period: Middle of April (P. 71).

8. 'Kyosarasa'

Origin: It was recorded in Edo era (17th-19th) in Japan (Ito, 1695).

Characteristics: Bark grey-brown, twigs green; variegated flower with white and pink, petal 5, 1.6cm long, 1.3cm wide, flower diameter 3.6cm, pollen viability is 95.7%; leaf blade green, elliptic-lanceolate, 12cm × 3.7cm, L/W ratio 3.3; petiole 1.1cm long (Pawasut *et al.*, 2004).

Blooming period: Middle of April (P. 71).

9. 'Bai Bitao'

Origin: An ancient Chinese cultivar record dating back to the Song Dynasty (960-1279). It is also called "thousand or cent-leaves white peach" in ancient poems.

Characteristics: Bark grey-brown, twigs green; flower bud round, petal ovate-oblong, white (155D), 2.4cm long, flower diameter 5.1cm; semi-double, petal (22-)27.7(-28); stamens 59.7, filament 1.03cm long, anthers yellow; pistil longer than stamens; medium flower density (flower bud 0.88/cm), calyx green, 2 round; filament and calyx petaloids; pedicle 0.48cm long; leaf blade green, elliptic-lanceolate, 11.4cm × 3.4cm, L/W ratio 3.35; leaf margin finely serrate; petiole 0.96cm long; 2-3 reniform glands; drupe green, 3.82cm × 3.58cm, round; stone 2.44cm × 1.48cm, obovate; smooth on stone surface.

Blooming period: Middle of April (P. 72).

10. 'Zi Qi'

Origin: An early blooming purple leaf cultivar derived from the seeds, which from open-pollinated 'Zi Ye Tao' (double red flower, purple leaf), irradiated by γ-ray (Shen *et al*., 2007).

Characteristics: Up to 4.5m tall (6 years old), 44.6 cm diameter. vigorous, upright; bark grayish-brown, lots of lenticels; one-year-old twig glabrous, red on exposed side; bright pink (62B); petal spatulate, flower diameter 3.9-4.4cm; semi-double, 12-18 petals; stamens 29-36, filament and anther all red, slight petaloids of filament; dense flower (flower bud 1.58/cm); calyx 5, reddish, pubescent outside; 1 pistil, pedicle 0.5cm long; leaf blade dark purple in spring, fainted in summer; ovate-lanceolate, 9.1cm × 3.2cm, L/W ratio 2.84; leaf smooth, leaf apex acuminate, leaf base cuneate, leaf margin serrate; petiole 1cm long, 2-4 reniform glands; drupe green, round, 3.09cm × 3.06cm, pubescent.

Blooming period: Early and long, the end of March in Tai'an, Shandong province. Blooming date 18-21d. Zone 5-9. No obvious disease and insects (P. 73).

11. 'Harrow Frostipink'

Origin: Cold-hardy and disease resistant, late blooming cultivar bred in Agriculture Canada Research Station, Harrow Ontario (Layne, 1981). Derived from the cross ('Harrow Blood' × NY555036) × open-pollinated.

Characteristics: Vigorous, upright to spread tree; large flower, light pink (55C), stamens from initially white, becoming light pink (56D) at anthesis and darker pink to red (53C) after full bloom; dense flower (flower bud 1.25/cm); leaf blade green (147A), elliptic-lanceolate, 9.5cm × 3.0cm, L/W ratio 3.16; leaf margin crenulate, no gland; petiole 0.5-1cm long; cold-hardy, could be grown successfully without protection in Zone 6b, with protection may be in 6a; a good level of field tolerant to brown rot, powdery mildew and bacterial spot.

12. 'Harrow Candifloss'

Origin: See 'Harrow Frostipink'.

Characteristics: Vigorous, upright to spread tree; large flower, light pink (52C); medium flower density (1/cm); leaf blade green (147A), elliptic-lanceolate, 9.0cm × 2.5cm, L/W ratio 3.6; leaf margin crenulate; several reniform glands; cold-hardy, could be grown successfully without protection in Zone 6b, with protection may be in 6a; a good level of field tolerant to brown rot, powdery mildew and bacterial spot.

13. 'Jerseypink'

Origin: Originated at New Jersey Agr. Expt. Sta. New Brunswick. Selected in 1985 and introduced in 1990 by J.C. Goffreda, A.M. Voordeckers, and S.A. Mehlenbacher for ornamental qualities (Goffreda *et al*., 1992).

Characteristics: Vigorous, prolific, flower diameter 5.5-6cm. rich pink, double, anthers yellow, reniform glands, late blooming; leaf blade green; reniform glands; drupe green, 5.5-6cm oblong-round, tolerant to bacterial spot and cytospora canker.

Blooming period: Average bloom date at Cream Ridge, N.J. is April 26. This late blooming habit allows the flowers to escape spring frost injury.

14. 'Hehuan Ersetao' ('Yaguchi', 'Yingchun')

Origin: An ancient Chinese cultivar record dating back to the Song Dynasty (960-1279).

Characteristics: Vigorous, bark grayish-brown, twigs red on exposed side; from early bloom pink (65B) to darker pink (62B) after full bloom; flower bud oval, petal oval, 2.2cm × 1.4cm, flower diameter 4.6cm; semi-double, petal (19-)21.3(-25); stamens 84; filament 1.1cm long, petaloids, anthers orange red; pistil apparently longer than stamens; dense flower (flower bud 1.2/cm); calyx 5, reddish; pedicle 0.63cm long; leaf blade green, elliptic-lanceolate, 14.5cm × 3.7cm, L/W ratio 3.9; leaf margin crenulate; petiole 1.0cm long; drupe green, 3.98cm × 3.74cm, round; stone 2.2cm × 1.64cm, elliptic, smooth on stone surface.

Blooming period: Around April 10. It is the first pink cultivar blooming except David Group.

'Yaguchi' was recorded since Edo era (17th-19th) (Ito, 1695). Morphologically, it shares similar characteristics to the 'Hehuan Ersetao'. According to the Article10.1 of *International Code of Nomenclature for Cultivated Plants* (ICNCP) (Brickell *et al.*, 2004), we suggest that 'Yaguchi' should be combined and the accepted name is 'Hehuan Ersetao', because it was established earlier. Supplementary characteristic, pollen viability 85.3% (Pawasut *et al.*, 2004). The same resolution should also be applied to 'Yinchun' (Wang and Zhuang, 2001) and the accepted name should be 'Hehuan Ersetao'. Supplementary characteristics, chilling requirement 600h (P. 75).

15. 'Yaezaki bantou'

Origin: Unknown. Japanese cultivar.

Characteristics: Upright, medium size, twigs red on exposed side; flower bud ovate-oblong, petal oval, pink (62B), 1.96cm long, flower diameter 4.65cm; semi-double, petal (23-)26.5(-29); stamens 68.6, filament 1.24cm long, anthers orange red; pistil is little bit longer than stamens; medium flower density (flower bud 0.93/cm); calyx 10; filament petaloids; pedicle 0.73cm long; leaf blade green, elliptic-lanceolate, 12.42cm × 3.65cm, L/W ratio 3.7; leaf margin crenulate; petiole 0.98cm long; 2-3 reniform glands; drupe green; oblate, 5.24cm × 3.02cm; stone 1.16cm × 1.76cm, oblate, rough on the surface.

Blooming period: Middle of April (P. 76).

16. 'Akabana bantou'

Origin: It is a selection from an open-pollination seedling population of 'Yaezaki Bantou' (Yoshida *et al.*, 2000).

Characteristics: Upright, medium size, twigs red on exposed side; flower bud ovate-oblong, petal oval, red (65D); 1.97cm × 1.78cm; flower diameter 4.3cm; semi-double, petal 18-20, stamens 70.3, filament 0.97cm long, anthers white, pistil longer than stamens; sparse flower density (flower bud 0.53/cm); calyx 2 round, reddish; filament and calyx petaloids, pedicle 0.53cm long; leaf blade green; elliptic-lanceolate, 13cm × 2.98cm, L/W ratio 4.36; leaf margin crenulate; petiole1.08cm long; 2 reniform glands; drupe green, oblate, 4.33cm × 2.33cm; stone 1.16cm × 1.6cm, oblate; rough on the surface.

Blooming period: Middle of April (P. 76).

17. 'Jiang Tao'

Origin: An ancient Chinese cultivar records dating back to 5 Dai eras (907-979).

Characteristics: Upright, medium size, bark grayish-brown, twigs red on exposed side; flower bud oval, petal ovate, 2.2cm long, red(65D); flower diameter 4.73cm, semi-double, regular, petal (13-)14.7(-18); stamens 39.7, filament 1.23cm long, reddish, close to petal, filament petaloids; anthers orange red; pistil nearly the same length as stamens; medium flower density (flower bud 0.75/cm); calyx 2 round, reddish; pedicle 0.9cm long; leaf blade green, elliptic-lanceolate, 13.3cm × 3.7cm, L/W ratio 3.59; leaf margin crenulate, petiole 1.0cm long; 2-4 reniform glands; drupe green, 3.46cm × 3.26cm, round; stone 2.88 cm × 1.96cm, oval, rough on the surface.

Blooming period: Middle of April (P. 77).

18. 'Harrow Rubirose'

Origin: See 'Harrow Frostipink'.

Characteristics: Vigorous, upright to spread tree; large flower, red (52A), then fades to dark pink(52B); semi-double, petal 15-20; dense flower (1.75/cm); leaf blade green, lanceolate, 11cm x 3.5cm; leaf margin crenulate; several reniform glands; cold-hardy, could be grown successfully without protection in Zone 6b, with protection may be in 6a; a good level of field tolerant to brown rot, powdery mildew and bacterial spot (P. 78).

19. 'Zi Ye Tao'

Origin: An ancient Chinese cultivar record dating back to the Song Dynasty (960-1279).

Characteristics: Vigorous, upright, one-year-old twig glabrous, red on exposed side; flower bud oval, petal ovate, red (65D), 1.93cm long; flower diameter 4.73cm, semi-double, regular, petal (25-)27(-30); stamens 46, filament 1.07cm long; anthers orange red; 1-2 pistils, longer than stamens; medium flower density (flower bud 0.85/cm); calyx reddish, 2 round; filament and calyx petaloids; pedicle 0.53cm long; leaf blade red, elliptic-lanceolate, 13.8cm × 3.8cm, L/W ratio 3.65; leaf margin crenulate, petiole 1.08cm long; drupe red on exposed side, 4.4cm × 3.93cm, oval; stone 2.92cm × 1.98cm, obovate; rough on the surface.

Blooming period: Middle of April (P. 78).

20. 'Zuoshuang'

Origin: It derived from bud mutation of red leaf cultivar 'Zi Ye Tao' in Beijing Botanical Garden (Hu *et al*., 2005a).

Characteristics: Vigorous, upright, one-year-old twig red on exposed side; flower bud oval, petal ovate, red (65D), 1.7cm long, flower diameter 4.2cm, semi-double, regular, petal (24-)25.3(-27); stamens 47, filament 1.09cm long; anthers orange red; pistil longer than stamens; medium flower density (flower bud 0.83/cm); calyx reddish, 2 round; filament and calyx petaloids ; pedicle 0.43cm long; leaf blade red, or green (137C), or red spots or strips (187A) on green; elliptic-lanceolate, 13.8cm × 3.68cm, L/W ratio 3.75; leaf margin crenulate, petiole 0.75cm long; 2-4 reniform glands; drupe green, 4.34cm × 3.72cm, oval; stone 2.92cm × 1.9cm, elliptic, deep furrows on the stone surface.

Blooming period: Middle of April (P. 79).

21. 'Peppermint Stick'

Origin: Introduced by W.B. Clarke nursery of San Jose, CA. in 1933 (Jacobson, 1996).

Characteristics: Vigorous, large tree size; semi-double flower, white with red stripes, petal (20-) 24.4 (-32); filament petaloids, anthers yellow, 1 pistil, longer than stamens (P. 79) .

22. 'Ningxia Zi Ye'

Origin: It was derived from the mutation of 'Zi Ye Tao' (Zhang, 1993).

Characteristics: Vigorous, upright, twig red on exposed side; flower bud oval, petal ovate,1.9 cm long, mostly red (65D), with pink (68C) petals; flower diameter 4.5cm, semi-double, regular, petal (24-)26(-30); stamens 47, filament 1.1cm long, anthers orange red; pistil 2-6, longer than stamens; medium flower density (flower bud 0.95/cm); calyx reddish, 2 round; filament and calyx petaloids; pedicle 0.52cm long; leaf blade red, elliptic-lanceolate, 13.6cm × 3.5cm, L/W ratio 3.89; leaf margin crenulate, petiole 1.01cm long; drupe green with purple hue, 3.88cm × 3.70cm, oval; stone 2.80cm × 1.78cm, obovate; rough on the surface.

Blooming period: Middle of April (P. 80).

23. 'Renmian Tao'

Origin: An ancient Chinese cultivar records dating back to 5 Dai eras (907-979).

Characteristics: Vigorous, upright, bark grey; twig yellow-brown; flower bud round; petal ovate, 2.19 cm long, pink (65C); flower diameter 4.88cm, semi-double, petal (33-)36(-39); stamens 56, filament 1.22cm long, white; anthers orange red; pistil 2-5, nearly the same length as stamens; medium flower density (flower bud 1 /cm); calyx reddish, 2 round; filament and calyx petaloids; pedicle 0.75cm long; leaf blade green, elliptic-lanceolate, 15.24cm × 4.19cm, L/W ratio 3.64; leaf margin finely serrate' petiole 0.73cm long; 2-3 reniform glands; drupe green, 4.83cm × 4.03cm, oval; stone 3.1cm × 1.98cm, obovate, rough on the surface.

Blooming period: Middle to the end of April (P. 80).

24. 'Erse Tao'

Origin: An ancient Chinese cultivar record dating back to the Song Dynasty (960-1279).

Characteristics: Vigorous, upright, bark grey; twig yellow-brown; flower bud round; petal ovate, 2.23 cm long, light pink(62D) with pink (64D) in the same branch; flower diameter 5cm, semi-double, petal (35-)37.3(-42); stamens 53, filament 1.27cm long, white; anthers orange red; pistil 2-6, nearly the same length as stamens; medium flower density (flower bud 1.1/cm); calyx reddish, 2 round, filament and calyx petaloids, pedicle 0.8cm long; leaf blade green, elliptic-lanceolate, 15.34cm × 4.3cm, L/W ratio 3.57; leaf margin finely serrate; petiole 0.73cm long; 2-3 reniform glands; drupe green, 5.1cm × 4.1cm, oval; stone 3.3cm × 2.0cm, obovate, rough on the surface.

Blooming period: Middle of April (P. 81).

25. 'Hanhong Tao'

Origin: First named in 1993 (Zhang).

Characteristics: Vigorous, upright, bark grayish-brown; twig glabrous, yellow-brown; flower bud oval; petal ovate, 1.87 cm long, flower diameter 4.37cm, semi-double, 1.7cm tall; petal (36-)44.3(-48); stamens 43.3, filament 1cm long; flower heart white; filament light red; anthers orange red, pistil 1-5, nearly the same length as stamens; sparse flower density (flower bud 0.6/cm); calyx reddish, 2 round; pubescent in margin, filament and calyx petaloids, pedicle 0.73cm long; leaf blade green, elliptic-lanceolate, 11.84cm × 3.62cm, L/W ratio 3.27; leaf margin crenulate; petiole 1cm long; 2 reniform glands; drupe green, 4.62cm × 4.08cm, round; stone 3.16cm × 1.9cm, elliptic, rough on the surface.

Blooming period: It is the earliest red cultivar among the ornamental peach. The Blooming period is usually in the early April (P. 82).

26. 'Kikoumomo'

Origin: It was recorded in Qing Dynasty (1644-1911) in China. Also it was recorded in Edo era (17th-19th) in Japan (Ito, 1695).

Characteristics: Medium tree vigor, upright; bark dark grey; twig slender, yellow-brown; longer internodes than normal upright type cultivars, internodes length about 2.3cm; flower bud ovate; petal lanceolate, irregular waved, 2.13 cm long, 0.6 cm wide; pink (65A); flower diameter 4.53cm, semi-double, chrysanthemum type, 2cm tall; (22-)29(-32); stamens 32.7, filament 1cm long, curled; filament petaloids, anthers yellow; pollen viability 88.2% (Pawasut *et al.*, 2004); pistil little bit longer than stamens; medium flower density (flower bud 0.75/cm); calyx reddish green, 2 round; pedicle 0.43cm long; leaf blade green, little curled in margin; elliptic-lanceolate, 10.2cm x 2.82cm, L/W ratio 3.62; leaf margin finely serrate; petiole 0.92cm long; 2-3 reniform glands; drupe green, conoid, 4.02cm × 3.05cm; stone 2.92cm × 1.56cm, elliptic; rough on the surface.

Blooming period: Middle of April (P. 83).

27. 'Kyoumaiko'

Origin: It was discovered as a mutation of 'Kikumomo' (Yoshida, 1985).

Characteristics: Medium tree vigor, upright; bark dark grey; twig slender, yellow-brown; flower bud ovate, petal lanceolate, 1.7cm long, 0.5 cm wide; red (52B); flower diameter 4.4cm, semi-double, chrysanthemum type; 2.1cm tall; petal (33-)36 (-40); stamens 38; filament 0.67cm long, white; turn to pink after pollination; anthers orange red; pollen viability 95.2% (Pawasut *et al.*, 2004); 2-3 pistil, shorter than stamens; medium flower density (flower bud 0.83/cm); calyx reddish, 2 round; filament and calyx petaloids; pedicle 0.73cm long; leaf blade green, elliptic-lanceolate, 8.1cm × 2.2cm, L/W ratio 3.68; leaf margin finely serrate; petiole 1.2cm long; 1-2 reniform glands; drupe green, conoid, 4.76cm × 3.56cm; stone 3.12cm × 1.6cm, elliptic, smooth on the surface.

Blooming period: Middle to the end of April (P. 84) .

28. 'Wanbai Tao'

Origin: First published in 1991 (Zhang and Chen).

Characteristics: Medium tree vigor, bark grey; twig glabrous, yellow-brown; flower bud oblate, petal ovate-oblong, 2.2cm long; white(155D); flower diameter 5.46cm; double, peony type, 2.23cm tall; petal (55-)63(-68); stamens 50.7, filament 1.17cm long, white, anthers yellow; pistil2-5, longer than stamens. medium flower density (flower bud 0.7/cm); calyx green, 2 round; filament and calyx petaloids; pedicle 0.9cm long; leaf blade green, elliptic-lanceolate, 13.9cm × 4.06cm, L/W ratio 3.42, leaf margin crenulate; petiole 0.95cm long, 2-3 reniform glands; drupe green, 3.62cm × 3.56cm, round; stone 2.52cm × 1.54cm, obovate, rough on the surface.

Blooming period: The end of April to early May. Late blooming cultivar (P. 85).

Blooming period is later than 'Baibitao'; peony flower type is also different from rose type of 'Baibitao'.

29. 'Fei Tao'

Origin: An ancient Chinese cultivar records dating back to 5 Dai eras (907-979).

Characteristics: Medium tree vigor, bark dark grey; twig green; flower bud oblate, petal ovate-oblong, 2.43 cm long, red (59D); flower diameter 5.63cm, double, peony type, round, 2.48cm tall; petal (58-)68.2(-83), fully doubled, 120 petals have been ever found; stamens 44, filament 1.22cm long, white, turn to red after pollination; anthers yellow; pistil 2-5, longer than stamens, sparse flower density (flower bud 0.55/cm); calyx reddish, 2 round; filament and calyx petaloids; pedicle 0.65cm long; leaf blade green, ovate-lanceolate, 12.86cm × 4.26cm, L/W ratio 3.02; leaf margin crenulate; petiole 0.97cm long; 3-4 reniform glands; drupe green, 5.44cm × 4.33cm, oval, stone 3.8cm × 1.7cm, elliptic; rough on the surface.

Blooming period: The end of April. Sometimes could last to early May (P. 85).

It is a red flower cultivar with green twig, which is the apparent characteristic to identify 'Feitao'.

30. 'Hong Bitao'

Origin: An antient Chinese cultivar.

Characteristics: Medium tree vigor, bark dark grey; twig grayish-brown; flower bud round; petal ovate-oblong, 2.43cm long, light red (66D); flower diameter 5.63cm; double, peony type, round, loose, 2.33cm tall; petal (45-)55.4(-64); stamens 32.8, filament 1.33cm long, white, turn to red after pollination; filament petaloids, anthers orange red; pistil 2-3, longer than stamens; medium flower density (flower bud 0.65/cm); calyx reddish, 2 round; filament and calyx apparently petaloids, pedicle 0.73cm long; leaf blade green, elliptic-lanceolate, 14.14cm × 3.76cm, L/W ratio 3.76; leaf margin crenulate; petiole 1.28cm long; 1-2 reniform glands; drupe green, 4.3cm × 3. 78cm, oval; stone 2.7cm × 1.75cm, obovate, rough on the surface.

Blooming period: Middle to the end of April (P. 87).

The flower color and type are apparent different from 'Hong Bitao' to 'Feitao', which the former is bright red and loose flower, whereas the latter is dark red with round flower.

31. 'Meizi'

Origin: A mutation found in Beijing Botanical Garden in 1998 (Hu and Zhang, 2000).

Characteristics: Medium tree vigor, bark dark grey; twig grayish-brown; flower bud round; petal ovate-oblong, 2.17 cm long, pink red (73B); flower diameter 4.93cm, double, semi-round, 1.7 cm tall; petal (54-)60.7(-65); stamens 86.67, filament 1.07cm long, white, extend to flower budding pink stage; anthers white; pistil 2-5, nearly the same length as stamens; medium flower density (flower bud 0.75/cm); calyx reddish, 2 round; filament and calyx apparently petaloids, pedicle 1.15cm long; leaf blade green, ovate-lanceolate, 12.74cm × 4.02cm, L/W ratio 3.17; leaf margin crenulate; petiole 0.8cm long; 2 reniform glands. No fruit through several years' observation.

Blooming period: Middle to the end of April (P. 87).

The unique flower color is the apparent character to identify 'Mei Zi'.

32. 'Bi Tao'

Origin: An ancient Chinese cultivar records dating back to 5 Dai eras (907-979).

Characteristics: Medium tree vigor, bark grayish-brown; twig purple with green hint; flower bud oval; petal ovate, 2.87cm long, light pink (62B); flower diameter 6.25cm; double, peony type, round, 2.10 cm tall; petal (42-)50.8(-63); stamens 32.67, filament 1.53cm long, white; anthers yellow; pistil 2-, nearly the same length as stamens; medium flower density (flower bud 0.8/cm); calyx green, 2 round; filament and calyx apparently petaloids; pedicle 0.75cm long; leaf blade green, elliptic-lanceolate, 13cm × 3.5cm, L/W ratio 3.72; leaf margin finely serrate; petiole 1.22cm long; 2-4 reniform glands; drupe green, 5.1cm ×

4.1cm, oval, stone 3.55cm × 1.9cm, obovate, rough on the surface.

Blooming period: Middle to the end of April (P. 88).

33. 'Zan Fen'

Origin: First published in 1993 (Zhang).

Characteristics: Spreading, bark grey, twig grayish-brown; flower bud oblate, petal ovate-oblong, 2.47cm long, purple pink (73C); flower diameter 5.63cm, double, peony type, semi-round, 2.03cm tall; petal (53-)62.67(-68); stamens 60.67, filament 1.2cm long, filament petaloids; anthers orange red; pistil 3-5, longer than stamens; medium flower density (flower bud 0.65/cm); calyx reddish, 2 round; filament and calyx apparently petaloids; pedicle 0.86cm long; leaf blade green, elliptic-lanceolate, 13.68cm × 3.8cm, L/W ratio 3.6; leaf margin finely serrate; petiole 1.06cm long; 1-3 reniform glands; drupe green, 4.65cm × 4.48cm, round; stone 3cm × 1.9cm, obovate, rough on the surface.

Blooming period: Middle to the end of April (P. 89).

34. 'Sahong Tao'

Origin: An ancient Chinese cultivar.

Characteristics: Medium tree vigor; spreading, bark grayish-brown; twig green with purple strips or spots; flower bud oblate; petal ovate-oblong, 3.02cm long, pink (66D), and red (59D) on white (155D) flower; flower diameter 6.08cm, double, peony type, round, 2.1cm tall; petal (58-)63.3(-66); stamens 35.3, filament 1.17cm long, anthers yellow; pistil 2-4, longer than stamens; medium flower density (flower bud 0.7/cm); calyx color related to petal color, while white flower with green calyx, bicolor flower with bicolor calyx; 2 round, filament and calyx apparently petaloids; pedicle 0.86cm long; leaf blade green, elliptic-lanceolate, 15.5cm × 3.95cm, L/W ratio 3.92; leaf margin crenulate; petiole 1.22cm long; 1-4 reniform glands; drupe green, 4.86cm × 4.12cm, conoid, stone 3.08cm × 1.84cm, elliptic, rough of the surface.

Blooming period: Middle to the end of April (P. 89).White is the leading color of the tree, but few pure white flowers, always with pink or red strips or dots.

35. 'Wubao Tao'

Origin: An ancient Chinese cultivar.

Characteristics: Vigorous; upright, bark grayish-brown; twig yellow-brown; flower bud oblate; petal ovate-oblong, 2.57cm long, pink (65D) with light pink (66D) and red (59D) strips in one flower; flower diameter 5.33cm, double, peony type, round, 2.1cm tall; petal (50-)59.2(-74); stamens 46.5, filament 1.11cm long; anthers orange red; pistil 4-6, nearly the same length as stamens; medium flower density (flower bud 0.65/cm); calyx reddish, 2 round, filament and calyx apparently petaloids; pedicle 0.57cm long; leaf blade thick and paperish; more green than other cultivars; ovate-lanceolate, 11.23cm × 3.78cm, L/W ratio 2.98; leaf margin coarsely serrate; petiole 1cm long; 2-3 reniform glands; drupe green, 5.44cm × 4.22cm, oval; stone 3.02cm × 1.62cm, obovate; rough on the surface.

Blooming period: The end of April, sometimes could last to early May (P. 90).

36. 'Er Qiao'

Origin: It was a mutation of 'Hong Bitao' found in Beijing Botanical Garden in 1997 (Zhang and Dai) (P. 91).

Characteristics: Medium tree vigor; upright; twig yellow-brown; flower bud oblate; petal ovate-oblong, 2.47cm long, bicolor on one flower, pink (65B) petal on red flower (63A); flower diameter 4.17cm, double, peony type, semi-round, 1.67 cm tall; petal (50-)53.67(-59); stamens 40.33, filament 1.2cm long, petaloids; anthers orange red; pistil 2-3, longer than stamens; medium flower density (flower bud 0.68/cm); calyx reddish, 2 round; pedicle 0.9cm long; leaf blade green, elliptic-lanceolate, 15cm × 3.67cm, L/W ratio 4.0; leaf margin crenulate; petiole 1.04cm; 2-4 reniform glands; drupe green, 5.32cm × 4.11cm, oval; stone 3cm × 1.68cm, obovate; rough on the surface.

Blooming period: The end of April, sometimes could last to early May (P. 91) .

Dwarf Group

Specimens in the dwarf group are, arborescent shrubs, small in stature with internodes usually shorter than 10mm. They commonly have a dense, compact shape and exhibit long narrow-leaves.

Records of dwarf peach varieties date back to the Ming Dynasty (960-1279). All dwarf peaches in western countries originate from China. The earliest examples of dwarf as a bonsai were found in 1911 in the USA (Hedrick, 1917). Pink, red, and white dwarf peaches were introduced from China by W.B. Clarke nursery (Jacobson, 1996). Cultivar names of the group are primarily adapted from the original Chinese names. A key to the dwarf group based on flower type, color, leaf and resistance to adversity is detailed below:

Table 5.2   Key to ornamental peach cultivars of Dwarf group

1. Short internodes, single flower form, green or red leaves
  2. White flowers, green leaves
    3. Fruit surface pubescent...'Danban Shoubai'
    3. No pubescent on fruit surface..'You Shoubai'
  2. Pink or red flowers, green or red leaves
    4. Pink or red flowers, green leaves
      5. Pink flowers.. 'Danban Shoufen'
      5. Red flowers.. 'Danban Shouhong'
    4. Pink flowers, red leaves
      6. No specific tolerance to disease.. 'Red Dwarf'
      6. Tolerant to bacterial spot .... 'Bonfire'
1. Short internodes, semi-double flower, green or red leaves
      7. White flowers
        8. Petal ovate, regular 3-4 layer flat petals .... 'Shoubai'
        8. Petal lanceolate-ovate, chrysanthemums like flower ....'Saiwai Howaito'
      7. Pink, red or variegated flowers
        9. Pink flowers.. 'Shoufen'
        9. Red or variegated flowers
          10. Red flower .... 'Shouhong'
          10. Variegated flowers
            11. Pink and white flowers ....'Xiayu Shouxing'
            11. Pink and red flowers ....'Erqiao Shouxing'

37. 'Danban Shoubai'

Origin: Unknown.

Characteristics: Dwarf, bark grey, twigs green; flower bud oval, petal oval, 1.5cm long, red (155D); flower diameter 3.4cm, single, petal 5-6; stamens 32, filament 1.2cm long, white; anthers yellow; pistil longer than stamens; dense flower (flower bud 1.7/cm); calyx green, 5; pedicle 1.4cm long; leaf blade green, elliptic-lanceolate, 14.16cm × 2.66cm, L/W ratio 5.32; leaf margin finely serrate, wavy; petiole 0.88cm long; 2-3 reniform glands; drupe, green, pubescent, 4.42cm × 3.7cm, oval; stone 2.88cm × 1.88cm, elliptic, rough on the surface.

Blooming period: Middle of April (P. 93).

38. 'You Shoubai'

Origin: Mutation from 'Shoubai' in Beijing Botanical Garden in 1997 (Zhang and Dai, 1997).

Characteristics: The distinguish difference from 'Danban Shoubai' is that 'You Shoubai' is nectarine. Fruit is about 3.88cm long, 3.44cm wide, smaller than 'Danban Shoubai'; mostly pit on the stone of 'You Shoubai' is different from pit and furrow of 'Danban Shubai' (P. 93).

39. 'Danban Shoufen'

Origin: Unknown.

Characteristics: Dwarf, bark grey, twig reddish brown; flower bud oval, petal oval, 2.23cm long, pink (65B), flower diameter 4.57cm, single, petal 5-6; stamens 41.3, filament 1.47cm long, whitish pink, anthers orange red; pistil nearly the same length as stamens; medium flower density (flower bud 1.8/cm); calyx reddish brown, 5; pedicle 0.43cm long; leaf blade green, elliptic-lanceolate, 15.88cm × 2.9cm, L/W ratio 5.47; leaf margin finely serrate, wavy; petiole 0.92cm long; drupe green, 4.2cm × 4.14cm, round; stone 2.5cm × 1.74cm, elliptic, smooth on the surface.

Blooming period: Middle of April (P. 94).

40. 'Danban Shouhong'

Origin: Unknown.

Characteristics: Dwarf, bark grey, twig reddish brown; flower bud oval, petal ovate, 2.1cm long, red (59D); flower diameter 4.23cm, single, petal 5-6, stamens 32.7, filament 1.17cm long, white to light red, flower heart white; anthers orange red; anthers orange red; pistil nearly the same length as stamens; medium flower density (flower bud 1.6/cm); calyx reddish brown, 5; pedicle 0.4cm long; leaf blade green, elliptic-lanceolate, 14.2cm × 2.8cm, L/W ratio 5.14; leaf margin finely serrate, wavy; petiole 0.88cm long; 2-3 reniform glands; drupe green.

Blooming period: Middle of April (P. 95).

41. 'Red Dwarf'

Origin: It was a red-leaved dwarf cultivar, released by Yoshida and Seike (1974) in Japan. This plant originated from F_2 of a red-leaved, upright growth type and a green leaved dwarf type peach.

Characteristics: Dwarf, bark grey, twig reddish brown; internodes length 0.53cm; flower bud narrow ovate; petal ovate-oblong, 2.cm long; pink (69C); flower diameter 4.13cm, single, 5; stamens 40.7, filament 1.17cm long, anthers orange red; pistil longer than stamens; dense flower (flower bud 1.9/cm); calyx reddish brown, 5; pedicle 0.33cm long; leaf blade dark red, elliptic-lanceolate, 19.8cm × 3.7cm, L/W ratio 5.35; leaf margin finely serrate, wavy; petiole 0.84cm long; drupe green with red hint, 4.2cm × 3.6cm, oval; stone 2.8cm × 2.1cm, oval, rough on the surface.

Blooming period: Middle of April (P. 95).

42. 'Bonfire': Tom Thumb™

Origin: The cultivar 'Bonfire' was a selection from seedling population of F₂ of 'Tsukuba No.2' open-pollination (Moore et al. 1993). Untied States Patent: 8509, for the name of TOM THUMB.

Characteristics: Dwarf, bark glabrous, dense lenticels; internodes length 8mm; large flower, pink (62D), single, 5; flower diameter 3.5cm; leaf blade dark red (187A), elliptic-lanceolate, 18.45cm × 3.36cm, L/W ratio 5.49; 2-4 reniform glands; petiole 1.22cm long; drupe small, greenish-yellow, pubescent, elliptic; stone 3.1cm × 2.51cm, deeply furrowed surface cold-hardy to -23°C.

Blooming period: Blooming at mid-March in Arkansas, full bloom in early April (P. 96).

The fruit of 'Bonfire' is purple than 'Red Dwarf'. The leaves have good resistance to bacterial leaf spots.

43. 'Shoubai'

Origin: Unknown.

Characteristics: Dwarf, bark rough, dark grey; twig green; flower bud round, petal oval, 2.17 cm long, 1.8cm broad; white (155D); flower diameter 4.6 cm, semi-double, petal (25-) 25.2 (-29); stamens 51.8, filament 1.05cm long, white, anthers yellow; pistil longer than stamens; dense flower(flower bud 1.91/cm); calyx green, 2 round; filament and calyx apparently petaloids; pedicle 0.77cm long; leaf blade green, elliptic-lanceolate, 14.93cm × 2.9cm, L/W ratio 5.2; leaf margin finely serrate, wavy; petiole1.43cm; 3-4 reniform glands; drupe green, 3.84cm × 3.68cm, round; stone 2.7cm × 1.94cm; elliptic, rough on the surface.

Blooming period: Middle to the end of April (P. 97).

44. 'Saiwai Howaito'

Origin: It is the selected seedling from a cross between 'Kikumomo' and double white dwarf (Maeda *et al.*, 2003).

Characteristics: Dwarf, very small size; internodes very short; medium tree vigor; thick twigs, green, petal ovate-lancolate; white; semi-double, chrysanthemum-like flowers; filament petaloids.

Blooming period: Early April in Tokyo (P. 97).

45. 'Shoufen'

Origin: Unknown.

Characteristics: Dwarf, bark dark grey; twig reddish green; flower bud oval, petal oval, 1.98cm long, 1.67cm broad, pink (65B); flower diameter 4.5cm, semi-double, petal (22-) 23.8 (-25); stamens 43.8, filament 1.18cm long, white, anthers orange red; pistil longer than stamens; dense flower (flower bud 1.47/cm); calyx reddish green, 2 round; filament and calyx petaloids; pedicle 0.57cm long; leaf blade green, elliptic-lanceolate, 15.6cm × 2.69cm, L/W ratio 5.2; leaf margin finely serrate, wavy; petiole1.1cm; 2 reniform glands; drupe green, 4.15cm × 4.15cm, round; stone 2.42cm × 1.47cm, oval, rough on the surface.

Blooming period: Middle of April (P. 98).

46. 'Shouhong'

Origin: Unknown.

Characteristics: Dwarf, bark rough, grayish black, twig yellow brown; flower bud round, petal oval, 2.07cm long, pink (59D); flower diameter 4.33cm, semi-double, (16-) 18.3 (-29); stamens 32.7, filament

1.3cm long, light red, flower heart white; anthers yellow; pistil nearly the same length as stamen; dense flower (flower bud 1.57/cm); calyx reddish brown, 2 round; filament and calyx petaloids; pedicle 0.27cm long; leaf blade green, elliptic-lanceolate, 16.5cm × 3.1cm, L/W ratio 5.3; leaf margin finely serrate, undulate; petiole 0.98cm long; 1-3 reniform glands; drupe green, 4.22cm × 3.92cm, oval; stone 2.74cm × 1.94cm, elliptic, rough on the surface.

Blooming period: Middle to the end of April (P. 99).

47. 'Xiayu Shouxing'

Origin: First published in 1993 (Zhang).

Characteristics: Dwarf, bark dark grey; twig green or yellow-brown; flower bud ovate, petal ovate, 2.22 cm long; white flower (155C) with pink (62B) strips or dots; flower diameter 4.65cm, semi-double, petal (23-) 23.5 (-24); stamens 54, filament 1.53cm long, anthers color related to petal, yellow anther on pure white flower or white flower with strips or dots, pure pink flower with orange red anther; pistil nearly the same length as stamens; dense flower (flower bud 1.8/cm); calyx color related to petal color, while white flower with green calyx, bicolor flower with bicolor calyx, 2 round; filament and calyx apparently petaloids, pedicle 0.3cm long; leaf blade green, elliptic-lanceolate, 15.8cm × 2.5cm, L/W ratio 6.3; leaf margin finely serrate; petiole1.28cm; 2-3 reniform glands; drupe green, 4.78cm × 4.16cm, oval; stone 2.84cm × 1.86cm, elliptic, rough on the surface.

Blooming period: Middle to the end of April (P.100).

48. 'Erqiao Shouxing'

Origin: It was a mutation found in Beijing Botanical Garden in 1997 (Zhang and Dai, 1997).

Characteristics: Dwarf, bark rough, grayish black; twig green or yellow-brown; flower bud narrow-ovate, petal ovate-oblong, 2.12cm long, 1.8cm wide; pink (63D) and red (59D) in one or different flowers; flower diameter 4.13cm, semi-double, petal (12-)16.2 (-19); stamens 28, filament 1.6cm long; filament petaloids, anthers yellow; pistil longer than stamens; dense flower (flower bud 1.8/cm); calyx reddish brown, 2 round; pedicle 0.77cm long; leaf blade green, elliptic-lanceolate, 14.83cm × 2.57cm, L/W ratio 5.77; leaf margin crenulate, wavy; petiole 0.92cm long; 4 reniform glands.

Blooming period: Middle to the end of April (P. 101).

Pillar Group

Pillar group specimen exhibit upright branches with a narrow crotch angle as a typical characteristic. They have a narrow crown and pillar shape that enables cultivars in this group to be attractive ornamental plants for landscape use, not only in the garden, but also for highway, driveway, avenues and street tree planting.

Kadan-chikin-shio, written by Ihee Ito (1656-1739), described a pillar type peach, named 'Houkimomo'. This is one of Japan's greatest contributions to the world. Most 'modern' pillar type cultivars are descendents of this ancient cultivar.

A key to the pillar group based on flower type, color, leaf and resistance to adversity is detailed below:

Table 5.3 Key to ornamental peach cultivars of Pillar group

1. Upright twig, narrow crotch angle, variegated flowers .... 'Houki momo'
1. Upright twig, narrow crotch angle, single flower color
  2. White flowers, green leaves.. 'Teruteshiro'
  2. Pink or red flowers, green or red leaves
    3. Pink or red flowers, green leaves
      4. Pink flowers
        5. Light pink flowers (62C), petal 25-32 .... 'Terutehime'
        5. Pink flowers (73B), petal 18-24 .... 'Terutemomo'
      4. Red flowers.. 'Terutebeni'
    3. Pink flowers, red leaves
      6. Light pink flowers, red leaves .... 'Corinthian Pink'
      6. Dark rose pink flowers, dark purple leaves .... 'Corinthian Rose'

49. 'Houkimomo' (Pillar、Pyramid、Fastigiata)

Origin: An ancient cultivar from Japan (Yoshida *et al.*, 2000), which was first recorded in Edo Era (17th to 19th) (Ito, 1695).

Characteristics: Pillar type, bark yellow-brown, twigs in narrow branch angle, grayish-yellow; flower bud ovate, petal oval, 2.07cm long, 1.6cm wide; pink (65A) petals on white (155D) flower, or white and pink flower in one branch, mostly white flower; flower diameter 4.5cm, semi-double, petal (28-) 29 (-30), stamens 35.7, filament 1.1cm long, anthers color related to flower color, yellow anther on pure white flower or bicolor flower, pure pink flower with orange red anther; pistil nearly the same length as stamens; medium flower density (flower bud 0.8/cm); calyx color related to petal color, while white flower with green calyx, bicolor flower with bicolor calyx, 2 round; filament and calyx apparently petaloids; pedicle 0.83cmlong; leaf blade green, elliptic-lanceolate, 12.6cm × 3.5cm, L/W ratio 3.6; leaf margin crenulate, petiole 0.9cm long; drupe green; stone elliptic, smooth on the surface.

Blooming period: Middle of April (P. 103).

Houkimomo originated from Japan. Although there are different names in different countries for this cultivar, here the suggested name is still using the original name 'Houkimomo', instead of former "Pillar", "Columnar", "Pyramid" or "Fastigiata", etc.

50. 'Terutemomo' ('Corinthian Mauve')

Origin: It is a selection of F_2 generation of 'Houkimomo' and 'Akashidara' (Yamazaki *et al.*, 1987).

Characteristics: Pillar type, bark yellow-brown, twigs in narrow branch angle, yellow-brown; flower bud round, petal ovate-oblong, 2cm long, peach pink (73B); flower diameter 4.7cm, semi-double, petal (18-) 22.1 (-24), stamens 43, filament 1cm long, white, filament petaloids, anthers orange red; pistil nearly the same length as stamens; medium flower density (flower bud 0.8/cm); calyx reddish green, 2 round; pedicle 0.67cm long; leaf blade green, elliptic-lanceolate, 12.1cm × 3.5cm, L/W ratio 3.5; leaf margin crenulate; sometimes 2-5, usually 3 reniform glands; petiole 0.9cm long, drupe green, 3.62cm × 3.58cm, round; stone 2.36cm × 1.64cm, elliptic, smooth on the surface.

Blooming period: Middle of April (P. 104、105).

'Corinthian Mauve' was the selection from F_2 generation of 'NC174RL' nectarine and a selection of Japanese 'Pillar' peach (Werner *et al.*, 2000a). AFLP data showed that 'Corinthian Mauve' and 'Terutemomo' have little genetic difference and share similar morphological characteristics (Hu *et al.*, 2005). According to the Article10.1 of ICNCP (Brickell *et al.*, 2004), we suggest that 'Corinthian Mauve' should be combined and the accepted name is 'Terutemomo' because it was established earlier. 'Corinthian Mauve' could be used as the same effect as 'Terutemomo' in the North America.

51. 'Terutehime'

Origin: It is an open pollinated seedling selected from 'Houkimomo' (Horikoshi *et al.*, 1992).

Characteristics: Pillar type, bark yellow-brown, twigs in narrow branch angle, grayish-yellow; flower bud oval; petal ovate-oblong, 2.1cm long, 1.4cm wide; light pink (62C); flower diameter 4.7cm, semi-double, petal (25-)28.6(-32), stamens 49.7, filament 1.17cm long, white, anthers orange red; pistil nearly the same length as stamens; medium flower density (flower bud 0.8/cm); calyx reddish green, 2 round; filament and calyx petaloids, pedicle long 0.63cm。 Leaf blade green, should be elliptic-lanceolate, 12.3cm × 3.4cm, L/W ratio 3.6; leaf margin finely serrate to crenulate, petiole 0.9cm long; drupe green, 3.92cm × 3.78cm, oval; stone 2.54cm × 1.66cm, elliptic, rough on the surface.

Blooming period: Middle of April (P. 105、106).

52. 'Terutebeni'

Origin: It is a selection of F_2 generation of 'Akashidara' and 'Houkimomo' (Yamazaki *et al.*, 1987).

Characteristics: Pillar type, bark yellow-brown, twigs in narrow branch angle, grayish-yellow; flower bud ovate; petal ovate, 1.9cm long, 1.7cm broad; red (59D); flower diameter 4.6cm, semi-double, petal (17-) 21.5 (-24); stamens 43.5, filament 1.17cm long, red, anthers orange red; pistil nearly the same length as stamens; medium flower density (flower bud 0.65/cm); calyx reddish green, 2 round; filament and calyx apparently petaloids; pedicle 0.75cm long; leaf blade green, elliptic-lanceolate, 12.3cm × 3.4cm, L/W ratio 3.6; leaf margin crenulate, petiole 0.9cm long; drupe green, 3.76cm × 3.66cm, round; stone 2.52cm × 1.76cm, elliptic, rough on the surface.

Blooming period: Middle of April (P. 107).

53. 'Teruteshiro' ('Corinthian White')

Origin: It is a selection of F_2 generation of 'Houkimomo' and 'Zansetsushidare' (Yamazaki *et al.*, 1987).

Characteristics: Pillar type, bark yellow-brown, twigs in narrow branch angle, grayish-yellow; flower bud ovate, petal round, 1.8cm long, 1.67cm wide; white (155D); flower diameter 4.5cm, semi-double, petal (31-)34 (-37), stamens 45, filament 1.1cm long, white; anthers yellow; pistil longer than stamens; medium flower density (flower bud 0.6/cm); calyx green, 2 round; filament and calyx apparently petaloids; pedicle 0.67cm long; leaf blade green, elliptic-lanceolate, 13.8cm × 3.7cm, L/W ratio 3.7; leaf margin crenulate, petiole 0.84cm long; drupe green, 4.22cm × 3.47cm, oval; stone 2.42cm × 1.48cm, obovate, smooth on the surface.

Blooming period: Middle of April (P. 108).

'Corinthian White' was the selection from F_2 generation of 'NC174RL' nectarine and a selection of Japanese 'Pillar' peach (Werner *et al.*, 2000b). AFLP data showed that 'Corinthian White' and 'Teruteshiro' have little genetic difference and share similar morphological characteristics (Hu *et al.*, 2005). According to the Article10.1 of ICNCP (Brickell *et al.*, 2004), we suggest that 'Corinthian White' should be combined and the accepted name is 'Teruteshiro' since it was established earlier. 'Corinthian White' could be used as the same effect as 'Teruteshiro' in the North America.

54. 'Corinthian Pink'

Origin: It is the selection from F_2 generation of 'NC174RL' nectarine and a selection of Japanese 'Pillar' peach (Werner *et al.*, 2001). Patent number in the US: PP11902.

Characteristics: Pillar type, large size; trunk medium to rough, gray-green (197-B); twigs in narrow branch angle about 5°-20°;vigorous, annual growth 90-120cm; 4 years-old tree about 4.27m tall, 1.37mbroad; bark medium rough; twigs red on exposed side; flower pink, from early time (56D) to mature (56A); double, petal 31.4; flower diameter 3.8cm; stamens numerous, anthers bright yellow; pistil longer than stamens; calyx light brown (183-A); pedicle 1.4cm long; leaf blade purple, from young leaf (183A) to mature (178B), a little change; leaf blade, 14.1 cm × 4.8 cm, elliptic-lanceolate, L/W ratio 2.93; leaf margin crenulate, leaf apex acute; average 3 reniform glands, varies from 2 to 5; petiole 1.19cm long; drupe pubescent, green; small, tolerant to bacterial leaf spot, susceptible to peach borer; chilling requirement 950h; ovary fertile, low fruit set, less than 1% flower could get fruit set; drupe purple, 4.05 cm × 3.75cm, elliptic; stone 2.13cm × 1.78cm, obovate, rough on the surface.

Blooming period: Date of first bloom in North Carolina is March 15 to March 30. Date of full bloom. --March 25 to April 10. Varies yearly due to weather conditions. Bloom duration is typically 10-14 days, and individual flowers last about 7-10 days, depending on temperature during bloom (P. 109).

55. 'Corinthian Rose'

Origin: It is the selection from F_2 generation of 'NC174RL' nectarine and a selection of Japanese 'Pillar' peach (Werner *et al.*, 2000c). Patent number in the US: PP11564.

Characteristics: Pillar type, medium tree vigor; rapid growth, annual growth 90-120cm; 4 years-old tree could reach 4.3-4.88 m tall, branch smooth (new) to medium rough (old), one-year-old shoots gray-red (178-B), two-year-old shoots gray-brown (199-A) ; twigs in narrow branch angle about 5-20°; flower bud ovate, large, flower from pink (56C) in early time to red in mature (68C); double, petal 27.4; flower diameter 3.8cm; stamens numerous, anthers bright yellow; 1 pistil longer than stamens; calyx light brown (183A), pedicle 1.5cm long; leaf blade purple, large, 11.1cm × 3.8cm, L/W ratio 2.92, ovate-lanceolate; 3 reniform glands; drupe reddish green, small, <3cm, pubescent, elliptic.

Blooming period: Blooming period in North Carolina is the mid-March, Blooming period could last about 14 days (from 25 March to 10 April). The individual Blooming period could last 7-10d. Chilling requirement 950h (P. 111).

The difference from 'Corinthian Pink' is darker flower leaf color and lighter fruit color. It is also tolerant to bacterial leaf spot.

Weeping group

The pendulous twigs of this type give it an attractive umbrella like form. This form makes a delicate ornamental statement, making cultivars of this group an attractive choice for tranquil ornamental planting.

Based on flower type, color, and leaf color, the key to weeping group is as follows.

Table 5.4   Key to ornamental peach cultivars of Weeping group

1. Pendulant twigs, single or semi-double, green leaves
 2. Single flowers
   3. White flowers
    4. Fruit pubescent .... 'Shiroshidare'
    4. Fruit without pubescent, nectarine ....'White Glory'
   3. Pink, red, or variegated flowers
    5. Pink flowers.. 'Danfen Chuizhi'
    5. Pink-red, red, or variegated flowers
     6. Red flowers... 'Akashidare'
     6. Variegated flowers
      7. Mostly white flowers with sparse pink flowers or pink strips on white flowers ....
      ....'Genpeishidare'
      7. Variegated of white and pink, no pure color in any flowers......................... 'Sachishidare'
 2. Semi-double flowers
    8. White flowers
     9. Petal less than 25 .... 'Lü E Chuizhi'
     9. Petal about 28-42 .... 'Zansetsushidare'
    8. Red, pink, or variegated flowers
     10. Red flowers
      11. Petal less than 15 .... 'Hongyu Chuizhi'
      11. Petal about 18-30... 'Sagamishidare'
     10. Pink or variegated flowers
      12. Variegated flowers with white and pink ....'Yuanyang Chuizhi'
      12. Pure pink flowers with light or dark hue
       13. Pure light pink flowers without mix color................................... 'Daiyu Chuizhi'
       13. Different pink flowers on the branch
        14. Light pink with few pink flowers ....'Wubao Chuizhi'
        14. Rose pink with few light pink flowers...................................'Hagoromoshidare'
1. Pendulant twigs, semi-double flowers, red leaves
     15. Pink flowers... 'Pink Cascade'
     15. Red flowers .... 'Crimson Cascade'

56. 'Shiroshidare'('Danbai Chuizhi')

Origin: It was an open-pollinating seedling selected from 'Akashidare' population (Yoshida *et al.*, 2000).

Characteristics: Medium tree vigor; twigs pendulous, green; flower bud ovate-oblong, petal ovate, 1.97cm long; white (155D); flower diameter 3.6cm, single, petal 5-10; stamens 34, filament 1.63cm long; anthers yellow; pistil 1, nearly the same length as stamens; medium flower density (flower bud 0.78/cm); pedicle 0.57cm long; calyx 5, green; leaf blade green, elliptic-lanceolate, 13.7cm × 3.6cm, L/W ratio 3.8;

leaf margin crenulate, petiole 0.9cm long; drupe green, 3.3cm × 3.02cm, round; stone 2.28cm × 1.68cm, elliptic, smooth on the surface.

Blooming period: Middle of April (P. 113).

57. 'White Glory'

Origin: It resulted from open-pollination of S-37 peach and released by D.J. Werner (Werner *et al.*, 1985) in the North Carolina State University Arboretum in Raleigh, N.C. It has been registered in CRUWO ("Cultivar Registration of Unassigned Woody Ornamentals").

Characteristics: Vigorous; twigs pedulous, green, pure white flower, single, petal mostly 5, 5% semi-double, petal 1.5-1.9cm long, 1.2-1.8cm wide; filament white, 1.0-1.8cm long, stamens 41-45, petaloids, anthers yellow; style 1.4-1.6cm long, 20% to 30% of flowers do not form pistils; calyx green; pedicle 1cm long; leaf blade green, elliptic-lanceolate to ovate-lanceolate, (13-)17.2 (-20)cm long, (3.5-) 4.3 (-5) cm wide; L/W ratio 4; 2 reniform glands; petiole 1-1.5cm long; drupe glabrous, subglobose, about 3-4cm diameter; have survived —18°C with no injury; chilling requirement 900h (P. 114).

58. 'Danfen Chuizhi'

Origin: Unknown.

Characteristics: Medium tree vigor; twigs pendulous, reddish brown; flower bud ovate-oblong; petal oval, pink (65D), single, petal 5; flower diameter 3.7cm, filament white, anthers orange red; medium flower density; calyxes reddish brown, 5; Leaf blade green, elliptic-lanceolate, 11.8cm × 3.0cm, L/W ratio 3.9; leaf margin finely serrate, petiole 1.1cm long.

Blooming period: Middle of April (P. 114).

59. 'Akashidare' ('Danhong Chuizhi')

Origin: Unknown.

Characteristics: Medium tree vigor; twigs pendulous, reddish brown; flower bud ovate-oblong; petal oval, 1.53cm long, red (68B); flower diameter 3.9cm, single, petal 5; stamens 38; filament 1.27cm long, anthers reddish-brown; pistil 1, nearly the same length as stamens; medium flower density (flower bud 1.5/cm); pedicle 0.35cm long; calyx reddish brown, 5; leaf blade green, elliptic-lanceolate, 11.3cm × 3.1cm, L/W ratio 3.6; leaf margin crenulate, petiole 0.8cm long; drupe green, 3.2cm × 3.2cm, round; stone 2.32cm × 1.64cm, elliptic, rough on the surface.

Blooming period: Middle of April (P. 115).

60. 'Genpeishidare'

Origin: An ancient Japanese cultivar since Edo era (17th-19th) (Ito, 1695).

Characteristics: Medium tree vigor; twigs pendulous, flower bud ovate-oblong; petal ovate-oblong, 1.9cm long; flower white(155D) and pink (65B); flower diameter 3.5cm, single, petal 5-8; stamens 35, filament 1.27cm long; anthers yellow; pistil 1, nearly the same length as stamens; medium flower density (flower bud 0.7/cm); pedicle 0.33cm long; calyx 5, the color varying the petal color; filament and calyx petaloids; leaf blade green, elliptic-lanceolate, 13.6cm × 3.5cm, L/W ratio 3.8; leaf margin finely serrate; petiole 0.9cm long; drupe green.

Blooming period: Middle of April (P. 115).

61. 'Sachishidare'

Origin: Unknown. Japanese cultivar (Yoshida *et al.*, 2000).

Characteristics: Medium tree vigor; twigs pendulous, green; single flower, bicolor, pink strips and dots on white flower, no pure white flowers (P. 116).

62. 'Lü E Chuizhi'

Origin: Unknown.

Characteristics: Medium tree vigor; twigs pendulous, green; flower bud oval; petal ovate, 2.7cm long; white (155D); flower diameter 4.5 cm; semi-double, petal 22-24; stamens 39, filament 1.3cm long; filament petaloids, anthers yellow; pistil 2, longer than stamens; medium flower density (flower bud 0.65/cm); pedicle 0.56cm long; calyx green, 2 round; leaf blade green, elliptic-lanceolate, 13.2cm × 3.7cm, L/W ratio 3.6; leaf margin finely serrate; petiole 1.0cm long; drupe green, 4.3cm × 3.5cm, oval; stone 2.75cm × 1.83cm, elliptic; rough on the surface.

Blooming period: Middle of April (P. 116).

63. 'Zansetshidare'

Origin: An ancient Japanese cultivar since Edo era (17th-19th) (Ito. 1695).

Characteristics: The major difference from 'Lü E Chuizhi' is more petals (28-42) and ovate-lanceolate leaf, pollen viability 98.7% (Pawasut *et al.*, 2004).

64. 'Hongyu Chuizhi'

Origin: Unknown.

Characteristics: Medium tree vigor; twigs pendulous, reddish brown; flower bud oval; petal oval, 1.53cm long; pink red (63B); flower diameter 4.3cm, semi-double, petal 12-16, stamens 40, filament 1.1cm long, anthers orange red; pistil longer than stamens; medium flower density (flower bud 0.65/cm); calyx reddish-brown, 2 round, filament and calyx apparently petaloids; pedicle 0.57cm long; leaf blade green, elliptic-lanceolate, 12.8cm × 3.6cm, L/W ratio 3.5; leaf margin crenulate, petiole 0.8cm long; drupe green, 3.68cm × 3.44cm, round; stone 2.42cm × 1.47cm, elliptic, smooth on the surface.

Blooming period: Blooming began in 10th April (P. 117).

65. 'Sagamishidare'

Origin: An ancient Japanese cultivar since Edo era (17th-19th) (Ito, 1695).

Characteristics: The major difference from 'Hongyu Chuizhi' is more petals (18-30) and bigger size of flower (flower diameter 4.7cm, with 2.1cm long, 1.6cm wide); pollen viability 97.7%; leaf ovate-lanceolate (Pawasut *et al.*, 2004).

66. 'Yuanyang Chuizhi'

Origin: First named in 1991(Zhang and Chen).

Characteristics: Medium tree vigor; bark grey, twigs pendulous, yellow-brown; flower bud oval, petal ovate, 1.8cm long, pink (69B) and white (155C) flower in one branch; flower diameter 4.2cm, semi-double, petal 19-22; stamens 35, filament 1.53cm long, anthers color related to petal, white flower with strips or dots, pure pink flower with orange red anther; pistil nearly the same length as stamens; medium flower density (flower bud 0.75/cm); calyx color related to petal color, while white flower with green calyx, bicolor flower with bicolor calyx, 2 round; filament and calyx apparently petaloids; pedicle 0.6cm long; leaf blade green, elliptic-lanceolate, 12.1cm × 3.2cm, L/W ratio 3.8; leaf margin finely serrate; petiole 0.98cm long; drupe green, 3.68cm × 3.04cm, elliptic; stone 2.54cm × 1.64cm, elliptic, rough on the surface.

Blooming period: The end of April (P. 118).

67. 'Daiyu Chuizhi'

Origin: A bud mutation found in Beijing Botanical Garden in 1995 (Zhang and Dai, 1997).

Characteristics: Medium tree vigor; twigs pendulous, reddish brown; flower bud oval; petal ovate, 1.7cm long; light pink (69C), almost white; flower diameter 4.1cm, semi-double, petal 21; stamens 35, filament 1.23cm long, filament petaloids, anthers yellow; pistil longer than stamens; medium flower density (flower bud 0.65/cm); calyx green, 2 round; pedicle 0.63cm long; leaf blade green, elliptic-lanceolate, 13.9cm × 3.4cm, L/W ratio 4.1; leaf margin finely serrate; petiole 0.9cm long; drupe green, 3.46cm × 3.14cm, oval; stone 2.4cm × 1.48cm, elliptic, rough on the surface.

Blooming period: Mid to end of April (P. 119) .

68. 'Wubao Chuizhi'

Origin: Unknown.

Characteristics: Medium tree vigor; twigs pendulous, green; flower bud round, petal ovate, 1.9cm long; light pink (65D) and pink (65B) flowers in one branch, some with purple hues (62B); flower diameter 4.2cm, semi-double, petal 18-23; stamens 33, filament 1.4cm long; anthers color related to petal, yellow anther on light pink flower, pink or pink with purple hue flower with orange red anther; pistil longer than stamens; medium flower density (flower bud 0.65/cm); calyx green and reddish, 2 round; filament and calyx apparently petaloids; pedicle 0.68cm long; leaf blade green, elliptic-lanceolate, 14.5cm × 3.6cm, L/W ratio 4.1; leaf margin crenulate, petiole 0.97cm long; drupe green, 3.96cm × 3.36cm, elliptic; stone 2.72cm × 1.72cm, elliptic, rough on the surface.

Blooming period: Middle to the end of April (P. 120).

69. 'Hagoromoshidare'('Zhufen Chuizhi')

Origin: An ancient Japanese cultivar since Edo era (17th-19th) (Ito, 1695).

Characteristics: Flower diameter 4.7cm petal 2cm long, 2cm wide; leaf 13.1cm × 4.2cm, L/W ratio 3.1; petiole 1.2cm long, pollen viability 97.9% (Pawasut *et al.*, 2004).

The major characteristics of 'Zhufen Chuizhi', which published in 1991(Zhang), is as below. Twigs pendulous, bark grayish black, reddish green; flower bud oval, petal ovate, 2cm long; mostly pink red flower (64D), sparsely light pink (62D); flower diameter 4.0cm, semi-double, petal 22-29; stamens 35, filament 1.27cm long, filament petaloids, anthers orange red; pistil longer than stamens; medium flower density (flower bud 0.75/cm); calyx reddish green, 2 round; pedicle 1.53cm long; leaf blade green, elliptic-lanceolate, 13.6cm × 3.7cm, L/W ratio 3.6; leaf margin finely serrate to crenulate, petiole 1.2cm long;; drupe green, 4.2cm × 3.56cm, oval; stone 2.73cm × 1.78cm, elliptic; rough on the surface.

Blooming period: Middle to the end of April (P. 121).

There is no significant difference from these two cultivars. Since 'Zhufen Chuizhi' is published later, accordin to the Article 10.1 of *International Code of Nomenclature for Cultivated Plants* (ICNCP) (Brickell *et al.*, 2004), we suggest that 'Zhufen Chuizhi' should be combined and the accepted name is 'Hagoromoshidare'.

70. 'Pink Cascade'

Origin: It resulted from crossing an unknown named green-leaf, weeping, double flower, New Jersey peach selection with 'Rutgers Red leaf'. The F_2 of this cross was self-pollinated in 1978. This weeping cultivar was selected from a population of 70 F_2 seedlings in 1981. It was released in 1992 by J.N. Moore, R.C. Rom, S.A. Brown, and .L. Kingaman (Moore *et al.*, 1993) (P. 122).

Characteristics: Extreme double flower, pink, leaves red fading to greenish-red by midsummer, small fruit.

71. 'Crimson Cascade'

Origin: The same origin as 'Pink Cascade' (Moore *et al.*, 1993).

Characteristics: Weeping growth habit, vigorous; rapid growth habit, 4 years-old tree 2.5m broad, 8 m tall; twigs pendulous, reddish, flower bud oval, petal oval, 1.63cm long; red (65B); flower diameter 4.0cm, semi-double, 22-29; stamens 35, filament 1.27cm long, anthers orange red, filament petaloids, pistil longer than stamens; medium flower density (flower bud 0.55/cm); calyx reddish brown; pedicle 0.63cm long; leaf blade purple, elliptic-lanceolate, 12.2cm × 3.4cm, L/W ratio 3.6; leaf margin finely serrate; petiole 0.85cm; drupe green, with reddish hue; 3.27cm × 3.23cm, round; stone 2.33cm × 1.73cm, elliptic, smooth in the surface, most strips, fewer pit (P. 122).

Twisted Group

The twisted group presents a unique growth habit with twisted branches. Although there is currently only one cultivar that exhibits single pink flowers, cultivars with semi-double and double flowers with red, white, or even bicolor palettes are anticipated.

72. 'Unriumomo'

Origin: It was selected from a mutation of nectarine seedling (Yoshida *et al.*, 2000).

Characteristics: Upright, medium size; bark grey-brown, twigs green; flower bud ovate-oblong, petal 5, oval, light pink (65D), 1.9cm long, flower diameter 4.0 cm, single form, stamens 42.3, filament 1.33cm long, anthers orange red, pistil shorter than stamens; medium flower density (flower bud 0.65/cm), pedicle 0.57cm long; calyx 5, reddish brown; leaf blade green, elliptic-lanceolate, curled, 10.6cm × 2.8cm, L/W ratio 3.8; leaf margin finely serrate to coarsely serrate, petiole 1.1cm long; nectarine, oval, green with red on exposed side, 2.9cm long, 2.6cm wide; stone 2.5cm long, 1.7cm wide, elliptic, surface of stone smooth.

Blooming period: Middle of April (P. 123).

Although the single pink flower of 'Unriumomo' is not as striking as other ornamental peach cultivars, the interesting twisted branch illustrates a novel variety with potential for promising new cultivars.

David Group

The David group is hybrids of *P. persica* and *P. davidiana*, which possess characteristics inherited from their parents. The early flower and smooth bark are related to *P. davidiana*. Undeveloped pistil is a common characteristic of cultivars in this group.

The key of this group based on flowers is as below.

Table 5.5   Key to cultivars in David group

1. White double flower with abortion pistil, no fruiting, green twig
   .... 'Baihua Shanbitao'
1. Pink or rose flowers, reddish twigs, normal developed or slightly reduced pistil
   2. Light pink flowers .... 'Fenhua Shanbitao'
   2. Rose pink flowers .... 'Fenhong Shanbitao'

73. 'Baihua Shanbitao'

Origin: Supposed natural hybrid between *P. davidiana* and *P. persica*.

Characteristics: Large tree size, spreading; bark smooth, grey to red-brown; slender twigs, yellow brown; flower bud ovate, petal ovate, 1.83cm long, white (155D), flower diameter 4.3cm; semi-double, 16-23 petals, stamens 73.5, filament 1.83cm long, anthers yellow, stamens nearly as long as petals; no pistil; dense flower (flower bud 0.83/cm), pedicle 0.53cm long; calyx green, two round, triangular ovate; filament and calyx petaloids; leaf blade green, elliptic-lanceolate, 12.8cm × 3.2cm, L/W ratio 4; leaf margin finely serrate, petiole1.5cm long.

Blooming period: Early April. The earliest among all ornamental peach cultivars (P. 125).

74. 'Fenhua Shanbitao' ('Tanchun')

Origin: From a seedling population of 'Baihua Shanbitao' with 'Hehuan Erse Tao' crossing (Hu and Zhang, 2001).

Characteristics: Vigorous, large size, bark grayish-brown; slender twigs, glabrous; flower bud ovate, petal ovate, 1.97cm long, light pink (69A); flower diameter 4.1cm; semi-double, petal 17-23; stamens, filament 1.43cm long, anthers orange red; pistil apparently shorter than stamens; medium flower density (flower bud 0.83/cm); pedicle 1.07cm long; calyx reddish, 2 round, filament and calyx petaloids; leaf blade green, elliptic-lanceolate, 12.8cm × 3.2cm, L/W ratio 4; leaf margin finely serrater; petiole1.4cm long; drupe green, 3.4cm × 3.2cm, round; stone 2.55cm × 1.82cm, elliptic, smooth on the surface.

Blooming period: The early bloom stage around the April 6th based on 4 years' observation.

'Tanchun' was selected from seedling populations of 'Baihua Shanbitao' with 'Yingchun' (Fang et al., 2008). Morphologically, it shares similar characteristics to 'Fenhua Shanbitao'. According to the Article 10.1 of *International Code of Nomenclature for Cultivated Plants* (ICNCP) (Brickell *et al.*, 2004), we suggest that 'Tanchun' should be combined and the accepted name should be 'Fenhua Shanbitao' because it was established earlier. The supplementary characteristic, chilling requirement 400h (P. 126).

75. 'Fenhong Shanbitao'

Origin: From a seedling population of 'Baihua Shanbitao' with 'Jiang Tao' crossing (Hu and Zhang, 2001).

Characteristics: Vigorous, large size, spreading; bark grayish-brown; slender twigs, glabrous, reddish; flower bud ovate, petal oval, 2.1cm long, rose pink (55B), flower diameter 4.27cm, semi-double, petal 22-33; stamens 53, filament 1.43cm long, filament petaloids, anthers orange red; pistil shorter than stamens; medium flower density (flower bud 0.78/cm); calyx reddish, pedicle 0.93cm long; leaf blade green, elliptic-lanceolate, 13.2cm × 3.3cm, L/W ratio 4; leaf margin finely serrate; petiole 1.5cm long; drupe green, 4.24cm × 3.84cm, oval; stone 2.48cm × 1.65cm, oval, smooth on the surface.

Blooming period: The early bloom stage around April 8th (P. 127).

'Fenhua Shanbitao' and 'Fenhong Shanbitao' are two early blooming cultivars bred in Beijing Botanical Garden. They obtained from the advantages of parents, early blooming, large tree size and glabrous bark of the pollen parent and bright color with double flowers of the female parent. The Blooming time just fills the vacancy between *P. davidiana* and all other cultivars of ornamental peach. After several years' observation, the percentage of fruit set of both cultivars are very low, 7.5% and 6.4%, respectively. The neat green leaf blades without messy fruit make it striking among all other ornamental peach cultivars, making them good breeding resources for sterile cultivars.

The Criteria For Selecting Cultivars' Described in This Chapter

All cultivars described above are referenced in botanical literature with detailed characteristic descriptions. Those cultivar appearing in studies that did not verify results through systematics study are not selected in this chapter, though they could be set up as independent cultivars once systematics study and verification has taken place. Cultivar represented by name only and ambiguous insufficient characteristic descriptions have also been omitted.

For reference, here are 20 cultivars of ornamental peach, which are common used names in the western countries. All are omitted from the key in this chapter as their systematics studies are incomplete.

'**Alboplena**' (double, white): Introduced to England in 1849 (Jacobson, 1996). Flowers semi-double, large snow-white, yellow stamens. Not able to define the difference from this and existing cultivars since petal number is not recorded.

'**Alboplena Pendula**' (weeping, semi-double, and white): Flowers pure white; densely double. Introduced to England in 1850 (Krussmann, 1986). No petal number recorded.

'**Aurora**' (dark pink, double, flower diameter 3cm): Cultivated and introduced (in 1937) by W.B Clarke nursery of San Jose, CA. (Jacobson, 1996). Registered in CRUWO (Cultivar Registration of Unassigned Woody Ornamentals). Semi-double flowers, open-flared with stamens evident, very soft pastel pink, early blooming. Rapid vigorous growth. Without detail description of flower color and petal number, there is no conclusive verification of type is possible.

'**Atropurpurea**' (purple leaf, single, pink): characteristics quite similar to 'Beijing Zi', and 'Follis Rubris' (Leaves rich purplish-red when young, becoming bronze-green. Single flowers). These may all be a single cultivar developed in different countries.

'**Cardinal**' (pink-red, semi-double): flowers rosette form, semi-double, intensely pink-red (Krussmann, 1986). The simple description seems similar to 'Magnifica' or 'Russel's Red' or 'Jiangtao', though no detailed data exists.

'**Camelliaeflora**' (large flower, double, pink): Introduced from Japan by Siebold before 1845. Vigorous growth habit, flowers double, crimson (Notcutt, 1926). The flowers are very similar to those of Clara Meyer camelliaeflora, with the exception of the rich color of the blossom. Its rich color makes it one of the most attractive spring-flowering shrubs. No detailed data of flower color or petal number is recorded.

'**Chrysanthemum**' (large flower, light pink, chrysanthemum-like petal): Introduced ≤ 1906 by Fruitland nursery of Augusta, GA. Later offered by W.B. Clarke nursery of San Jose, CA. (Jacobson, 1996). Flowers large, light pink, the center quilled like a chrysanthemum. There is no evidence to support whether this cultivar originated from the orient, however, its date of introduction; it is no earlier than the same cultivar in Japan and China.

'**Dianthifolia**' (large flower, semi-double, petals narrow and acuminate, dark red stripe): Introduced from Japan by Siebold before 1845 (Krussmann, 1986). There is no detailed data regarding flower color or petal numbers exists.

'**Duplex**' (light pink, semi-double, small flower): Recorded in France since 1636 (Krussmann, 1986). No detailed data of flower color or petal numbers to enable definition of distinction from cultivars detailed in my key.

'Foliis Rubris' (purple leaf, flower single, pale pink): Cultivated commercially in the USA since 1871. After being exported to Europe in 1873 it became known as 'Royal Redleaf' (Jacobson, 1996). Morphologically it is quite similar to 'Atropurpurea'.

'Helen Borchers' (semi-double, pink, flower diameter 6.23cm, late blooming) (Dirr, 1998): Introduced by W.B Clarke nursery of San Jose, CA. prior to 1949 (Krussmann, 1986). Strong growth. No detailed data on flower color or petal numbers exists.

'Iceberg' (semi-double, white, flower diameter 4cm): Introduced by W.B Clarke nursery of San Jose, CA. in 1938 (Jacobson, 1996).

'Klara Mayer'(flower bright pink, double, diameter 4cm wide, late bloom)(Jacobson, 1996): Introduced around 1890 by Spath (Krussmann, 1986). Medium vigor with abundant flower. May also be spelled 'Clara Mayer'.

'Magnifica' (semi-double, bright carmine red or crimson): Introduced not later than 1894 by Veitch nursery of England (Krussmann, 1986). Spreading growth pattern. Similar to 'Cardinal' and 'Russel's Red'. Not definition of color or petal number.

'Rubro-plena' (semi-double or double, red): Exported from China to Europe in the 1840s. Described in 1854 (Jacobson, 1996). No more detailed description of morphological characteristics exists.

'Versicolor' (flower semi-double or double, white with pink or red stripes): Present in France before 1863 (Krussmann, 1986). Slow growth, leaves light green. Flowers numerous, sometimes totally white or with red or rust colored stripes. Characteristics quite similar to 'Sahong' mentioned above, however, without detailed data on this cultivar reliable definition is impossible.

'Windle Weeping' (weeping, semi-double, pink or crimson flower): Vigorous growth, spreading umbrella form, flowers cupulate, numerous (Krussmann, 1986). Introduced by W. B. Clarke nursery in San Jose, CA before 1949.

'Pink Peachy', **'Red Peachy'** and **'White Peachy'**: dwarf cultivars with differing flower color. All approximately 1-1.5 m high. Semi-double flower, pink, red and white, respectively. No detailed flower color or petal number data recorded.

BREEDING AND IMPROVEMENT

Major Mendelian Characteristics of Ornamental Peach

Prunus genome is one of the smallest among all cultivated species, with an estimated length of 290 Mbp (Baird *et al.*, 1994), approximately twice the size of Arabidopsis. The small genome, short juvenile phase and ease of hybridization all contributed to the peach being selected as a model species for of study genomics in Rosaceae.

A total of 42 morphological characteristics of simple Mendelian inheritance were discovered during the last century (Dirlewanger and Arus, 2008). They now form the basis for all future breeding of ornamental peach cultivars. The following chapter details traits or characteristics regarding growth habit, flower, leaf and hardiness that are considered of important ornamental value.

Growth habit

Growth habit is one of the greatest rewards of the ornamental peach giving it immense ornamental value over other spring flowers.

The term growth habit refers to characteristics of branch and shape. A unique growth habit to the ornamental peach species is the twisted variety. Branch angle, orientation and intensity of branch nodes compose the diversity of growth habits and therefore the many guises in which the ornamental peach exists.

1. Weeping

The weeping form is the distinctive type of ornamental peach, whose pendulant branches create a shape much like an umbrella. The extended branch angle can be of more than 90° (Monet *et al.*, 1988: Shen, *et al.*, 2008). The weeping trait (*pl*) is incomplete dominance (Dirlewanger and Bodo, 1994).

2. Pillar, Columnar, Broomy or Fastigiate

The pillar form, which is also known as columnar, broomy, or fastigiated is marked by vertical branches growing at a narrow crotch angle (no more than 35°-40°) (Layne and Bassi, 2008), giving a column like appearance. This wonderful form was contributed to the world by Japan with the earliest recorded description of the columnar peach dating back to the Edo era during the 17th-19th centuries (Ito, 1695). The application of columnar peaches as ornamental plants also began in Japan (Yamaguchi *et al.*, 1987a).

Broomy trait (*br*) is incomplete dominance. Phenotype is upright when *Br* is heterozygous with the alleles for the standard, dwarf, weeping, or compact growth habits (Scorza *et al.*, 1989; Scorza *et al.*, 2002). Yamazaki *et al.* (1987a) confirmed that Broomy trait (*br*) is dominant to weeping (*pl*) by crossing and back-crossing between columnar and weeping. Werner and Chaparro (2005) showed that columnar (*brbr*) was epistatic to the expression of weeping (*plpl*).

3. Twisted

'Unriumomo' is the only cultivar that presents twisted branches (Yoshida *et al*., 2000). Okie described this growth habit in 1998 and concluded that this trait seemed monogenic and recessive.

4. Dwarf

Dwarf peach trees have been studied since 1786 the first with the first recorded description by Monceau (in Monet and Salesses, 1975). In 1867, Strong (1867) and Hooper (1867) described dwarf peach tree phenotypes. Connors (1922) observed semidwarf trees in progeny from self-pollinated 'Elberta' and concluded that the trait was recessive. Since then, three dwarf genotypes have been identified. Lammerts (1945) obtained dwarf trees in F_2 populations from the cross of 'Chinese Dwarf' and a standard growth form and concluded that this dwarf type is monogenic and recessive. This dwarf, know as the "brachytic" dwarf (*dwdw*) type is characterized by short internodes (usually less than 10 mm in length) and large leaves. Lammerts (1945) reported a semidwarf peach trait ("bushy") that was inherited as a double recessive (*bu1bu1/bu2bu2*). A semidwarf mutant, that differs phenotypically and genetically from the commonly recognized "brachytic dwarf" described by Lammerts (1945); 'A72' was described by Monet and Salesses (1975) The 'A72' dwarf trait (*nn*) is incompletely dominant with the heterozygote expressing as a semidwarf tree. Gradziel and Beres (1993) reported an apparent mutant phenotype in a seedling of an open pollination of a clingstone breeding line. This tree, ('SD22–59'), that resembled the 'A72' dwarf and the segregation of growth habits in progeny of 'SD22–59' standard growth habit varieties (1 semi dwarf: 1 standard growth), was also suggestive of the phenotype produced by the 'A72' mutation. Hu and Scorza (2009) evaluated this growth habit and verified the inheritance of 'A72' in a population of 220 progeny derived from self-pollination. Detailed tree and branch measurements revealed a unique forked-branch (FBR) character in the 'A72' (*Nn*) phenotype. The progeny segregated into 1 *NN*: 2*Nn*: 1*nn*. *NN* trees were indistinguishable from standard peach trees, *Nn* were FBR, and *nn* were dwarf. Hybrids between 'A72'and columnar (*brbr*) peach trees confirmed that FBR is inherited as a monogenic trait that appears to express incomplete dominance.

5. Standard

Tilted branches with crotch angles of 40°-70° may be summarized as standard varieties of ornamental peach, though are classified as diverse types of fruit peach.

6. Combination growth habit

Interaction between different growth habits may result in new types and formations. Dwarf pillar (Scorza *et al*., 2002) (Fig. 7.2, P. 135) and compact-columnar trees (Scorza, 2003) distinctly show new forms produced through a combination genes, which not only confirm the plasticity in ornamental peach tree growth (Scorza *et al*., 2002; Werner and Chaparro, 2005), but also demonstrate great potential for development of predicted novel growth habits.

Both compact (*Ct*) (Mehlenbacher and Scorza, 1986) (Fig. 7.3, P.135)and globe types (*GL*) (Scorza *et al*., 1989) have intense branches with dense flowers. Both of these types are yet to be applied within landscaping. The compact tree size and profuse flowers exhibit tremendous potential for the future of ornamental peaches.

Flower

The flower is the most important part of the ornamental peach species. Most flower traits studied are controlled by single genes. When considering flower size the non-showy bell-shaped flower is dominant to the showy flower (Connors, 1920; Bailey and French, 1942; Lammerts, 1945). When considering color; pink to red, pink to white, red to white and dark pink to light pink are dominant, respectively (Lammerts, 1945; Yamaguchi *et al.*, 1987a). Regarding petal numbers single is absolutely dominant to double (Lammerts, 1945) and regarding the fertility or sterility of pollen two types exist: *ps/ps* (Connors, 1926; Blake and Connors, 1936; Scott and Weinberger, 1944) and *ps2/ps2* (Blake, 1932; Werner and Creller, 1997). Male sterility is described as monogenic and recessive (Bailey and French, 1949).

Leaf

Leaf color is monogenically controlled (*Gr/gr*; Black, 1937). Red is incompletely dominant over green in leaves (Black, 1937; Yoshida *et al.*, 2000). The gene controlling leaf color may be linked with the trait that decides flower color (Yamaguchi *et al.*, 1987a). Gland shape is associated with resistance to powdery mildew (Connors, 1921). Usually there are three types of peach presented: eglandular (*e/e*; recessive; Connors, 1921), globose (*E/e*; heterozygous) and reniform (*E/E*; homozygous dominant). The eglandular phenotype is associated with a strong susceptibility to powdery mildew. Cultivars with globose glands are more susceptible to powdery mildew than those with reniform glands (Saunier, 1973). Most ornamental peach cultivars have reniform glands, which are not susceptible to powdery mildew.

The wavy blade is recessive and monogenic (*Wa/wa*; Scott and Cullinan, 1942) and is often reported as being associated with the dwarf growth habit.

Narrow leaf (willow-like shape) (Chaparro *et al.*, 1994; Okie and Scorza, 2002) is approximately half the width of ordinary peach (Glenn *et al.*, 2000), which enables increased sunlight absorbtion and reduced susceptibility to disease. Okie and Scorza (2002) reported that narrow leaves show higher water efficiency than standard.

Disease resistance

Inheritance studies of diseases and insect resistance in fruit peach are well reported. In ornamental peach 'Harrow Frostipink', 'Harrow Candifloss' and 'Harrow Rubirose' are noted for their resistance to brown rot and powdery mildew (Layne, 1981). 'Bonfire' is noted as being resistant to bacterial spot (Moore *et al.*, 1993).

Breeding Methods

It is very easy to mutate ornamental peach through climatic effect and cultivation conditions. Conventional techniques such as selection (sport, clonally selection, and seedling collection), cross breeding (self-pollination and controlled cross) are still the most effective methods to breed new ornamental peach cultivar.

Segregation ratios of seedlings from controlled hybridization, either within the same or different species are markedly high. Controlled hybridization between different growth habits results in novel growth habits that show great potential for ornamental peach cultivation. Dwarf pillar is the result of the segregation of seedlings of pillar, for example (Scorza *et al.*, 2002).

Besides the diversity within cultivated ornamental peach cultivars abundant genetic resources exist in wild varieties of the species, including *P. davidiana*, which confers resistance to powdery mildew and leaf curl, sharka; *P. kansuensis*, which has a blooming date one week earlier than other peach; *P. ferganensis* which is early blooming and *P. mira* which has a long life and is tolerant to dry and cold plateau conditions. New genes found in these wild types of species are great breeding resources for early or late blooming cultivars with wide adaptability to adversity. Distance hybridization also reduces fertility which shows potential for the breeding of desired sterile cultivars of ornamental peach.

Artificially induced breeding is an important breeding technique for the development of new ornamental peach cultivars. It is possible the irradiation (X-rays, gamma rays, thermal neutrons), and polyploid (by colchicines) breeding could be utilized on a larger scale than its current application. Besides the usual radiation of branches seeds can also be used as alternative test materials.

Besides selection and controlled crossing, molecular breeding with identified genes may be applied in order to breed new cultivar with high disease resistance, cold hardiness and additional strengths and desired characteristics. Marker assisted gene selection in peach for resistance to sharka disease for instance, has been isolated (Pascal *et al.*, 2005). This will aid the selection of parental lines and with the planning and scheduling of crosses that are faster, less costly and specifically targeted.

Breeding Trend for Ornamental Peach

Immense ornamental value is found in the growth habit, flower and leaf of ornamental peach cultivars. Future breeding should address the improvement of ornamental value, extension of blooming time, improved disease resistance, adaptability to climate and environment and sterility (Hu and Zhang, 2008).

Improving ornamental value

As an ornamental tree, new ornamental peach cultivar should address ornamental values firstly, for instance, growth habits, flower color and unique foliage etc.

The more combinations of growth habit, flower form, flower color and leaf color that can be created the more cultivars with unique ornamental value for landscape and private garden application are available.

Novel growth habits combined with single growth habit looks to be one of the future trends of ornamental peach breeding.

According to the relationship that exists between different growth habits (Zhou *et al.*, 1998), all flower types and colors presented in the standard type should appear in all other growth habits. In the twisted type, for example, although only one cultivar exists that presents single pink flowers, cultivars exhibiting semi-double, double with red, white or even bicolor flowers will potentially be bred or selected in the near future.

There is no single commercial ornamental peach cultivar with yellow blooms in existance so far. However, creamy blooms with a hint of yellow have been found within a peach breeding program (Fig. 7.7, P. 145), which implies the possibility of breeding pure yellow color flowering cultivar eventually. The bell-shaped flower has not received much attention within recent or current breeding though semi-double and double flowering bell-shaped cultivars are to be expected in different growth habits.

Extending period of appreciation all year around

The extension of blooming period in ornamental peach cultivars is very important for the enhancement of good ornamental application within gardens.

The anthesis of one individual ornamental peach tree is about 8-10 days. However, with the application of early and late blooming cultivars it is possible to extend the total blooming period of ornamental peach beyond a month, making an entire spring season bloom possible.

Besides the attraction of flowers, red (or purple) leaved cultivars also provide striking ornamental displays and extend ornamental value and period of 'appreciation' significantly beyond the flowering season. Flowers during spring, leaves throughout summer, transition and fall during autumn and structure during winter, all of these should be considered future breeding targets of ornamental peach.

Enhancing adversity resistance

Enhancement of resistance to insects and disease, cold and drought hardiness are important issues that must be addressed as significant breeding goals.

Resistance to the Plum Pox Virus (PPV) in *P. davidiana*, has been characterized (Pascal *et al*., 2005) and can be used in ornamental peach breeding lines to breed new virus resistant cultivar. Cold hardiness and frost (early spring) resistant cultivars should be considered for breeding in the northern hemisphere. Great potential exists for new cultivars with better disease resistance with resistance to peach leaf curl and brown rot already confirmed though distance hybrid breeding (Komar-Tyomnaya, 1998).

Breeding sterile cultivars

The breeding of new fruitless ornamental peach cultivar is new challenge to ornamental peach breeding work. Contrary to the common fruit peach, where fruit is the focus of the species; ornamental peach cultivars focus entirely ornamental value. Fruit produced after blooming, which not only consumes the nutrients of a tree but also increase chances of infection presents significant difficulty; preventing the development ornamental peach in landscape and private gardens on a large scale. Cultivars with outstanding ornamental value that do not set fruit after blooming are the perfect objective of new ornamental peach developments. Sterile cultivars already exist, such as 'Baihua Shanbitao' and the hybrid between *P. persica* and *P. mira* (ref. personal conversation with Dr. Komar-Tyomnaya in Ukraine) and may be used as a breeding line. Meanwhile, the haploid also provides potential for breeding new and desirable sterile cultivars of ornamental peach.

Cultivars that possess both ornamental value and produce high quality fruit seem perfect for home garden use. However, in order to get desirable fruit great professional skill is needed in addition to constant attention which is difficult for the ordinary family to provide. Therefore when considering the particular purpose of ornamental peach fruit is never the focus of breeding.

Germplasm conservation is important for new cultivars' selection and breeding. With the aid of conventional and modern breeding techniques and collaboration among ornamental peach breeders; under careful observation and prudent selection many new ornamental peach cultivars can be expected in the near future.

PROPAGATION AND CULTIVATION

Propagation

Qi Min Yao Shu (an encyclopedia for agriculture), which was written by Jia Sixie (386-534), is the earliest record of grafting technique.

Depending on the purpose, different propagation methods are applied to ornamental peach. Seed propagation is usually for rootstock and hybridization, whereas most propagation methods for ornamental peach cultivars are vegetative methods, for instance, grafting, budding and cutting, etc.

The techniques of sowing, budding and grafting ornamental peach are basically the same as those for fruit peach and are discussed here only in the broadest sense.

Seed Propagation

Seed propagation is usually used for rootstock and hybridization.

Seeds should be collected in August and pits separated from fruit flesh promptly. Seeds should then be washed, dried and stored until needed. Seeds can be stored at 5°C for up to one year.

Peach seeds require 60-100d cold-moist stratification at a temperature of 3-5°C to break dormancy.

Sowing time in Beijing is usually in the end of March or early April.

Vegetative Propagation

1. Grafting

Grafting is still the most commonly practiced methods of propagation in nurseries breeding ornamental peach cultivars.

Rootstock: For ornamental peach propagation in China there is not yet particular rootstock cultivar. In northern China, *Prunus davidiana* is still the most common rootstock, valued for its cold-hardiness, whereas *P. persica*, for its barren tolerance, wet resistance and longer life than *P. davidiana* is a well-adapted as rootstock used in southern China.

Grafting technique: Budding and grafting are two major techniques adopted for propagating ornamental peach cultivars.

(1) Budding On account of its simplicity and low cost budding is becoming the most popular practiced technique in nurseries. Traditionally budding is done when rootstocks are actively growing so that the cambium divides and the bark separates readily from the wood.

Budding is normally performed during three periods during the year. In the Northern Hemisphere these periods are March to early May (spring budding), May to early June (June budding) and mid-July to early September (summer budding).

Summer budding is still the most important time for the budding of peach. A one-year scion/ two-year rootstock is generally produced.

June budding produces a smaller budded plant with a one-year scion/ one-year rootstock.

Spring budding is done as soon as new seasonal growth occurs. A one-year scion/ two-year rootstock is generally produced.

In budding, it is important to use vegetative buds rather than flower buds. The best buds to use are usually on the middle and basal portions of the branch.

(2) Grafting   Early spring is the best time for grafting, just when the buds of the rootstock are beginning to swell but before active growth has started. This method is usually used in order to keep cultivars with fewer scions.

The height of rootstock should be taken into consideration and 'ideals' very depending on growth characteristics. For example, weeping cultivars should be grafted on a higher rootstock, usually more than two years rootstock, in order to keep the pendulant branches and umbrella shape of the tree. To raise the height of dwarf peach higher rootstocks were also recently adapted on dwarf peach (Fig. 8.1, P.150).

2. Cuttings

Although it is well-known that cuttings provide more uniformity and a longer life than grafting; this method is still not popular within ornamental peach production, especially in China.

In preliminary experiments of semi-hardwood cuttings taken on 'Kikoumomo' and 'Terutehime' in 2005-2006, 20 sec quick dip in 600-1000 mg/L KIBA, under the mist with 10s/20m in the media of perlite, a better rooting rate was found (Fig. 8.2, P.151).

Cultivation

Temperature, light and water are the most important factors in a plant's development. The ornamental peach thrives in full sun and is tolerant to drought with relative cold-hardiness. They enjoy well-drained, fertile soil. An open area with good drainage is the best planting site for ornamental peach. In a congenial climate or growing condition the life expectancy of ornamental peach may be extended beyond the norm.

Cold-hardiness

The peach is one of the least cold-tolerant of the commercially grown *Prunus* species (Quamme *et al.*, 1982). Ornamental peaches can be extensively cultivated throughout temperate and subtropical zones (USDA Zone 5-9). Fully acclimatized peach flower buds can survive $-30°C$ and vegetative buds $-35°C$ (Layne, 1984). The most congenial temperature of peach is 18-23°C. Lower than $-18°C$, flowering buds will suffer cold injury (Qu, 1990), and lower than $-27°C$, most flowering buds will be killed outright. Open flowers and young fruitlets are killed by brief exposure to $-2°C$ or below.

The *P. persica* f. *hui-chun-tao* Gu which is found in Hunchun county in Jilin province, can tolerate temperatures of $-30°C$. It is the most northerly distribution of peach and might be an important germplasm resource for the breeding and development cultivars with cold-hardiness (Yu, 1984) for use in colder zones.

According to the relative electric conductivity (REC) method and growing recovery test of 5 ornamental peach cultivars, the semi-lethal temperature (LT_{50}) is lower than $-30°C$ (Li, 2000) in all. Five cultivars in decreasing order of cold tolerance - 'Baihua Shanbitao', 'Hanhong Tao', 'Kikoumomo',

'Sahong Tao' and 'Erse Tao'.

Light and water

Replete sunlight is the most important condition for planting an ornamental peach. Contrary to the fruit peach, replete sunlight is crucial to keeping growth habits forming a good ornamental effect.

Peaches originated from plateaus and are drought resistant. They cannot tolerant water logging and are is one of the most sensitive fruit species to water logging and anaerobic soil conditions (Anderson *et al.*, 1984; Alvino *et al.*, 1986; Schaffer *et al.*, 1992). Three days of water logging will result in death. In any kind of water logged soil or badly drained site a specimen eventually declines or dies.

Early spring watering is crucial for the production of flowering buds that are well-developed with a good number of high quality flowers. 80cm should be the lowest level of each watering. Repetitive lower level watering reduces the temperature.

General opinion is that specimens should be watered before blooming where possible. It will replenish water, elongate blooming time, and keep bigger and brighter flowers.

It is requisite for winter watering, which means in the end of November. It must be replete to keep water demands for the winter.

'Erse Tao' has high drought resistance and is able to grow well even when the water content of the soil is 5.284% (Li, 2000).

Soil and planting

Peaches have shallow roots with horizontal distribution larger than vertical distribution. Roots range is around 0.5-1 times the crown of the tree, while the depth of root is about 1/4 of the tree height. The absorbing roots distribute within 50-60cm depth of soil (Layne and Bassi, 2008).

Peaches will thrive in almost any soil of pH 5-8 (Wang & Zhuang, 2001). The best peach land is light and sandy as the air permeability of soil is closely related to root development. Soil which is badly drained with poor air permeability restricts the growth of root. Choosing correct rootstock is the best way to develop the adaptability to poor soil. For instance, 'Harrow Frostipink', 'Harrow Candifloss', and 'Harrow Rubirose' (Layne, 1981) are three cold-hardy, late blooming patterns that are suitable for growth in Zone 6.

Disease and Insect

Disease

Peaches are susceptible to pathogenic fungi, bacteria, viruses, and virus-like organisms. Major diseases affect ornamental peach attack fruit, leaves and then branches.

1. Brown rot

American brown rot [*Monilinia fruticola* (Wint.) Honey]

European brown rot [*M. Laxa* (Aderh. & Ruhl.) Honey]

Brown rot is a serious fungal disease that attacks peaches worldwide. Infected flowers wilt and turn brown, sticking firmly to branches. Infected twigs and branches show gum formations. Mummified fruit can also be a typical symptom of infected trees. Fungicides integrated with the removal of infected parts from trees and the reduction of overwintering inoculums are an important brown rot management approach.

2. Peach leaf curl [*Taphrina deformans* (Berk.) Tul.]

Peach leaf curl mainly occurs on new leaves. Infected leaves become discolored then thicken, wrinkle and pucker, causing the leaves to curl and drop prematurely. Fungicide can be applied to prevent this in spring before the buds swell, or in the fall after 90% of leaves have fallen.

3. Powdery mildew [*Sphaerotheca pannosa* (Wallr.:Fr.) Lev.]

Powdery mildew attacks leaves, buds, young shoots and fruits. White fungal growth forms on the upper surface of leaves. Branches may be killed, foliage greatly reduced in size and efficiency, and the entire tree debilitated. Protective fungicide should be applied.

4. Fungal gummosis [*Botryosphaeria dothidea* (Moug. ex Fr.) Ces. & de Not.]

Blisters around lenticels on young bark are primary symptom of fungal gummosis. They enlarge, are sunken and show excessive gum exudation (Fig. 9.2, P.162). From trunk to scaffold, even small fruiting branches are gradually infected which reduces the vigor and health of the tree. The pruning and destruction of infected wood is the best management approach for the reduction of inoculums sources.

5. Perennial canker [*Leucostoma persoonii* (Nitts.) Hoehn.]

Leucostoma canker is one of the most serious peach diseases of northern growing areas, especially in colder climates. It debilitates trees by killing large and small branches resulting in dieback. Amber gum may exude from the canker with infected branches exposing blackened tissue beneath. Improving culture and pest management practices is an integrated approach for the prevention of *Leucostoma* canker.

6. Bacterial canker (*Pseudomonas syringae* pv. *syringae* van Hall)

The first sign for bacterial canker appears with the buds in early spring. They may fail to open, open late or open and then stop growing, wilt and die. Reddish-brown and water-soaked streaks in woods are typical symptoms, which look similar to cold injury. It is preferable to spray chemicals in spring before the buds swell than in late summer or early fall. Pruning infected branches in time and promoting optimum tree health is an effective strategy to prevent infection and disease development.

7. Bacterial spot [*Xanthomonas arboricola* pv. *pruni* (Smith)]

Bacterial spots affects leaves, branches and fruits which can cause severe defoliation and fruit blemish. Leaf lesions and chlorosis are a first and obvious symptom. Leaf and flower buds fail to open and infected branches have a black, greasy and wet appearance. Once bacterial spot is established it is difficult to control. Improved culture practices and soil conditions to enhance resistance; integrated with foliage targeted chemicals consisting of lime sulfur materials, applied in early spring are good defense methods.

Insect

Principal insect pests of ornamental peach are aphids, leafhoppers, mite, leaf-miner, borers, scale and peach leaf-miner.

1. Aphids (*Myzus* ssp.)

The green peach aphid (*M. persicae*), *M. momonis*, and *Hyalopterus amygdali* are three major aphids infesting the leaves on ornamental peach. Young leaves, new growth, and flower buds are crowded with aphids. More than ten generations are produced a year for most aphids which presents serious problems for the peach. Green lacewing, ladybug and drone fly are natural enemies of the aphid. Combined with winter pruning, removal of infected wood and the spraying of 0.3-0.5 degree lime sulfur to kill overwintered larva in spring before buds swell aphid infestation should be avoided. May and June are crucial times for spraying imidacloprid and nimbin - chemicals for protection and for the introduction of natural enemies (P.165).

2. Leafhopper [*Empoasca flavescens* (Fabricius)]

Leafhopper pierces and sucks the juices of new growths and leaves. They occur during spring time with the first symptom being a white spot on the leaves with the entire leaf gradually turning white and then falling. This greatly reduces the vigor of the tree and flower buds. Five generations of leafhopper are produced per year. Leafhopper overwinters as adult on the fallen leaves and grass and in the cracks of bark. Spraying chemicals such as imidacloprid and nimbin during hatching season, cleaning fallen leaves and grass and burning out infected materials to kill the adult should effectively halt the spread (P.165).

3. Mite (*Tetranychus viennonsis* Zacher)

Adults and nymph mites are crowded under leaves to suck juices within foliage. Injury begins with small yellow or white spots along the midrib which net, turn grey- brown and finally wilt and drop reducing the vigor of the tree and flower buds. These mites produce 5-13 generations a year. *Tetranychus viennonsis* overwinters as female in the fallen leaves and cleaves close the bottom of the trunk, peaking in July and August. Wrapping grass robes on the trunk to induce overwintered mites then burning out dry barks covering the trunk along with resident overwintering pests is an effective control method. Spraying should be carried out in first-instar, which is not resistant to chemicals. Spray with 0.1-0.3 degree lime sulphur or abamectin (Fig.9.7, P.167).

4. Borer (*Aromia bungii* Faldermann)

The peach tree borer burrows under the bark and feeds near or below the ground line. A tree can be girdled and killed in a single season. One life cycle occurs within 2-3 years. To prevent larvae, kill them by hook and covering the hole with poisonous mud. Brush white (lime10, sulfur 1, salt 0.2, and oil 0.2) on the trunk to prevent adults lying eggs. The adult can be caught on the tree in June to July (Fig.9.8, P.167).

5. Scale (*Didesmococcus koreanus*)

The adult and nymph scale use their piecing and sucking jaws to imbibe plant juices and kill the plant. They produce one generation a year. *Didesmococcus koreanus* overwinters as a second-instars nymph in the stem and moves to find permanent place to suck sacks during April. Trim off the branches with pests and burn out the materials. Spray imidacloprid during peak hatching and introduce natural enemies, *Cybocephlus nipponicus* and *Chilocorus rulidus* for instance (Fig.9.9, P.168).

6. Peach leaf-miner (*Lyonetia clerkella* Linn.)

The larvae of the peach leaf-miner sneak onto leaves and injure the mesophyll cells causing the leaves to collapse and become off color without breaking the leaf form. They produce seven generations per year. Peach leaf-miner adults overwinter in the cracks of bark and grass. Clear and burn out infected materials and spray chemicals to exterminate larvae. Preparation carbamide and methyl abamectin is best applied to newly-hatched larvae (Fig.9.10, P.168).

Pruning Techniques

Objectives of pruning for ornamental peach

The objectives of pruning for ornamental peach differ to prune of fruit peach. Pruning techniques aim to show aesthetic standards at their utmost by keeping and building well-distributed structures around the tree.

On the basis of satisfying the demand for growth, pruning should be applied in different ways according to growth habits, cultivar characteristics, age, vigor, planting site, and the environmental conditions of each tree.

The promotion of good light distribution on branches, good nutritional balance and good aeration are components important for keeping excellent appearance in ornamental peach specimens. The peach is sympodial branching, the apical buds inhibit the lateral buds to growth. The objective of pruning is to stimulate a different growth trend. Types of buds have different functions and different branches require different pruning methods. The different functions of branches such as the leading branch, stem branch, laterals and interior branch etc. require the application of different pruning techniques.

A well-distributed structure with dense flowers is the best aim for ornamental peach pruning.

Heading back and thinning out

Heading back consists of cutting back the terminal portion of a branch to a bud, whereas thinning out is the complete removal of a branch to a lateral or main trunk. Because the heading back of a stem destroys apical dominance it is usually followed by the stimulation of several lateral bud breaks depending on the cultivars and the distance from the tip to the cut.

Heading back is used for perennial branches. To encourage spreading growth the branch is usually cut back to an outward pointing bud, leaving at least one vegetative bud.

For trees with over-exuberant growth, heading back is usually done during winter with a third to half of annual growth removed. Heading-in branches always tends to make a thick-topped tree.

Thinning, in contrast to heading, encourages longer growth of remaining terminal branches. Its net result is a reduction of laterals. Thinning of weak growth tends to "open up" the tree. The dead wood, watersprouts, diseased growth, closely spaced and intertwining branches and the small unprofitable wood from the center of the tree should be thinned out to promote good light conditions in the inner part of the tree enhancing photosynthetic efficiency and accumulating nutrients, better for flowering.

The rejuvenation of older trees by reducing and thereby stimulating remaining growing points is accomplished by thinning.

A moderate to severe dose of annual pruning is a must with the thinning out 1/2 to 2/3 of 1-year-old shoots and heading-back shoots longer than 60cm.

For ornamental purposes the fruit is never a desirable aspect. Thinning fruit is most easily done during the early stages of fruiting rather than late in the season. It should usually be done just after flowering when fruits are still tiny. This not only eases to cooperation but also reduces nutrient loss during fruit setting.

Timing of pruning

In contrast to summer pruning which is usually done for fruit peach varieties, winter pruning (dormancy prune) combined with post bloom pruning is most effective in the ornamental peach. These two pruning methods compliment to each other well.

The dormant prune is usually applied during December to the middle of February. This is best delayed until the severest weather has past in order to reduce winter injury to fresh cuts. It is crucial pruning for the tree structure. Training the tree form depends on different cultivar characteristics. Main branch direction, correct orientation of the bud, position and angle should be taken into account in order to keep abundant flowering feathers and laterals. There is no need to cut every branch, try to maintain the natural growth and shape by using more heading back. To take off the upward bud on retaining part of heading back branch while doing winter or post bloom pruning could save energy and labor next year once it developed strong branches.

Pruning after flowering completes the problems remaining after the winter prune, in addition to thinning fruit. Pruning after flowering produces a perfect tree structure when done well.

Pruning is a continual process. Perfect pruning within one or two years is impossible to achieve. Patience and skill are needed to achieve ideal pruning through annual adjustments. There is no absolutely right or wrong in pruning. The proper and best way, depends on the individual tree. Pruning by following each branch is suitable for ornamental peach varieties.

Principles for different growth habits

1. Pruning for Standard Type

The standard type is the most common growth habit of ornamental peach. They exhibit tilted branches and an upright growth habit.

The standard peach is commonly pruned to an open center and usually composes of two or three widely spaced main scaffold limbs. It is not as strict as fruit peach. To keep intense flowers that are on well-distributed (Fig. 9.11, P.173) tree structure and outward pointing buds that encourage spreading growth are basic principles to follow. The selective removal of buds, flowers or fruits to increase the size of the remaining parts must be considered a specialized aspect of pruning.

Different cultivars have their own characteristics, so pruning should not be blindly followed overall character. Heading height is usually about 30-40cm which allows the best view for most flowers.

'Kikumom' and 'Kyoumaiko' have relatively slender branches. Heading back for each branch should be avoided. Simply remove dead wood and diseased growth, closely spaced branches and keeping the natural growth habit is when pruning these type cultivars. 'Bai Bitao' and 'Wanbai Tao' do not easily form new branches and so require cutting in strong buds to promote new growth. To thin out and maintain good light penetration reduce susceptibility to infection. 'Erse Tao' and 'Jiang Tao', which form new branches easily, require the cutting in of relatively weak buds to promote flowering and relieve tree vigor. 'Sahong Tao' and 'Wubao Tao', which are late blooming cultivars, have the natural characteristics of dense flowering bud distribution. Moderate heading back can be done according to the age of the tree and should be done gradually. Avoid heading back too strongly.

2. Pruning for Dwarf Type

Dwarf type cultivars have short internodes and small tree size. They grow slowly generally do not need much pruning except to take remove diseased growth, dead wood, crossed branches and to keep plump and attractive shaped.

3. Pruning for Pillar Type

Pillar type cultivars have an upright tree form with a narrow crown compared to standard trees. Pillar type trees have a canopy that is 60% narrower than those of "standard type" specimen and required only 50% of the cuts of standard trees (Bassi *et al.*, 1994). Pillar trees require only light branch thinning and no major cuts besides those employed to remove diseased growth, dead wood, crossed branches and watersprouts. Since pillar trees have lower branch density the keeping more first order branches on the scaffold is crucial to keeping the tree form. Cutting in the leading center must be avoided as once the leading center is cut, laterals will be stimulated and whose growth will totally destroy the pillar shape.

4. Pruning for Weeping Type

Due to its pendulous branches that hang in conflict with the "normal" direction of nutrient transfer the weeping type has the narrowest trunk diameter and slowest growth. Strong wind in winter and early spring in Beijing create serious death and injury of young shoots. Dead wood, diseased growth and overlapping branches should be thinned in a timely fashion. Since unpruned trees produce the same amount of new growth as pruned weeping trees (Bassi *et al.*, 1994), the major purpose for the pruning of a weeping cultivar is in an attempt to keep a plump weeping tree form by heading back weak branches, upward pointing buds and to make new outward growth spreading and overhanging.

5. Pruning for Twist Type

The twist type has natural twisted branches which do not generally need to be pruned. Effective thinning out of overlapped twigs and branches will allow light to distribute well.

6. Pruning for David Type

The David type has apparent characteristics of *P. davidiana*, i.e. a large tree size with slender branches and smooth bark. To keep these characteristics do not prune strongly, simply maintain natural growth through the thinning out of dead wood and diseased growth. Any more is unnecessary.

Pruning is supply an assistant to improving the growth of a tree and cannot deal with the problems of a specimen. Good maintenance, sufficient water, adequate fertilizer supplies and good disease and pest control integrated with proper pruning should produce excellent specimens.

APPLICATION OF ORNAMENTAL PEACH IN LANDSCAPES

Peaches have been grown in gardens as ornamental plants since the Han Dynasty (206 BC -220 AD). Since then, this exquisite early blooming spring plant has attracted great attention in landscape gardening for the endless aesthetic possibilities presented through integration of its brightly colored flowers and diverse shapes. As a symbol of luck and happiness in China peach blossoms have always featured in poems and songs. They are present in ancient and modern landscapes, imperial, private and public including street and park planting. Feature throughout society as bonsai, flower arrangements and principal plants in themed gardens. Ornamental peaches are available everywhere, the species is far more than fruit tree in a farm or orchard.

Landscape Application Forms

The great diversity of ornamental peach cultivars make it possible to utilize the species as an ornamental plant extensively in a variety of landscapes.

In the garden

Ornamental peaches can be planted as features in gardens where they display their bright color and individual features. In classical gardens peaches were traditionally planted in front houses.

Cultivars with different growth habits provide a variety of peach scenes. Dwarf peach is the best choice for a small garden, besides having colorful double flowers in the springtime, dense green or purple leaves also bring a muted but flourishing scene during hot summer. Weeping peach, planted next to the lake or pond not only displays an interesting reflection in the water but the wavering pendulant branches bring a charm and romance whirling into a landscape.

Differently colored ornamental peach clustered together is a striking way to display the prosperity of spring, delivering a warm, bright and vivid ambience, widely displayed in modern gardens. Colorful flowers increase the visual impact a brilliant spring scene.

Placing ornamental peach with other ornamental plants creates varying scenes and ambience. The most famous of all traditional styles is the planting of ornamental peach with weeping willows on river and canal banks. The blooming period of peach and foliage bursting time of willow are well matched. This overlapping phenological phenomena and striking color contrast symbolize the coming of spring, well regarded as a scene of paradise on earth. West Lake in Hangzhou and Slender West Lake in Yangzhou are the most famous places in the world for the successful use of this planting model.

Pines and bamboos are also excellent partners of ornamental peach. The bright color and vivid flowers of ornamental peach demonstrate a contrast against the seriousness of the evergreen pine tree. Flexible bamboo branches with dynamic ornamental peach flowers boasts a glorious view full of temperament and interest, strong as well as graceful.

Street trees

Traditionally the peach plus willow planting model is regarded as an early example of a tree lined street.

Pillar type ornamental peach cultivars which have a narrow crown have great potential as foundation plants. They are the best choice for street tree planting. The unique pillar tree form coupled with red leaf color and plump tree structures in the winter provide an ornamental appearance that lasts well beyond the flowering season. 'Terutebeni', 'Teruteshiro', and 'Terutemomo' as planted around Yoyoki stadium in Tokyo is an excellent example of pillar type ornamental peach, street tree planting. In Chongqing, 'Terutebeni' is beginning to be planted along highways, driveways, walkways and as partitions in the street (P.183).

Bonsai, potted and cut flowers

The ornamental peach is not commonly potted as bonsai because of its short life span. However, dwarf plants are often produced by nurseries and trained into curious shapes for special holidays, especially Chinese New Year -- Spring Festival. These are sold at plant markets or distributed by itinerant plant sellers. In northern China peaches are usually planted in pots after leaf falling and placed greenhouses, blooming phases are adjusted to fit the Chinese New Year. In southern China, peach trees are cut off at the ground level and sold in a similar fashion to freshly-cut Christmas trees during December in western countries. Cut peach trees represent a wish to bring home good fortune and are a fabulous holiday decoration during the Spring Festival (P.186).

Themed gardens

Peach themed gardens can be traced back in the Tang Dynasty (618--907). In imperial gardens a thousand years ago, a peach garden was established especially for the Emperor's enjoyment.

Themed gardens best display the diversity of ornamental peach cultivars. In addition to simple tourist sites displaying single flower fruit peaches, Beijing Botanical Garden, Kunming outskirts park and Shanghai Nanhui park are three major ornamental peach themed gardens where the ornamental peach is absolutely the leading point in the garden, with a very diverse display.

Beijing Botanical Garden (BBG) has celebrated the Ornamental Peach Festival since 1989 and began collecting ornamental peach from all over the world in 1991. BBG is the largest ornamental peach germplasm collection in China. Besides ornamental peach cultivar application BBG explores the long-standing flower culture associated with this species. Stone inscriptions of peach poems, painting and calligraphy exhibitions and horticultural deposition are integrated with the "planting" displays of the Ornamental Peach Festival. These add artistic atmosphere and contextualize the cultural importance of the species. Spring bulbs, such as tulips, hyacinth and fritillary are planted together with ornamental peaches constructing a colorful scene in the springtime. All of these aspects have made the Ornamental Peach Festival one of the most famous festivals of spring in Beijing.

Ornamental Peach and Tourism

Peach blossom is not only a symbol of spring but is also symbolic of longevity and good fortune. The appreciation peach blossom during spring has a long history in China. Peach has a combined value for food, medicine, wine, tea, cosmetics and blossom which, when integrated with its rich culture makes it a wonderful resource for local tourism. In order to develop and expand peach culture and to promote local tourism annual peach blossom festivals have been held at various sites in China during the month of peach bloom since the 1990s. Besides the themed gardens mentioned above, Hunan Taohuayuan, Shanghai Nanhui, Chengdu Longquan, Lanzhou Anning, Guangxi Liuzhou, Guangdong Qingyuan, all hold peach blossom themed activities, integrating them with sight-seeing, recreation and gourmet. The development of peach blossom related industry has provided brand new experiences to people the world over, continuing the importance of peach blossom culture and enhancing its popularity within gardens.

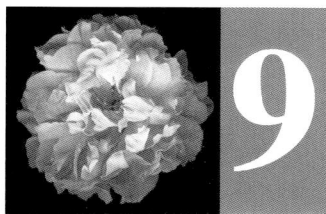

THE HAPPINESS AND LUCK FLOWER
——PEACH BLOSSOM IN ORIENTAL ART AND MYTHOLOGY

Legends of Peach

In ancient Chinese folklore the peach is believed to have originated from the far west of the country, in the Kunlun Mountains that are traditionally considered to be the home of the deity Xiwangmu - the Queen Mother of the West. It was said that in her peach orchard the flowers only bloomed once every 3000 years and the fruits of these flowers would bestow immortality on any who ate them. This tale retains its place in Chinese culture; shaped and colored peach fruit with sweetmeats called "Shou Tao" are always considered the perfect gift at a birthday party, especially for senior people, as a blessing of their longevity. "Shou Xing", the Chinese god of longevity, is typically depicted as an old man with characteristically prominent cranium, holding a staff his left hand and a peach in his right.

Peach tree elements such as bark, wood, branches, seeds, gum, leaves, roots and flowers have been used medicinally for centuries (*Shen Nong Ben Cao Jing*, 2 AD, earliest recorded Chinese pharmacopoeia). Peach seed is an effective medicine in the promotion of good blood circulation and often used in traditional Chinese medicine (or TCM) for the treatment of high blood pressure. Peach tree gums have been employed in and as medicines for diabetes and diarrhea and peach leaves are excellent at aiding the recovery of boils and abscesses, tinea manuum, and dispersing pathogenic wind. Peach root pounded with brown sugar and applied externally can treat osteomyelitis, peach bark helps to relieve chronic bronchitis, coughs and asthma and peach flower is often applied as a cosmetic by those striving for a delicate facial appearance.

For centuries peach woods were considered to possess magical properties and branches and carved peach stones were considered to ward of evil spirits. All of these medicinal indications capacitate functional immortality and the magical properties of the peach tree have been immortalized in abundant tales, legends and superstitions.

Legends of Peach Blossom

Peach has been grown in gardens as an ornamental plant since the Han Dynasty (206 BC -220 AD), this beautiful early blooming spring plant has attracted a great deal of attention in China and become an important part of Chinese culture as a symbol of good luck and happiness. There are multitudinous Chinese legends connected with peach blossoms.

Symbol of Luck and Happiness

In the ancient Chinese tale: "The tale of the Sun", a Giant hero fought with the Sun; before he died, he laid down his weapon and transformed into a forest of peach, a gift of hope for the people of the future. This tale demonstrates how peach is traditionally regarded as a holy plant and has transformed from a common fruit to a symbol of happiness, perfection, luck and jubilance.

One of the most charming legends of peach flower lore is that of the *Peach Blossom Fountain*, an allegory written by Tao Yuan Ming between 365-427 AD; describing a fisherman who became lost whilst boating along a stream. Entering from a cave, from which a faint light emanated; he found himself at a creek bordered by a dense grove of blossoming peach trees. No tree of any other kind stood. Walking through the cave he emerged into a land frozen in the times of the First Emperor (five centuries earlier). Farming families in this land where time stood still continued to wear the garb of olden days and were all living a cheerful and contented life take care to be kind and courteous to each other. The fisherman returned from his sojourn in this magical place and told his fellow villagers of his wonderful discovery. They followed him back and searched for the entrance to the cave, but, needless to say, the fisherman never found his way back and never again saw the peach blossom paradise or spent days amidst its rosy dreams and ideals that came but once in a lifetime. In homage to this famous tale, places filled with peach blossom trees in China are always considered places of ease and calm - paradise on earth and "Peach Blossom Fountain" in Chinese implies the same earthly paradise and utopia as Shangri-La does in a great many English speaking countries.

Since *Peach Blossom Fountain* a great many legends relating to peach blossom appeared. Places named "Peach Blossom Fountain" can be found in Hunan, Anhui, Zhejiang, Jiangxi and Sichuan to name a few. Even Japan has a "Peach Blossom Fountain", indicating that the peach is popular and considered as fondly in other oriental countries as it is in China.

For the Bai ethnic minority of Yunnan province, China; small branches of peach-blossom play an important role in their most important popular rural folk dancing festival that takes place every year (Chen, 2004).

Symbol of Female Beauty

The ordinary name for pink in Chinese means "peach flower colour". In ancient Chinese prose and poems bright pink peach-blossom always refers to a beautiful woman. The most famous poem of all was written by a poet named Hu Cui during the Tang Dynasty (618-907), a very simple translation may sound as follows: "There was a beautiful girl with a branch of peach blossom when the door opened last year. But the girl is not there any more this year, while the peach blossoms are still bright. " The peach-blossom represents female beauty in this poem, not only because of its bright colour but also for its delicate flower form, reminiscent of the beautiful face of a girl. Since this poem was written a great many more poems and songs have been composed that depicts the beauty of a woman through the beauty of peach blossoms. Even now in China many people still like to call beautiful girls "peach blossom".

In Japan peach-blossom is not quite as popular as their flowering cherry, though they still hold a traditional festival named "Girl's Day" (also called Peach-blossom Day) every springtime when peach trees are in full bloom. All girls wear their most beautiful kimonos and have a wonderful day under the peach trees with their close girl friends.

Chinese "Christmas Tree"

In additions to the cultural importance of peach blossom already mentioned the blooms hold special importance for people in China, particularly southern China (including Hong Kong, Macau, Guangdong and the Zhujiang Triangle region) during the lunar New Year. During Chinese New Year people like to buy a tree filled with blossoms to take home. The pronunciation of red peach-blossom is similar to the meaning of "great prosperity" in Cantonese and the taking home of blossoms brings with it hopes for happiness, prosperity, good luck and good fortune for the coming year. Not unexpectedly, the red flowered cultivars are the most popular at this time, traditionally these trees are placed in a tall vase tied with a red ribbon. People enjoy hanging many holiday presents and decorations on the tree making it, in a way; very similar to the Christmas trees of western countries. Hence the peach tree is often called the "Chinese Christmas Tree". Peach trees for the Chinese New Year are cultivated in specialist nurseries by specialist households. Around 70-80 thousand ornamental peach trees are sold during each Chinese New Year in Guangdong province alone. The Chinese Spring Festival is the most important festival in many south-eastern countries including Singapore and Malaysia, for example. Peach flowers are a popular and important cultural artefact wherever the Chinese New Year or Spring Festival is celebrated.

As a symbol of the springtime, female beauty, good fortune and happiness, the spectacular peach blossom and this charming plant have attracted wondrous attention throughout thousands of years of oriental culture.

参考文献
Reference

[1] 陈丹维, 周明芹, 高晓芹, 陈龙清. 武汉地区观赏桃花品种资源及其应用的初步调查[J]. 中国观赏园艺研究, 2006 : 60-62.

[2] 陈淏子. 花镜 (1688) (伊钦恒修订本) [M]. 北京：农业出版社重印本, 1985 : 172-175.

[3] 陈梦雷. 古今图书集成. 草木典·桃部：215-219[M]. 北京：中华书局影印本, 1934.

[4] 陈耀华. 桃花新品种的选育及对桃花品种分类的再认识[J]. 广东园林, 1997(4) : 29-30.

[5] 陈重明等. 民族植物与文化[M]. 南京：东南大学出版社, 2004, 25 : 89-91.

[6] 程中平, 陈志伟, 胡春根, 邓秀新, 罗正荣. 利用RAPD技术对碧桃种质资源的分析[J]. 西南农业大学学报, 2003, 25(1) : 4-6.

[7] 程中平, 陈志伟, 胡春根, 邓秀新, 罗正荣. 寿星桃种质资源的RAPD分析[J]. 北京林业大学, 2002, 24(3) : 74-77.

[8] 董国正. 西藏光核桃的调查[J]. 中国林副特产, 1991, 3(18) : 44-45.

[9] 段木干. 中外地名大辞典[J]. 台北：人文出版社, 1982.

[10] 方伟超, 朱更瑞, 王力荣. 短低温早花观赏桃新品种——探春的选育[J]. 果树学报, 2008. 25(6) : 957-958.

[11] 高国藩. 植物民俗比较研究. 中日民俗比较研究学术讨论会论文集[M]. 北京：北京大学出版社, 1993 : 88.

[12] 高玲. 长江三峡风景名胜区宜昌景区桃花品种资源的初步调查[J]. 花木盆景, 2002, 8 : 7.

[13] 高玲, 周凤英. 观赏桃主要病虫害调查与防治对策初探[J]. 湖北林业科技, 2004, 1:25-28.

[14] 葛洪 (晋). 西京杂记[M]. 北京：中华书局, 1985.

[15] 谷向阳. 中国楹联大观[M]. 长春：吉林文史出版社, 1993.

[16] 顾振华, 王燕, 步洪年, 柳小年. 常德地区桃花品种资源调查[J]. 湖北农业科学, 2009, 48(2) : 376-378.

[17] 韩振海. 落叶果树种质资源学[M]. 北京：中国农业出版社, 1995 : 269-271.

[18] 何浩, 何晓平. 合肥地区观赏桃花品种资源调查[J]. 安徽农业科学, 2007, 35(35) : 11472-11473, 11495.

[19] 赫西 (C.O. Hesse), 威因伯格 (J.H. Weinberger), 贝莱 (C.H. Bailey) & 豪格 (L.F. Hough), 福戈尔 (H.W. Fogle). 沈德绪等译. 桃、李、杏、樱桃育种进程[M]. 北京：中国农业出版社, 1980.

[20] 胡东燕, 俞思佳. 桃花常见病虫害及防治[J]. 花木盆景, 2002, 3 : 26-27.

[21] 胡东燕, 张秀英. 早花桃花品种的育种[J]. 北京林业大学学报, 2001, 23 : 39-40.

[22] 胡东燕, 张佐双. 桃花品种资源多样性的保护[M]. 中国园艺学会70周年优秀论文集. 北京：科学出版社, 2000. 518-520.

[23] 胡东燕. 桃花品种资源多样性调查及其分类研究[D]. 北京林业大学硕士论文, 1999.

[24] 胡东燕. 分子标记技术在桃花品种系统分类中的应用研究[D]. 北京林业大学博士论文, 2004.

[25] 黄岳渊, 黄德邻. 花经[M]. 上海：上海书店, 1985.

[26] 江雪飞. 观赏桃花设施栽培的花期调控及花芽分化特性的研究[D]. 西北农林科技大学硕士论文, 2003.

[27] 凯尔曼[美] (M.C.Kellman). 张仲等译. 植物地理学[M]. 北京：高等教育出版社, 1987.

[28] 李防 (宋) 等撰. 太平御览[M]. 北京：中华书局, 1960.

[29] 李世访, 陈策. 褐腐病的发生和防治[J]. 植物保护, 2009, 35(2) : 134-139.

[30] 李时珍. 本草纲目 (1578)[M]. 北京：人民卫生出版社, 1982.

[31] 李书文. 几种桃花品种抗旱、抗寒性初步研究[D]. 北京林业大学硕士论文, 2000.

[32] 卢戈编. 果树修剪图说[M]. 成都：四川科学技术出版社, 1990.

[33] 李亚蒙. 桃垂枝性状生理特性及SSR分子标记研究[D]. 山东农业大学硕士论文, 2006.

[34] 刘佳梦, 王虞英, 宋婧一. 北京地区两用桃育种研究进展[J]. 北京农业科学, 2000, 18(6)：23-26.

[35] 鲁振华, 宋银花, 牛良, 刘淑娥, 王志强. PermutMatrix软件及其在观赏桃形态性状聚类分析中的应用[J]. 果树学报, 2009, 26 (4)：577-580.

[36] 吕文彦, 娄国强, 张育平, 秦雪峰. 观赏桃园昆虫群落结构及多样性的研究[J]. 贵州农业科学, 2008, 36 (1)：90-92.

[37] 孟元老等. 东京梦华录[M]. 上海：古典文学出版社, 1957：374.

[38] 渠红岩. 中国古代文学桃花题材与意象研究[D]. 南京师范大学博士论文, 2005.

[39] 曲泽洲. 北京果树志·桃志[M]. 北京：北京出版社, 1990.

[40] 茹考夫斯基. 育种的世界植物基因资源[M]. 北京：科学出版社, 1974.

[41] 沈向, 毛志泉, 胡艳丽, 王丽琴, 陈学森, 束怀瑞, 李晓磊. 桃花新品种 '紫奇'[J]. 园艺学报, 2007, 34 (5)：1336.

[42] 史丽华. 中日传统节日的比较. 中日民俗的异同和交流[M]. 北京：北京大学出版社, 1993：88.

[43] 孙旭武. 低需冷量桃童期及其花芽分化期的生理生化特性研究[D]. 甘肃农业大学硕士论文, 2003.

[44] 汪灏. 广群芳谱 (1708)[M]. 上海：上海书店, 1985年影印本.

[45] 王冰[唐]注. 黄帝素问[M]. 北京：人民卫生出版社, 1982.

[46] 王珂. 桃属不同种花芽分化特性及遗传倾向的研究[D]. 河南农业大学硕士论文, 2005.

[47] 王力荣, 朱更瑞, 方伟超, 左覃元. 桃品种需冷量评价模式的探讨[J]. 园艺学报, 2003, 30(4)：379-383.

[48] 王仁裕 (五代) 撰, 丁如明辑校. 开元天宝遗事[M]. 上海：上海古籍出版社, 1985.

[49] 王象晋纂辑. 群芳谱[M]. 北京：农业出版社, 1994.

[50] 王焰安. 桃文化研究[M]. 北京：中国档案出版社, 2003.

[51] 汪祖华, 庄恩及. 中国果树志·桃卷[M]. 北京：中国林业出版社, 2001.

[52] 汪祖华, 陆振祥, 郭洪. 李、杏、梅亲缘关系及分类地位的同工酶研究[J]. 园艺学报, 1991. 18 (2)：97-101.

[53] 汪祖华. 桃种质资源亲缘演化关系研究——花粉形态分析[J]. 园艺学报, 1990, 17 (3)：161-168.

[54] 星川清亲. 栽培植物的起源与传播[M]. 郑州：河南科学技术出版社, 1931：200-201.

[55] 许士捷. 养花技艺[M]. 兰州：甘肃人民出版社, 1982.

[56] 严绍璗. 实物信仰与"桃崇拜". 中日民俗比较研究学术讨论会论文集[M]. 北京：北京大学出版社, 1993：150.

[57] 杨恭毅. 杨氏园艺植物大名典Ⅷ[M]. 台北：中国花卉杂志社, 1984.

[58] 俞德浚. 落叶果树分类学[M]. 上海：上海科学技术出版社, 1984. 82-95.

[59] 俞德浚. 中国果树分类学[M]. 北京：中国农业出版社, 1979.

[60] 藏得奎, 于东明, 杨美铃, 于文胜, 张明欣. 山东省观赏桃花品种资源的初步调查[J]. 山东农业科技, 1998 (4)：1-6.

[61] 札林库伯. 伊朗民族史 (波斯文)[M]. 德黑兰：阿米尔卡迪尔出版社, 1985：355-356.

[62] 张春英. 桃花种质资源的数量学分类研究[J]. 北京林业大学学报, 1999, 21 (3)：41-45.

[63] 张春英, 林同香, 戴思兰. 桃花种质资源亲缘关系的研究——RAPD分析[J] 北京林业大学学报, 1999, 21 (5)：26-31.

[64] 张秀英. 桃花[M]. 上海：上海科学技术出版社, 2001.

[65] 张秀英. 桃花品种分类第二报[J]. 广东园林, 1993 (3)：33-38.

[66] 张秀英, 王雁, 王桂萍. 桃花种质资源花粉形态的观察与比较[J]. 北京林业大学学报, 1997, 19 (2)：57-62.

[67] 张秀英, 陈忠国. 北京市桃花品种调查与分类初探[J]. 园艺学报, 1991, 18 (1)：67-74.

[68] 周建涛, 李惠芬. 观赏桃主要性状遗传规律与国外育种实践[J]. 北京林业大学学报, 1998, 20 (2)：100-103.

[69] 周师厚. 洛阳花木记 (1082), 说郛[M]. 上海：上海古籍出版社, 1985.

[70] 周文华. 汝南圃史 (1620) (抄本) 卷三, 六—八[M]. 北京：农业出版社, 1982：6-8.

[71] 朱更瑞, 方伟超, 王力荣. 观赏桃品种需冷量的研究[J]. 植物遗传资源学报, 2004, 5 (2)：176-178.

[72] 竹内理三等. 日本历史辞典[M]. 天津：天津人民出版社, 1988.

[73] 宗学普, 段玉春. 光核桃的分布及其类型初探. 西藏作物品种考察文集[M]. 北京：中国农业科学技术出版社, 1987：184-185.

[74] 邹喻萍, 葛颂, 王晓东. 系统与进化植物学中的分子标记[M]. 北京：科学出版社, 2001.

[75] Fleur. 花の百科全集46, もも·菜の花(日文) [M]. 东京：讲谈社, 1996：6.

[76] 吉田雅夫. 桃圖鑑(日文) [M]. 东京：世界文化社, 1985.

[77] 吉田雅夫, 清家金嗣. 赤葉のわい性モモ‘レッド·ドワ フ’について(日文) [J]. 果树试报I, 1974, 25—32.

[78] 前田克夫, 萩原勲, 志村勲, 箱田直纪, 石川俊二. もも 幸ホワイト(さいわいホワイト)[J]. 果树种苗, 2005, vol. 99：34-35.

[79] 伊藤伊兵卫三之丞. 花坛地锦抄 (日文) [M]. 农山渔村文化协会发行, 1995：142-144.

[80] 塚洋太郎等. 花坛纲目. 园艺植物大事典(日文) [M]. 小学馆出版社, 1994：2480-2488.

[81] 山崎和雄, 冈部诚, 高桥荣治. ハナモモ新品种‘照手红’.‘照手桃’.‘照手白’の育成经过と特性 (日文) [J].神奈川园试研报, 1987, 34：54-56.

[82] 堀越祯一, 佐藤嘉子, 冈部诚, 山崎和雄, 高桥荣治. ハナモモ新品种‘照手姬’の育成经过と特性 (日文) [J]. 神奈川园试研报, 1992, 42：29-31.

[83] Alvino, A., V. Maliulo, and G. Zerbi. Problems of peach (*Prunus persica*) tolerance to anaerobic conditions due to excess soil water [J]. *Rivista* Ortoflorofrutticolura Italiana，1986，70:263-270.

[84] Anderson, P.C., P.B. Lombard and M.N.Westwood. Leaf conductance, growth, and survival of willow and deciduous fruit tree species under flooded soil conditions. [J]. J. Amer. Soc. Hort. Sci，1984，109:132-138.

[85] Aranzana, M.J., M.C. Vicente and P. Arus. Comparison of fruit and leaf DNA extracts for AFLP and SSR analysis in peach (*Prunus persica*) [C]. Acta Hort，2001，546:297-300.

[86] Bailey, J.S. and A.P. French. The inheritance of blossom type and blossom size in the peach[C]. Proc. Amer. Soc. Hort. Sci，1942，40:248-250.

[87] Bailey, J.S., and A.P. French. The inheritance of certain fruit and foliage characters in the peach[R]. Massachusetts Agri. Expt. Sta. Bul，452，1949.

[88] Baird, W.V., A.S., Estager, and J., Wells. Estimating nuclear DNA content in peach and related diploid species using laser flow cytometry and DNA hybridization [J]. J. Amer. Soc. Hort. Sci，1994，199:1312-1316.

[89] Ballinger, W.E., H.K. Bell and N.F. Childers. Peach nutrition. In: Childers, N.F. (ed.) Nutrition of Fruit Crops, Temperate, Sub-tropical, Tropical [M]. Somerset Press, Sommerville, New Jersey，1966，pp:276-390.

[90] Bassi, D., A. Dima, and R. Scorza. Tree structure and pruning response of six peach growth forms [J]. J. Amer. Soc. Hort. Sci, 1994, 119(3):378-382.

[91] Batra, L.R. World species of *Monilinia* (Fungi): Their ecology, biosystematics and control [M]. Mycol. Mem，1991，16: 1-246. J. Cramer, Berlin.

[92] Batsch, A.J.G.K. Beytrage und entwurfe zur pragmatischen genschichte der drey nature-reiche nach ihren, verwandschaften (Beytr. Entw. Pragm.Gesch.Natur) [M]. p.96; in 4. Weimar: im verlage des industrie comptoirs. 1801.

[93] Bean, W. J. Trees and shrubs, hardy in the British Isles. Vol.2 [M]. 1950, London: 564.

[94] Bentham, G. and J.D. Hooker. Genera plantarum. vol. 1 [M]. Lovell Reed and Co. London, 1865.

[95] Berman M.E. and T.M. Dejong. Water stress and crop load effects on fruit fresh and dry weights in peach (*Prunus persica*) [M]. Tree Physiology, 1996, 16:859-864.

[96] Biggs, A.R. and N.W. Miles. Suberin deposition as a measure of wound response in peach bark [J]. HortScience, 1985, 20:903-905.

[97] Blake, M.A. The J.H. Hale peach as a parent in peach crosses[C]. Proc. Amer. Soc. Hort. Sci, 1932, 29:131-136.

[98] Blake, M.A. Progress in peach breeding [C]. Proc. Amer. Soc. Hort. Sci, 1937, 35:49-53.

[99] Blake, M.A. and C.H. Connors. Early results of peach breeding in New Jersey [R]. New Jersey Agri. Expt. Sta, 1936, Bul:599.

[100] Bortiri, E., S. Oh, J. Jiang, S. Baggett, A. Granger, C. Week, M, Buckingham, D. Potter, and D.E. Parfiti. Phylogeny and systematics of Prunus (Rosaceae) as determined by sequence analysis of ITS and the chloroplast trnL-trnF spacer DNA[J]. Syst. Bot. 2001, 26:797-807.

[101] Bowen, H.H. Breeding peaches for warm climates [J]. HortScience 1971, 6:153-157.

[102] Brannen, P., D. Horton, B. Bellinger, and D. Ritchie. Southeastern peach, nectarine and plum pest management and cultural guide [R]. Georgia extension bulletin, 2007, 1171:Univ. Georgia cooperative extension service, College of Agri. And environmental Sci., Athens, Georgia.

[103] Brickell, C.D., B.R. Baum, W.L.A. Hetterscheid, A.C. Leslie, J. McNeill, P. Trehane, F. Vrugtman, and J.H. Wiersema. 2004. International Code of Nomenclature for Cultivated Plant. Acta Hort. 647.

[104] Brickell, C. and J. D. Zuk. A- Z Encyclopedia of garden plants [M]. DK Publishing, 1997, Inc.: 839.

[105] Byrne, D.H. Mechanisms of spring freeze injury avoidance in peach [J]. HortScience, 1986, 21; 1235-1236.

[106] Chang, L.S., A. Iezzoni, G. Adams, and G.S. Howell. *Leucostoma persoonii* tolerance and cold hardiness among diverse peach genotypes [J]. J. Amer. Soc. Hort. Sci, 1989, 114:482-485.

[107] Chen, D., M. Zhou, X. Gao, and L. Chen. Preliminary investigation of peach-blossom cultivar resources and, it's landscaping application in Wuhan city. Advances in ornamental horticulture of China [M] (in Chinese). Beijing: China Forestry Publishing House. 2006, 60-62. ▲

[108] Chen, H. Hua jing [M] (in Chinese). Beijing: Agricultural Press. 1985, 1688. ▲

[109] Chen, J. Quan fang bei zu [M] (in Chinese). Beijing: Agricultural Press. 1982, 1256.

[110] Chen, M. Gu jin tu shu ji cheng [M] (in Chinese) (548). Beijing: Chinese Book Bureau, (reprint in 1934), 1728. ▲

[111] Chen, Y. History study on evolution relationships between ornamental peach cultivars [J] (in Chinese).Guangdong gardening, 1993, (1):35-38. ▲

[112] Cheng, Z. RAPD Analysis of Red-leaf Peach Germplasm Resource [J]. J. Southwest Agri, (in Chinese).Colleg, 2003a, 18 (2): 85-87.

[113] Cheng, Z., Z., Chen, C., Hu, X., Deng, and Z, Luo. RAPD analysis of the germplasm of dwarf peach [J]. (in Chinese). J. Beijing For. Univ, 2002, 24(3):74-77. ▲

[114] Cheng, Z. RAPD analysis of the germplasm of *Amygdalus persica* var. duplex [J]. (in Chinese). J. Southwest Agri, Univ, 2003b, 25(1): 4-6. ▲

[115] Connors, C.H. Some notes on the inheritance of unit characters in the peach [C]. Proc. Amer. Soc. Hort. Sci, 1920, 16:24-36.

[116] Conners, C.H. Peach breeding. A summary of results [C]. Proc. Amer. Soc. Hort. Sci, 1922, 19:108–115.

[117] Connors, C.H. Inheritance of foliar glands of the peach [C]. Proc. Amer. Soc. Hort. Sci, 1921, 18:20-26.

[118] Connors, C.H. The sterility of 'J.H. Hale' [R]. New Jersey Agriculture Experimental Station Annual Report (1925), 1926, 46:90-91.

[119] Culham, A. and M.L. Grant. DNA markers for cultivar identification and classification [M]. p.183-198. In: S. Andrewsm, A. Lesile, and C. Alexander (eds.). Taxonomy of cultivated plants: third international symposium. Royal Botanic Garden, Kew. London, 1999, 183-198.

[120] Daane K.M., R.S. Johnson, T.J. Michailides, C.H. Crisosto, J.W. Dlott, H.T. Ramirez, G. Y. Yokota, and D.P. Morgan. Excess nitrogen raises nectarine susceptibility to disease and insects [R]. California Agri, 1995, 49:13-18.

[121] De Candolle, A. P. Prodromus systematis naturalis regni vegetabilis [M]. Vol. 2.Treuttel et Wurtz. Paris, 1825.

[122] Debener, T. Molecular markers as a tool for analysis of genetic relatedness and selection on ornamentals [M]. p. 329-345. In: A. Vainstein (ed.). Breeding for ornamentals: Classical and molecular approaches. Kluwer Academic Publishers. London. 2002.

[123] Debergh, P.C. Effect of agar brand and concentration on the tissue culture medium [M]. Physiologia plantarum, 1983, 59:270-276.

[124] Dirlewanger, E. and C., Bodo. Molecular genetic mapping in peach [J]. Euphytica, 1994, 77:101-103.

[125] Dirlewanger, E. and P. Arus. Markers in fruit tree breeding: improvement of peach [M]. In: Lorz, H. and Wenzel, G. (eds.) *Molecular Marker Systems in Plant Breeding and Crop Improvement*. Springer, Berlin, pp, 2008, 79-304.

[126] Dirr, M.A. Manual of woody landscape plants: their identification, ornamental characteristics, culture, propagation and uses (5th edition) [M]. Stipes Publishing, Champaign, IL, 1998.

[127] Downing, C. The fruits and fruit trees of America [M]. Wiley, New York, 1866.

[128] Dowing, A.J. Dowing's fruit and fruit trees of America [M]. 1869, 580-639.

[129] Dye, M. H., L. Buchanan, F.D. Dorofaeff and F.G. Beecroft. Boron toxicity in peach and nectarine trees in Otago [J]. N. Z. J. of Expt. Agri. 1984, 12:303-325.

[130] Edwards, G.R. Temperature in relation to peach culture in the tropics [C]. Acta Hort, 1987, 199:61-62.

[131] Erdtman. Handbook of palynology [M]. Munksgaard, Copenhagen,1969.

[132] Everett, T.H, New illustrated encyclopedia of gardening [M]. New York Botanical Garden Press, New York, 1967.

[133] Fang, W., G. Zhu, and L. Wang. Tanchun, a new early blosooming ornamental peach cultivar [J] (in Chinese). J. Fruit Sci, 2008, 25(6):957-958. ▲

[134] Fernald, M.L., Gray's manual of botany [M]. American Book Company, NY, 1987.

[135] Finch, C.R. D.H. Byrne, C.G. Lyons and H.D. Pennington. Sulfur nutrition requirements of peach trees [J]. J. Plant Nutrition, 1997, 20:1711-1721.

[136] Gao, L. A preliminary study on the investigation of peach-blossom cultivars in Yichang area of Three, Gorgers Resort of Yangtze River [J] (in Chinese). Flowers and Potted Landscape, 2002. 8:7. ▲

[137] Garcia, J.A. and M. Cambra. Plum pox virus and sharka disease [J]. Plant Viruses, 2007, 1:69-79.

[138] Ge, H. (1256). Xi jing za ji [M] (In Chinese). Chinese Book Bureau, Beijing (reprint in 1985). ▲

[139] Geruter, W., J. McNeill, F.R. Barrie, H. M. Burdet, V. Demoulin, T. S. Filgueiras, D.H. Nicolson, P.C. Silva, J.E. Skog, P. Trehane, N.J. Turland, D.L.Hawksworth (eds.). International Code of Botanical Nomenclature [M]. Koeltz Scientific Books, Konigstein. Germany, 2000.

[140] Gentry, C.R. and J.M.Wells, Evidence of an oviposition stimulant for peach tree borer *Synanthedon exitiosa* [J]. J. Chem. Ecol, 1982. 8:1125-1132.

[141] Giesberger, G. Climatic problems in growing deciduous fruit trees in the tropics and subtropics [C]. Tropical Abstracts, 1972. 27:1-8.

[142] Glenn, D.M., R. Scorza, and C. Bassi. Physiological and morphological traits associated with increased water use efficiency in the narrow-leaf peach [J]. HortScience, 2000. 35:1241-1243.

[143] Goffreda, J.C., A.M. Voordeckers, S.A. Mehlenbacher, 'Jerseypink' ornamental peach [J]. HortScience, 1992. 27(2):183.

[144] Gradziel, T.M. and W. Beres. Semidwarf growth habit in clingstone peach with desirable tree and fruit qualities [J]. HortScience, 1993. 28:1045–1047.

[145] Grant, M.L. and A. Culham. DNA fingerprinting and the identification of cult vars [J]. New Plantsman, 1997. 4 (3):157-168.

[146] Grasselly, C. Study of the possibilities of producing intra- and interspecific F₁ hybrids in the sub-genus *Amygdalus*. I. Heterosis in *Prunus persica* hybridized with *Prunus persica*, *Prunus davidiana* and *Prunus kansuensis* (in French) [J]. Ann. Amel. Planets, 1974. 24:302-315.

[147] Gu, Z., Y. Wang, H. Bu, and X. Liu. Investigation of peach blossom cultivar resources in Changde area [J] (in Chinese). Hubei Agri. Sci, 2009. 48(2):376-378. ▲

[148] Guerriero, R. and G. Scalabrelli. Effect of stratification duration on seed germination of several peach line rootstocks [C]. Acta Hort, 1984. 173:185-190.

[149] Gupta, U.C., Y.W. Jame, C.A. Campbell, A.J. Leyshon and W. Nicholaichuk. Boron toxicity and deficiency: a review [J]. Can. J. Soil Sci, 1985. 65:381-409.

[150] Harrison, R.E. Handbook of trees and shrubs for the Southern Hemisphere [M]. Angus & Robertson LTD, Sydney. 1963.

[151] Hartmann, H.T., D.E. Kester, F.T. Davies Jr. and R.L. Geneve. Plant propagation. Principles and practices [M]. 7th edition. Prentice-Hall, Upper Saddle River, New Jersey. 2002.

[152] Haw, S. Chinese flowering plums and cherries [J]. The garden, 1987. 112(5):224-228.

[153] Hedrick, U.P. The Peaches of New York [M]. J.B. Lyon Company Printers, Albany, New York. 1917.

[154] Heslop-Harrison, J. Pollen wall development [J]. Science, 1968. 161:230-237.

[155] Hesse, C.O. Peaches. In: J. Janick and J.N. Moore (*eds.*) Advances in fruit breeding [M]. Purdue Univ. Press, West Lafayette, Indiana. 1975, p.285-355.

[156] Hooper, J. The Italian dwarf peach [J]. Amer. J. Hort, 1867. 2:287–288.

[157] Horikoshi T., Y. Sano, M. Okabe, K. Yamazaki, and E. Takahashi. New broomy flowering peach cultivars, 'Terutehime' [J] (in Japanese). Bull. Kanagawa Hort Expt Sta, 1992. (42):29-31. *

[158] Hoshikawa, S. Origin and spreading of cultivated plants [M]. (in Chinese translation version). Zhengzhou: Zhengzhou Science and Technology Press, 200-201. 1981. ▲

[159] Hu, D. Studies on ornamental peach systematics using molecular markers [D] (In Chinese). Diss. Beijing For. Univ, 2004. ▲

[160] Hu, D. Studies on diversities investigation and classification of ornamental peach cultivars [D] (in Chinese). Thes. Beijing For. Univ, 1999. ▲

[161] Hu, D. and R. Scorza. Analysis of the 'A72' Peach Tree Growth Habit and Its Inheritance in Progeny Obtained from Crosses of 'A72' with Columnar Peach Trees [J]. J. Amer. Soc. Hort. Sci, 134 2009. (2):236-243.

[162] Hu, D., S. Zhang, and D. Zhang. 'New ornamental peach 'Zuoshuang' [J]. HortScience, 2005b, 40(2):500.

[163] Hu, D. and X. Zhang. Breeding of early ornamental peach cultivars [J] (in Chinese). J. Beijing For. Univ, 2001. 23:39-40. ▲

[164] Hu, D., Y. Li, and Y. Huo, Investigation of peach-blossom (ornamental peach) cultivars resources in Shanghai and Hangzhou (in Chinese) [J]. J. Beijing For. Univ, 1998. 20:114-117.

[165] Hu, D. and Z. Zhang. Breeding trends for ornamental peach [C]. Acta Hort, 2008. 769: 351-356.

[166] Hu, D. and Z. Zhang. The diversity conservation of peach-blossom cultivars [C]. Proc. Third conference of international association of botanical gardens Asian division:117-121. 2000. ▲

[167] Hu, D, Z. Zhang, X. Zhang, and Q. Zhang. The germplasm preservation of ornamental peach cultivars [C]. Acta

Hort, 2003. 620:395-402.

[168] Hu, D., Zhang, Z., Zhang, Q., Zhang D, and J. Li. Ornamental Peach and Its Genetic Relationship Revealing by Inter-Simple Sequence Repeat (ISSR) Fingerprints [C]. Acta Hort, 2006. 713:113-120.

[169] Hu, D., Z., Zhang, Q., Zhang, D., Zhang, and J., Li. Genetic Relationship of Ornamental Peach Determined Using AFLP Markers [J]. HortScience, 2005a. 40 (6): 1782-1786.

[170] Huang, Y. and D., Huang. Hua jing [M] (in Chinese). Shanghai: Shanghai Bookstore, 1985. ▲

[171] Huxley, A. and Griffiths M. The new Royal Horticulture Society dictionary of gardening. vol. 3[M]. 1999. 3:738.

[172] Illustrated Encyclopedia of Gardening in South Africa, Reader's digest [M]. Cape Town, 1984:204.

[173] International Union For The Protection of New Varieties of Plants [S/OL], TG/1/3, April 19, 2002

[174] Ito, I.S. 1695. *Kadan-chikin-sho* selection. Nousan gyozon culture association [M] (in Japanese). Tokyo (reprint in 1995), 142-144. *

[175] Jacobson, A.L. North American landscape trees [M]. Ten Speed Press, Berkeley, California:511-519. 1996.

[176] Jia, S. (533-544). *Qi Min Yao Shu* (in Chinese) [M]. Beijing: Reprinted in Beijing Agri. Press, 1963, 49-52.

[177] Jiang, X. Study on Blooming Date Regulation of Ornamental Peaches under Protected Culture and the Characteristics of Flower Bud Differentiation [D] (in Chinese). Thes. Northern Agri. and For. Univ, 2003. ▲

[178] Johnson, R.S., H. Andris, K. Day and B. Beede. Using dormant shoots to determine the nutritional status of peach trees [C]. Acta Hort, 2006. 721:285-290.

[179] Kester, D.E. and C.O. Hesse. Embryo culture of peach varieties in relation to season of ripening [C]. Proc. of the Amer. Soc. Hort. Sci., 1955. 65:265-273.

[180] Koenhe, E. Deutsche dendrology [M]. Stuttgart: Verlag von Ferdinand Enke, 1893.

[181] Komar-Tyomnaya, L.D. 'Lel' - new varieties of ornamental peach [G]. Agrarian science to production, (in Russian) 2008.

[182] Komar-Tyomnaya L.D., The new perspective varieties of ornamental peach. (in Ukraine) Varieties studying and guard of the rights on plants varieties [J]. 2007. 6:102-108.

[183] Komar-Tyomnaya L.D., The breeding of ornamental peaches // Intensification and breeding of the fruit cultures (in Russian) [G]. Collected scientific works of the State Nikita Botanical Garden. Yalta. State Nikita Botanical Garden, 1999.

[184] Komar-Tyomnaya, L.D., Estimation to resistance of the peach to *Taphrina deformans* depending on species accessories. (in Russian) [G]. Materials of YI Inter. conf. young scientists, "Problems of dendrology, horticulture and floriculture" : 229-235. 1998.

[185] Komar-Tyomnaya, L.D. Main directions of selection of ornamental peaches [C]. 1st Inter. Meeting of Young Scientist in Horticulture. Lednice, Czech, Rep. 1998.

[186] Koornneef, M., L. Bentsinka and H. Hilhorstb. Seed dormancy and germination [G]. Current opinion in plant biology, 2002. 51:33-36.

[187] Krussmann, G. Manual of cultivated broad-leaved trees and shrubs [M]. Timber Press (Vol.3). Trans. M.E.Epp. Portland, Oregon. 1986.

[188] Lammerts, W.E. The breeding of ornamental edible peaches for mild climates. I. Inheritance of tree and flower characters [J]. Amer. J. Bot, 1945. 32:53-61.

[189] Larsh, H.W. and H.W. Anderson. Bacterial spot of stone fruit [R]. Univ. Ill. Agr. Expt. Sta. Bul, 530. 1948.

[190] Layne, D.R. and D. Bassi. The peach, botany, production and uses [M]. London: CABI Intl, 2008

[191] Layne, R.E.C. Breeding peaches in North America for cold hardiness and perennial canker (*Leucostoma* spp.) resistance-Review and outlook [J]. Fruit Var. J, 1984.38:130-136.

[192] Layne, R.E. C. Harrow Frostipink, Harrow Candifloss and Harrow Rubirose ornamental peaches [J]. Can. J. Plant Sci, 1981. 61:157-159.

[193] Layne, R.E.C. Peach canker research at harrow [C]. Proc. Natl. Peach Council 30th Ann. Conv, 1971. P:40-41.

[194] Leece, D.R., F.W. Cradock and O.G. Carter. Development of leaf nutrient concentration standards for peach trees in New South Wales [J]. J. Hort. Sci, 1971. 46:163-175.

[195] Li, H.L. The garden flowers of China [M]. Ronald Press, New York. 1959.

[196] Li, S.1596. Ben cao gang mu [M] (in Chinese). Beijing: Sanitation Press, (reprint in 1982). ▲

[197] Li, S. Preliminary studies on drought resistance and cold hardiness in several ornamental peach cultivars (*Prunus persica* Batsch.) [D] (in Chinese). Thes. Beijing For. Univ, 2000. ▲

[198] Li, Y. Physiological characteristics of weeping character and study on SSR molecular marker [DJ] (in Chinese). Thes. Shandng Agri. Univ, 2006. ▲

[199] Linnaeus, C. 1753. p.677. Species plantarum [M]. 2 vols, Stockholm. (Facsimile edition: London, 1957-1959).

[200] Lobit, P., S. Soing, M. Genard and R. Habib. Effects of timing of nitrogen fertilization on shoot development in peach (*Prunus persica*) trees [J]. Tree Physiology, 2001. 20:35-42.

[201] Loreti, F., R. Muleo and S. Morini. Effect of light quality on growth of *in vitro* cultured organs and tissues [C]. The International Plant Propagation Soc. Combined Proc, 1991. 40:615-623.

[202] Loreti, F., S. Morini and A. Grilli. Rooting response of BS B2 and G.F. 677 rootstocks cutting [C]. Acta Hort, 1985. 173:261-269.

[203] Maeda, K., I. Ogiwara, I. Shimura, N. Hakoda, and S. Ishikawa. 'Saiwai Howaito' [J]. Fruit Trees Seedling, vol. 2005. 99:34-35. *

[204] Malcolm, History of the peach tree [M/OL]. http://articles.directorym.net/History_of_Peach_Trees_Prunus_Persica_Wichita_KS-r861433-Wichita_KS. html.

[205] Massoné, G. Investigations on the resistance of peach (*Prunus persica*) to peach aphids (in French). Lutte biologique et integree contre les pucerons [G]. Collique franco-societtique, Rennes, 26-27 Sept. 1979 p. 1979. 73-79.

[206] Massonie, G., P. Maison, R. Monet, and C. Grasselly. Resistance to the green aphid *Myzus persicae* Sulzer (Homoptera Aphididae) in *Prunus persicae* (L.) Batsch and other *Prunus species* [J]. Agronomie, 1982. 2:63-70.

[207] Meader, E.M. and M.A. Blank. Some plant characteristics of the second generation progeny of *Prunus persica* and *Prunus kansuensis* crosses [C]. Proc. Amer. Soc. Hort. Sci, 1939. 37: 223-231.

[208] Mehlenbacher, S.A. and R. Scorza. Inheritance of growth habit in progenies of 'Com-Pact Redhaven' peach [J]. HortScience, 1986. 21:124-126.

[209] Messeguer, R., P. Arus, and M. Carrera. Identification of peach cultivars with pollen isozymes [J]. Scientia Hort, 1987. 31:107-117.

[210] Moing, A. J.L. Pöessel, L. Svanella-Dumas, M. Loonis, and J. Kervella. Biochemical basis of low fruit quality of *Prunus davidiana*, a pest and disease donor for peach breeding [J]. J. Amer. Soc. Hort. Sci, 2003. 128:55-62.

[211] Monet, Le pêcher. Geneticque et physiolgie (in French) [M]. Masson, Paris, France, 1983.

[212] Monet, R. and G. Salesses. Un nouveau mutant de nanisme chez le pecher [J] (in French). Annales de l' Amelioration des Plantes, 1975. 25:353–359.

[213] Monet, R., Y. Bastard, and B. Gibault. Etude genetique du caractere "port pleureur" chez le pecher [J] (in French).

Agronomie, 1988. 8:127–132.

[214] Moore, J.N., R.C. Rom, S.A. Brown, and G.L. Klingaman. 'Bonfire' dwarf peach, 'Leprechaun' dwarf nectarine, and 'Crimson Cascade' and 'Pink Cascade' weeping peaches[J]. HortScience, 1993. 28(8):854.

[215] Morini, S., P. Fortuna, R. Sciutti and R. Muleo. Effect of different light-dark cycles on growth of fruit tree shoots cultured *in vitro* [M]. Advances in Hort. Sci, 1990. 4:163-166.

[216] Naor, A., I. Klein, H. Hupert, Y. Grinblat, M. Peres and A. Kaufman. Water stress and crop level interactions in relation to nectarine yield, fruit size distribution, and water potentials [J]. J. Amer. Soc. Hort. Sci, 1999. 124:189-193.

[217] Notcutt, R.C., Flowering trees and shrubs. Edited by the Late W.R. DYKES [M]. Martin Hopkinson and Company LTD, London:128-133 1926.

[218] Oakenfull, J.C.1913. Brazil in 1912 [M]. Robert Atkinson（London）Limited. London, 1913.

[219] Okie, W.R. and R. Scorza. Breeding Peach For Narrow Leaf Widt [C]h. Acta Horticulturae, 2002. 592:137-141.

[220] Parfiti. Phylogeny and systematics of *Prunus* (Rosaceae) as determined by sequence analysis of ITS and thechloroplast trnL-trnF spacer DNA [J]. Syst. Bot, 2001. 26:797-807.

[221] Pascal, T., Sicard, O., Abernathy, D., Lambert, P., Foulongne, M., Schurdi-Levraud, V., Kervella, J., Decroocq, V. Genetic Mapping of Quantitative Trait Loci Involved in the Resistance to Plum Pox Virus in *Prunus davidiana* and Development of Markers Linked to Resistance to Sharka Disease in Peach [C]. Proc. VI Peach Symposium. Santiago, Chile, 2005.

[222] Paunovic, S.A., D. Ogasanovic, and R. Plazinic. The effect of breeding parents on the susceptibility of F_1 progeny of peaches towards powdery mildew [*Sphaerotheca pannosa* (Wallr.) Lev.] [J]. J. Yugoslav Pomology, 1976. 37-38: 331-340.

[223] Pawasut, A., K., Yamane, Y.T., Yamaki, K., Tsukahara, Y. Ijiro, and N., Fujishige, Studies of tree performance, morphology and phenology in ornamental peaches [J]. Bul. of the research farm faculty of Agri, Utsunomiya Univ, 2004. 21:1-12.

[224] Pawasut, A., N. Fujishige, K. Yamane, Y. Yamaki, and H. Honjo. Relationships between Chilling and heat Requirement for Flowering in Ornamental Peaches [J]. J. Japan. Soc. Hort. Sci, 2004. 73(6):519-523.

[225] Pisani, P.L. and G. Roselli. Interspecific hybridization of *Prunus persica* × *P. davidaina* to obtain new peach rootstocks [J]. Genetica Agraria, 1983. 1/2:197-198.

[226] Pusey, P.L. Availability and dispersal of ascospores and conidia of *Botryosphaeria* in peach orchards [J]. Phytopathology, 1989. 79:635-639.

[227] Qu, Z. Beijing fruit monograph-Peach flora [M]. Beijing: Beijing Press, (in Chinese). 1990. ▲

[228] Quamme, H.A., R.E.C. Layne, and W.G. Ronald. Relationship of super-cooling to cold hardiness and the northern distribution of several cultivated and native *Prunus* species and hybrids [J]. Can. J. Plant Sci, 1982. 62:137-148.

[229] Ramming, D.W. The use of embryo culture in fruit breeding [J]. HortScience, 1990. 25:393-398.

[230] Rehder, A., Manual of cultivated trees and shrubs--hardy in North America [M]. Macmillan Company, New York: 462. 1927.

[231] Ritchie, D.F. Sprays for control of bacterial spot of peach cultivars having different levels of disease susceptibility[J]. Fungicide & Namaticide tests, 1999. 54:63-64.

[232] Robinson, J.B., M.T. Treeby and R.A. Stephenson. Fruits, vines and nuts. in: Reuter, D.J. and J.B. Robinson, (eds.) Plant Analysis, An Intl. manual, 2nd edition [M]. CSIRO Publishing, Collingwood, Australia, 1997. pp:349-382.

[233] Roselli, G. and E. Bellini. Investigations on peach susceptibility to mildew *Sphaerotheca pannosa* (Wallr.) Lev. (in

Italian) [J]. Riv. Ortoflorofrutt, 1976. 60:1-16.

[234] Rouse, R.E. and W.B., Sherman, High night temperature during bloom affect fruit set in peach [C]. Proc. of the Florida State Hort. Sci, 2002. 115:96-97.

[235] Ruan, Y. 1150. Shi san jing zhu shu ben [M] (in Chinese). Beijing: Chinese Book Bureau, (reprint in 1985).

[236] Sanchez, E.E., S.A. Weinbaum and R.S. Johnson. Comparative movement of labeled nitrogen and zinc in 1-year-old peach *Prunus persica* (L.) Batsch. trees following late-season foliar application [J]. J. Hort. Sci. & Biotechnoloy, 2006. 81: 1667-1675.

[237] Sanchez, E.E. Nutricion mineral de frutale de peptia y carozo [R]. Publication del Instituto Nacional de Tecnologia Agropecuaria. Estacion Experimental Alto Valle de Rio Negro, Macrorregion Patagonia Norte, Argentina. 1999.

[238] Saunier, R. Contribution to the study of relationships between certain characteristics of simple genetic 1973. determination in the peach tree and susceptibility of peach cultivars to oidium, *Sphaerotheca pannosa* (Wllr.) Lev. (in French) [G]. Ann. Amel. Plantes, 23:235-243.

[239] Sciutti, R. and S. Morini. Effect of relative humidity *in vitro* culture on some growth characteristics of a plum rootstock during shoot proliferation and on plantlet survival [M]. Advances in Hort. Sci, 1993. 4:153-156.

[240] Schaffer, K.A., P.C. Anderson, and R.C. Ploetz. Responses of fruit crops to flooding [J]. Horticultural Reviews, 1992. 13:257-301.

[241] Scorza, R. Ornamental peach trees for landscape and garden [J]. Landscape Plant News, 2003. 14 (2):1-4.

[242] Scorza, R., D. Bassi, and A. Liverani, Genetic interactions of pillar (columnar), compact, and dwarf peach tree genotypes [J]. J. Amer. Hort. Sci, 2002. 127(2): 254-261.

[243] Scorza, R., G.W. Lightner, and A. Liverani. The pillar peach tree and growth habit analysis of compact × pillar progeny [J]. J. Amer. Soc. Hort. Sci, 1989. 114:991–995.

[244] Scorza, R. and W.B. Sherman. Peaches. In: Janick, J. and J.N. Moore (eds.) Fruit Breeding. Vol.1. Tree and Tropical Fruits [M]. Wiley, New York, pp:325-440. 1996.

[245] Scorza, R. and W.R. Okie. Peaches (*Prunus*). In: Moore, J.N. and J.R. Ballington, Jr. (eds.) Genetic resources of temperate fruit and nut crops [C]. Acta Hort, 1991. 290:177-231.

[246] Scorza, R., S.A., Mehlenbacher, and G.W. Lightner. Inbreeding and co-ancestry of freestone peach cultivars of the eastern United States and implications for peach germplasm improvement [J]. J. Amer. Soc. Hort. Sci, 1985. 110: 547-552.

[247] Scott, D.H. and F. Cullinan. The inheritance of wavy leaf character in the peach [J]. J. Heredity, 1942. 33:293-295.

[248] Scott. D.H. and J.H.Weinberger. Inheritance of pollen sterility in some peach varieties [C]. Proc. Amer. Soc. Hort. Sci, 1944. 45:229-232.

[249] Shen, X, Y, Li, L. Kang, Y., Zou, and H. Shu. Relationship between morphology and hormones during weeping peach (*Prunus persica* var. *pendula*) shoot development [J] (in Chinese). Acta Hort. Sinica, 2008. 35(3): 395-402. ▲

[250] Shen, X., Z. Mao, Y. Hu, L. Wang, X Chen, H. Shu, and X. Li. A new ornamental peach cultivar 'Ziqi' [J] (in Chinese). Acta Hort Sinica, 2007. 34(5)1336. ▲

[251] Smith, C.A., C.H. Bailey and L.F. Hough. Methods for germinating seeds of some fruit species with special reference to growing seedlings from immature embryos [R]. New Jersey Agri. Expt. Sta. Bul, 1969. 823.

[252] Strong, W.C. The van buren golden dwarf peach [J]. Amer. J. Hort, 1867. 2:171–172.

[253] Swofford, D. L. PAUP, phylogenetic analysis using parsimony [CP]. Version 4.0b10. Sinauer, Sunderland, Mass, 2002.

[254] Tagliavini, M., J. Abadia, A.D. Rombola, A. Abadia, C. Tsipouridis and Agronomic means for the control of iron

deficiency chlorosis in deciduous fruit trees [J]. J. Plant Nutrition, 2000. 23: 2007-2022.

[255] Tagliavini, M., D. Scudellari, B. Marangoni. and M. Toselli. Nitrogen fertilization management in orchards to reconcile productivity and environmental aspect [R]. Fertilizer research, 1996. 43:93-102.

[256] Tworkoski, T., S.S., Miller, R. Scorza. Effects of pruning on auxin and cytokinin levels and subsequent shoot regrowth among different growth habits of peach [C]. Proc. of Plant Growth Regulation Soc. Amer, 2005. 32:46.

[257] Wang, R. 950.Kai Yuan Tian Bao Yi Shi [M] (in Chinese). Beijing: Agricultural Press, (reprint in 1990) ▲

[258] Wang L., G. Zhu, W. Fang, Q, Zuo. Estimating models of chilling requirement for peach [J] (in Chinese). Acta Hort. Sinica, 2003. 30(4):379-383. ▲

[259] Wang, X. 1621. Qun fang pu [M] (in Chinese). Beijing: Agricultural Press, reprint in 1994. ▲

[260] Wang, Z. and E. Zhuang, Chinese fruit monograph-Peach flora [M] (in Chinese). Beijing: China Forestry Publishing House, 2001.▲

[261] Wang, Z. and J., Zhou. Pollen morphology of peach germplasm [J] (in Chinese). Acta Hort. Sinica, 1990. 17(3):161-168. ▲

[262] Weinberger, J.H. Characteristics of the progeny of certain peach varieties [C]. Proc. of the Amer. Soc. Hort. Sci, 1944. 45, 233-238.

[263] Weinberger, J.H. Chilling requirement of peach varieties [C]. Proc. Amer. Soc. Hort. Sci, 1950. 56:122-128.

[264] Weir, R.G and G.C. Cresswell. Plant Nutrient Disorders 1. Temperate and Subtropical Fruit and Nut Crops [M]. Inkarta Press, Melbourne, Australia, 1993.

[265] Werner, D.J. and J.X., Chaparro, Genetic interactions of pillar and weeping peach genotypes [J]. HortScience, 2005. 40(1):18-20.

[266] Werner, D.J.and J.X. Chaparro. Variability in flower bud number among peach and nectarine cultivars [J]. HortScience, 1988a. 23:578-580.

[267] Werner D.J., B.D. Mowrey, and E. Young. Chilling requirement and post-rest heat accumulation as related to difference in time of bloom between peach and western sand cherry [J]. J.Amer. Soc. Hort. Sci, 1988b. 113:775-778.

[268] Werner, D.J., Frantz, P.R., and Raulston, J.C. 'White Glory' weeping nectarine [J]. HortScience, 1985. 20(2):308-309.

[269] Werner, D.J., and M.A. Creller. Inheritance of sweet kernel and male sterility [J]. J. Amer. Soc. Hort. Sci, 1997. 122:215-217.

[270] Werner, D.J., S.M. Worthington, and L. K. Snelling, Peach tree named 'Corinthian Mauve' [P]. United States Patent: PP11576. 2000a.

[271] Werner, D.J., S.M. Worthington, and L. K. Snelling, Peach tree named 'Corinthian Rose' [P]. United States Patent: PP11504. 2000b.

[272] Werner, D.J., S.M. Worthington, and L. K. Snelling, Peach tree named 'Corinthian White' [P]. United States Patent: PP11493. 2000c.

[273] Werner, D.J., S.M. Worthington, and L. K. Snelling, Peach tree named 'Corinthian Pink' [P]. United States Patent: PP11902. 2001.

[274] Werner, D.J., W.R. Okie. A History and Description of the *Prunus persica* plant introduction collection [J]. HortScience, 1998. 33(5):787-793.

[275] Wickson, E.J. The California Fruits and how to grow them [M]. 308. 1889.

[276] Wu, Z.and P. H. Raven (eds.). Flora of China. Vol. 9 (Pittosporaceae through Connaraceae) [M]. Beijing: Science Press, and Missouri Botanical Garden Press, St. Louis. 2003.

[277] Yamazaki, K., M. Okabe, and E. Takahashi. Inheritance of some characteristics and breeding of new hybrids in flowering peaches [J] (in Japanese). Bull. Kanagawa Hort Expt Sta, 1987a. (34):46-53. *

[278] Yamazaki, K., M. Okabe, and E. Takahashi. New broomy flowering peach cultivars 'Terutebeni', 'Terutemomo', and 'Teruteshiro' [J] (in Japanese). Bull. Kanagawa Hort Expt Sta, 1987b. (34):54-56.

[279] Yoshida, M. Peach [M]. Horticulture in Japan. Asakura. Tokyo, 1994. 35-37.

[280] Yoshida, M. Peach Illustrated Book [M] (in Japanese). World Culture Publisher. Tokyo, * 1985.

[281] Yoshida, M. and K. Seike. 'Red Dwarf', the new red-dwarf peach [J] (In Japanese). Bull. FruitTree Res. Sta. A, 1974. (1):25-32. *

[282] Yoshida, M., K. Yamane, Y. Ijiro, N. Fujishige, M. Yamaguchi and E. Takahashi. Studies on ornamental peach cultivars [J] (in Japanese). Bul. College of Agri. Utsunomiya Uni, 2000. 17(3):1-14.

[283] Yu, D. Taxonomy on deciduous fruit trees [M] (in Chinese). Shanghai: Shanghai Science and Technology Press, 87. 1984. ▲

[284] Zang, D., D. Yu, M. Yang, W. Yu, and M. Zhang. A preliminary study on the investigation of peach-blossom cultivars in Shandong [J]. Shandong Agri. Sci. and Technology Press, 1998. (4):1-6. ▲

[285] Zhang C., T. Lin, S. Dai. Studies on evolution relationships between ornamental peach germplasm by using RAPD [J]. J. Beijing For. Univ, 1999. 21(5):26-31. ▲

[286] Zhang, X. and Z. Chen. Preliminary studies on peach-blossom (ornamental peach) classification [J] (in Chinese). Acta Hort. Sinica, 1991. 18(1):67-74. ▲

[287] Zhang, X., S. Dai, and L. Shi. Studies on peach-blossom (ornamental peach) germplasm diversity [J] (in Chinese). Chinese Gardening, 1997. 13(2):17-19. ▲

[288] Zhang, X. The second report in the classification of ornamental peach-blossom (ornamental peach) cultivars [J] (in Chinese). Guangdong Garden, 1993. 3:33-38. ▲

[289] Zhang, X., Y. Wang, and G. Wang. Observation and comparisons on the pollen morphology of peach-blossom (ornamental peach) cultivars germplasm resources [J] (in Chinese). J. Beijing For. Univ, 1997. 9(2):57-62. ▲

[290] Zhou, J. and H. Li. Inheritance of some characters in ornamental peaches and breeding of new varieties in foreign countries [J] (in Chinese). J. Beijing For. Univ, 1998. 20(2):100-103. ▲

[291] Zhou, S. 1082. Luoyang hua mu ji [M] (in Chinese). Shanghai: Shanghai Ancient Book Press, (reprint in 1985). ▲

[292] Zhou, W.1620. Ru nan pu shi [M] (in Chinese). Beijing: Agricultural Press, reprint in 1982. ▲

[293] Zhu, G., W. Fang and L. Wang. Chilling requirement of ornamental peach [J]. J. Plant Genet. Resource, 5(2):176-178. 2004. ▲

▲和*：均为已经出现在中文文献中的相关文献的英译名，以便英文版本检索。

致　谢

当这本书真的完成，就这么放在我面前的时候，出乎我自己的意料：竟然是如此的坦然，似乎只是完成了自己的一项使命而已！尽管这一切并不容易，尽管这一做起来竟花去了我20年的时间，但一切都是这么自然，似乎命中注定，这一天终将来临！

一直认为自己可以工作在北京植物园是多么的幸运！不仅仅是每天可以和花草打交道，更幸运的是还有那么多精心呵护这些植物的同事们！能够走到今天，特别是这本书得以最终呈现在大家面前，是和同事们对我的爱护和点点滴滴的帮助分不开的。感谢付俊秋女士在花粉萌发试验、电镜扫描试验、苗木繁殖、育种、观测，以及数据库编制等方面的辛勤付出；感谢郭翎博士一直以来在桃花品种的繁殖和推广方面所给与的支持和鼓励；感谢吴超然、曹颖女士和樊金龙先生在种子及苗木繁殖方面的合作；感谢俞思佳女士和熊德平先生在桃花病虫害防治方面给与的指导和帮助；感谢李菁博和周达康先生无私提供病虫害方面的照片；感谢穆志刚先生在桃花修剪方面的独到经验；感谢陈晨女士绘制的精美图画和在全书版式方面的建议……感谢北京植物园的桃花节！感谢支持我们桃花节一年年坚持办下来的所有同事们和来自各地的参观者！

更幸运的还有无论哪一个阶段我都能够遇到来自各方的老师和朋友们的指导和提携。感谢北京林业大学张秀英教授为我打开了桃花这一方天地，引领我认知了桃花的魅力；感谢北京林业大学张启翔教授和美国缅因大学张冬林教授为我提升了一个新的高度，开阔了桃花研究的领域；感谢日本宇都宫大学吉田雅夫教授一直为我提供日本桃花研究的最新动态，以及10多年来的无私指教；感谢美国农业部Ralph Scorza博士给我机会与他共同工作一年时间，分享他在桃研究及枝型育种方面近30年的积累，以及对桃花发展方向的独到认识，特别感谢他在本书出版之际欣然作序，沟通了东西方桃领域的研究；感谢乌克兰尼基塔植物园Larisa Komar-Tyomnaya博士热情的邀请和合作，开启了桃花育种的崭新前景。应该感谢的还有曾经在日本农林水产省工作的山口正己教授和曾经在美国J.C. Raulston树木园工作的Todd Lasagna博士提供桃花新品种接穗，日本的高桥荣治先生和宇都宫大学山根健治博士及时交流最新动向，美国哈佛大学李建华博士在分子生物学实验方面的指导和帮助……还有众多在桃花研究方面提供了大量宝贵资料的素未谋面的前辈们：美国的D.J. Werner，W.R. Okie，J.N. Moore先生，法国的R. Monet先生，加拿大的R.E.C. Layne先生……还有一位要特别感谢的是我的合著者——张佐双教授，是他的远见卓识为我提供了认识、了解、探索和研究桃花世界的保障，也是他20年来的鼓励、支持与合作才使得在桃花方面的工作得以一路坚持，并有机会完整地呈现在大家面前！

　　在本书出版之际，特别感谢中国工程院陈俊愉院士为本书欣然作序。陈先生关于书名的建议体现了他严谨的治学态度，令我们深深敬佩。长久以来，人们对于桃花的观赏性的认识一直仅仅局限在其艳丽的花朵上，然而作为以观赏为主要目的的观赏桃在枝型、叶色等花以外的很多方面均有着很高的观赏价值，因此，本着科学的态度，更为全面地反映出观赏桃的本来价值，特将本书定名为《观赏桃》。对于陈先生提出的其他建议，我们将在今后的工作中逐步落实。

　　最后要感谢我的家人长期以来为我付出的无怨无悔的爱与支持。

　　能够和桃花相伴走过20年，对于我，正像是桃花本身的寓意：那是一种幸运，更是一种幸福！

胡东燕　　2010年4月于北京

Acknowledgements

Bestow the people who appreciate the beauty of ornamental peach

Working with the ornamental peach for the 20 years has absolutely been my destiny. As a symbol of luck and happiness, peach blossom brought me so much···

Working with ornamental peach, or should say appreciating it for all these years, the peach has led me meet new friends, travel and understand the world through "peach pink tinted spectacles", full of optimism. The peach has changed my life and brought me where I am today and I offer it my sincere thanks!

The knowledge, enthusiasm and awe-inspiring efforts of many peach breeders and researchers the world over have provided me with support and guided me spiritually and academically through my persistence. Thanks my peach friends!

Each trip to collect and investigate ornamental peach varieties has helped me to gain knowledge, insight and information. Thanks for the memories!

Beijing Botanical Garden provided endless opportunities for study, travel and self-improvement and my twenty years of interest in the ornamental peach have allowed me stay here without change. Thanks to my garden and thanks to my passion!

With smiles and delight every day my parents and family give me love and support, especially when the road is long and hard. Thanks to my family!

To everybody who helps me and loves me in my life—thank you, it would never have happened without you.

Here you are!

Dongyan Hu

April 2010 in Beijing